Singular Perturbations and Hysteresis

Singular Perturbations and Hysteresis

Edited by
Michael P. Mortell
University College Cork
Cork, Ireland

Robert E. O'Malley
University of Washington
Seattle, Washington

Alexei Pokrovskii
University College Cork
Cork, Ireland

Vladimir Sobolev
Samara State University
Samara, Russia

siam Society for Industrial and Applied Mathematics
Philadelphia

10 9 8 7 6 5 4 3 2 1

Library of Congress Control Number: 2005927224
ISBN 0-89871-597-0

Royalties from the sale of this book are placed in a fund to help students attend SIAM meetings and other SIAM-related activities. This fund is administered by SIAM, and qualified individuals are encouraged to write directly to SIAM for guidelines.

List of Contributors

Eric Benoît
Laboratoire de mathématiques
Université de la Rochelle
La Rochelle, France

Viatcheslav Bykov
Department of Mathematics
Ben-Gurion University of the Negev
Israel

Rod Cross
Department of Economics
University of Strathclyde
Glasgow, Scotland, UK

Augustin Fruchard
Laboratoire de mathématiques
Université de la Rochelle
La Rochelle, France

Igor Goldfarb
Department of Mathematics
Ben-Gurion University of the Negev
Israel

Vladimir Gol'dshtein
Department of Mathematics
Ben-Gurion University of the Negev
Israel

Michael Grinfeld
Department of Mathematics
University of Strathclyde
Glasgow, Scotland, UK

Abdallah El Hamidi
Laboratoire de mathématiques
Université de la Rochelle
La Rochelle, France

Pavel Krejčí
Mathematical Institute
Academy of Sciences
of the Czech Republic
Prague, Czech Republic

Harbir Lamba
Department of Mathematical Sciences
George Mason University
Fairfax, USA

Nikolai Nefedov
Department of Mathematics
Faculty of Physics
Moscow State University
Moscow, Russia

Anne Pittock
Department of Philosophy
University of Glasgow
Scotland, UK

Alexei Pokrovskii
Department of Applied Mathematics
University College Cork
Cork, Ireland

Oleg Rasskazov
Department of Applied Mathematics
University College Cork
Cork, Ireland

Elena Shchepakina
Department of Differential Equations
and Control Theory
Samara State University
Samara, Russia

Vladimir Sobolev
Department of Differential Equations and Control Theory
Samara State University
Samara, Russia

Richard Studdert
Department of Computer Science
University College Cork
Cork, Ireland

Adelaida Vasilieva
Department of Mathematics
Faculty of Physics
Moscow State University
Moscow, Russia

Contents

Preface

Time relaxation and hysteresis are common strongly nonlinear phenomena which occur in many industrial, physical and economic systems. The wording 'strongly nonlinear' means that linearization will not encapsulate the observed phenomena. Often these two types of phenomena manifest different stages of the same or similar processes. A number of fundamental hysteresis models can be considered as limit cases of time relaxation processes, or admit an approximation by a differential equation which is singular with respect to a particular parameter.

However, the amount of interaction between practitioners of theories of systems with time relaxation and systems with hysteresis (and between the 'relaxation' and 'hysteresis' research communities) is quite low. Thus, the International Workshop on Relaxation Oscillations & Hysteresis was held at University College Cork, Ireland on April 1-6, 2002 as an initial attempt to address this situation. Among the aims of the workshop were

- to bring together leading experts in time relaxation and hysteresis phenomena in applied problems;
- to discuss important problems in areas such as reacting systems, semiconductor lasers, shock phenomena in economic modelling, fluid mechanics, etc. with the emphasis on hysteresis and singular perturbations;
- to learn and to share modern techniques in areas of common interest.

Further details concerning the workshop can found at

http://www.ucc.ie/ucc/depts/physics/ins/roh2002.htm

This book is based on results of the workshop. It is hoped that it will facilitate the cross-fertilization process between these two important topics, and the emergence of promising new areas of research.

The first chapter presents an introduction to a relationship between singularly perturbed differential equations and equations with hysteresis. The basic notions of hysteresis and singular perturbations theory are introduced. The chapter contains a number of very simple examples from both physics and mathematics.

The second chapter is entirely devoted to applications of models of hysteresis in economics. The authors introduce a simple model of human motivation and investigate its consequences for the original formulation of the El Farol bar problem of W. Brian Arthur.

In the third chapter a discontinuous hysteresis law is rigorously derived as a singular limit in differential equations with non-monotone nonlinearities, which arises, for example, in a model for instabilities of a fluid flow in a tube with a pump and a valve with uncertain parameters.

Various aspects of the asymptotic theory of singularly perturbed systems are presented in following three chapters.

A structured and synthetic presentation of Vasilieva's combined expansions is given in the fourth chapter. These expansions simultaneously take into account the limit layer and the slow motion of solutions of a singularly perturbed differential equation.

In the fifth and sixth chapters singularly perturbed partial differential equations are considered and typical problems of the asymptotic theory of contrast structures are discussed.

The last four chapters are concerned with the geometrical approach to an investigation of models with singular perturbations and hysteresis.

The integral manifold method is elaborated in Chapter 7, which is devoted to the study of different critical cases in the theory of singular perturbations. The presentation of mathematical results is combined with applications to problems of mechanics and of control.

A further development of this method, with applications to lasers, control and problems of chemical kinetics and combustion, is contained in the eighth chapter.

In the ninth chapter the integral manifold method is used to investigate the problem of pressure driven flames in inert porous media.

The last chapter is devoted to an extension of the geometric approach to systems with small hysteresis. The chaotic behavior of laser models is examined to demonstrate the validity of this approach.

We wish to record our thanks to the authors for their important contributions. We believe this book brings together many important recent developments in the analysis of singular perturbation and hysteresis phenomena in an accessible and reasonably comprehensive fashion. We want the reader to share the excitement of present day research in this rapidly growing field. We hope this book will not only be a useful and accessible introduction to techniques and the research literature, but will also generate new ideas for researchers, and attract new researchers into this vibrant and dynamic field.

Finally, we wish to gratefully acknowledge the support of the Boole Centre for Research in Informatics, University College Cork.

The Editors

Chapter 1

A Naive View of Time Relaxation and Hysteresis

A. Pokrovskii and V. Sobolev

Relationships between singularly perturbed equations and equations with hysteresis are discussed.

1.1 Introduction

1.1.1 Stop and Play nonlinearities

Singular perturbations[1] (including time relaxation) and hysteresis phenomena are among the most common strongly nonlinear elements which arise in almost any industrial, physical or economic system. Often these two types of phenomena manifest different stages of the same or similar processes. In this introductory section we illustrate the relationships between hysteresis nonlinearities and singularly perturbed equations by a very simple example.

Let us consider the mechanical device pictured in Fig. 1.1.

Here a small bead B can be moved along the string. We control a frame F which can also be moved along a string to the left or to the right. The position $x(t)$, $t \geq 0$ of the right edge of the frame is a varying input. The bead moves only if the left end of the frame pushes it to right, or if the right end of the frame pushes it to the left. The varying distance $y(t)$, $t \geq 0$ between the bead and the right edge of the frame is the output. This 'device' is called the *Stop nonlinearity* (with the limits 0 and h, where h coincides with the width of the frame). We suppose, that the size of bead is negligible compared with h. If, as the output, we measure the position $z(t)$ of the bead with respect to the string then the same device represents

[1]We recall that a system of differential equations $\dot{x} = f(\cdot)$, $\varepsilon\dot{y} = g(\cdot)$ is said to be *singularly perturbed* if ε is a small parameter.

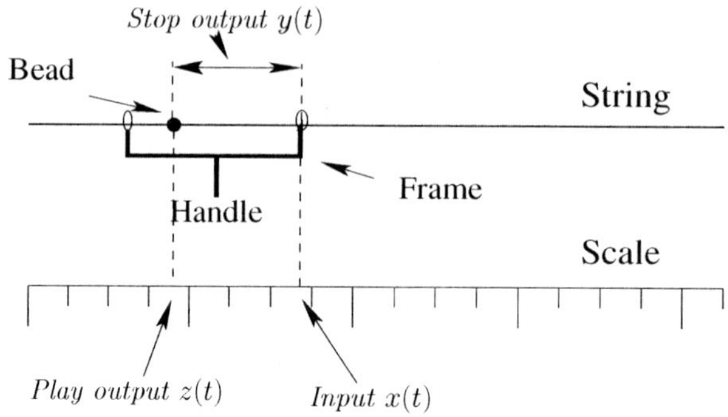

Figure 1.1. *The Play nonlinearity*

a *Play nonlinearity* (with the width h)[2]. The Play nonlinearity is sometimes called the *Backlash nonlinearity* [54].

The relationships between an input $x(t)$, $t \geq 0$, and the corresponding outputs of the Stop or Play nonlinearities are independent of the time scale: these nonlinearities are *rate independent.* This means that if $y(t), z(t)$ $t \geq 0$, are possible dynamics of the (relative for the Stop or absolute for the Play) bead positions for a given $x(t)$, $t \geq 0$, then the functions $v(t) = y(\alpha t), w(t) = z(\alpha t)$ describe possible dynamics of a bead for the input $u(t) = x(\alpha t)$. Thus, we can represent these relationships by a two-dimensional graph, see Fig. 1.2, where a typical trajectory $(x(t), y(t))$ is represented by the directed bold line. Fig. 1.1 is the analogue of Fig.

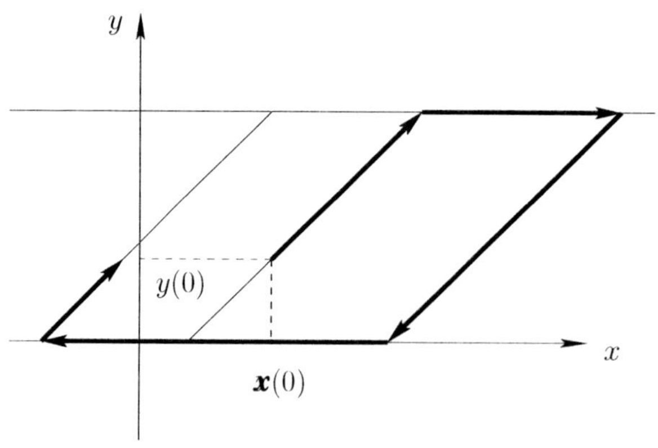

Figure 1.2. *A typical phase diagram for the Stop nonlinearity*

[2] The definition of a Play nonlinearity may vary slightly in different texts: e.g. in the monograph [28], p.8 the input is interpreted as the position of the *left* edge of the frame; in some other publications the input is interpreted as the position of the handle, etc.

1.2 for the Play nonlinearity; see Chapter 3 for more detail.

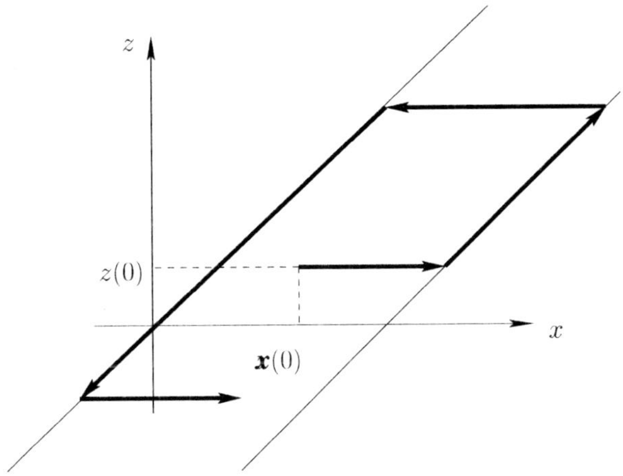

Figure 1.3. *A typical phase diagram for the Play nonlinearity*

The Stop and Play nonlinearities are simple but important examples of hysteresis relationships between scalar functions $x(t)$ and $y(t)$. We note, for example, that the Stop nonlinearity is the classical Prandtl model of an elastic element [45, 46] with Young's modulus 1 and with plasticity limits 0 and h. The same nonlinearity is sometimes referred to as the main nonlinearity in aircraft and rocket engineering [33, 34].

The relationship between $x(t)$ and $y(t)$ is not a function: for the same current position $x(t)$ of the frame the relative position $y(t)$ of the bead within the frame could be different. Note that the output $y(t)$ for $t \geq 0$ of the Stop (or the Play) nonlinearity depends only on the input $x(t)$, $t \geq 0$ and on the initial position of the bead. We can use the operator notations: $y(t) = S[y_0]x(t)$ for the Stop nonlinearity and $z(t) = P[z_0]x(t)$ for the Play nonlinearity. Here $y_0 = y(0)$ and $z_0 = z(0)$ play the role of parameters. Operators $S[y_0]$ and $P[y_0]$ obviously satisfy the formula:

$$S[y_0]x(t) + P[z_0]x(t) = x(t), \tag{1.1}$$

where the initial states satisfy the equality $y_0 + z_0 = x(0)$.

Although the device shown in Fig. 1.1 is extremely simple, as are typical phase diagrams in Fig. 1.2, 1.3, it is not easy to suggest a convenient description for the operators $S[y_0]$ and $P[y_0]$ for sufficiently general inputs, e.g. for all continuous $x(t)$, which is critically important in the investigation of differential equations with hysteretic terms. The 'naive' attempt to describe the operators $S[y_0]x(t)$ and $P[y_0]x(t)$ would be via the standard differential equations:

$$\dot{y} = f(x(t), y), \quad \text{or} \quad \dot{z} = g(x(t), z), \tag{1.2}$$

where $x(t)$ is a given input (that is $x(\cdot)$ plays the role of a functional parameter). This, however, is hopeless, because *the relationships between $x(t)$ and solutions of* *the equation (1.2) are not rate independent.* Indeed, the substitution $t = a\tau$ changes equation (1.2) to

$$a^{-1}\dot{y} = f(x(a\tau), y).$$

This obstacle is conceptual: in contemporary scientific language hysteresis is defined as *Rate Independent Memory* [56]. This instructive definition will be discussed in more detail later.

1.1.2 Netushil's representation

Now we come to one of the motivations of this paper. The operators $S[y_0]$ and $P[y_0]$ may be explicitly described in terms of *singular limits* of an auxiliary singularly perturbed scalar differential equation of the form

$$\varepsilon \dot{z} = f(x(t), z). \tag{1.3}$$

Here $x(t)$ is a given continuous function, which again plays the role of a functional parameter, and ε is a small positive parameter. Denote by $z(t; z_0, \varepsilon)$ the unique solution of equation (1.3) satisfying the initial condition

$$z(0) = z_0. \tag{1.4}$$

Recall that the singular limit $z(t, z_0)$ of the initial value problem (1.3), (1.4) is defined as

$$z(t, z_0) = \lim_{\varepsilon \to 0} z(t, z_0, \varepsilon). \tag{1.5}$$

We introduce an auxiliary function

$$G(u) = \begin{cases} -u, & if \quad 0 \le u, \\ 0, & if \quad -h \le u \le 0, \\ -(u+h), & if \quad u \le -h, \end{cases}$$

see Fig. 1.4.

Let us consider the following singularly perturbed equation

$$\varepsilon \dot{z} = G(z - x(t)). \tag{1.6}$$

The following important observation is due to Netushil [33, 34]:

> *The singular limit (1.5) exists for any continuous $x(t)$, and the function $x(t) - z(t, z_0)$ coincides with the output $S[y_0]x(t)$ of the Stop nonlinearity, where $y_0 + z_0 = x(0)$.*

Fig. 1.5 (compare with Fig. 1.2) demonstrates the effectiveness of using the singular perturbation approach suggested by Netushil in the approximation of the Stop nonlinearity.

We will make some comments on Netushil's observation. We return to a general singularly perturbed equation of the form

$$\varepsilon \dot{z} = f(x(t), z), \tag{1.7}$$

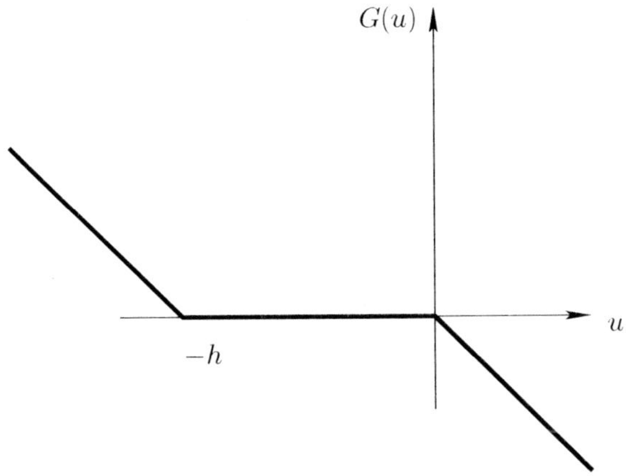

Figure 1.4. *Graph of the function $G(u)$*

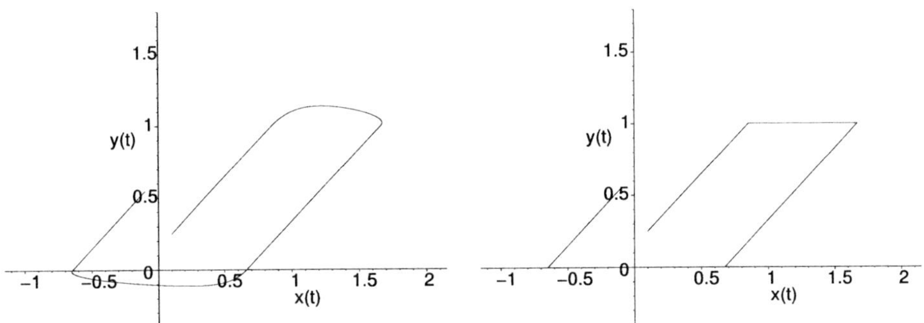

(a) $\varepsilon = 0.1$ (b) $\varepsilon = 0.001$

Figure 1.5. $h = 1, x(t) = e^{0.3t}(2.5 \sin t + 0.1 \cos t)$

and suppose that the function f is strictly positive for $z < R(x)$ and strictly negative for $z > R(z)$, where $R(x)$ is a continuous function on $-\infty < x < \infty$, see Fig. 1.6. This situation is usual in classical singular perturbation theory. It is well known that for $z(0) = R(x(0))$ the singular limit of equation (1.7) can be described by the functional relationship: $z(t) = R(x(t))$, [32]. It suffices to note that for 'infinitesimally' small $\varepsilon > 0$ the derivative $\dot{z}$ is 'equal to $+\infty$' for $z < R(x(t))$, is 'equal to $-\infty$' for $z > R(x(t))$.

The situation is different if the set of zeros of the function $f(x, z)$ has a more complicated structure: as the reader will see, in this case singular limits of the corresponding equations are operators of a hysteresis nature. Let us consider, in particular, the case when the set of zeros of the function $f(x, z)$ is a strip bounded by the graphs of two different continuous functions $z = R_-(x)$ and $z = R_+(x)$,

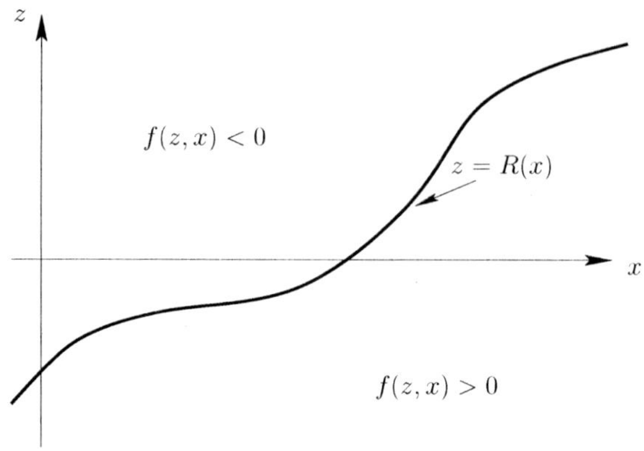

Figure 1.6. *Function $R(x)$*

satisfying the inequality $R_-(x) \le R_+(x)$. We also suppose that the functions $R_-(x)$ and $R_+(x)$ are continuous and monotone in x, and that f is positive for $z < R_-(x)$, and is negative for $z > R_+(x)$, see Fig. 1.7.

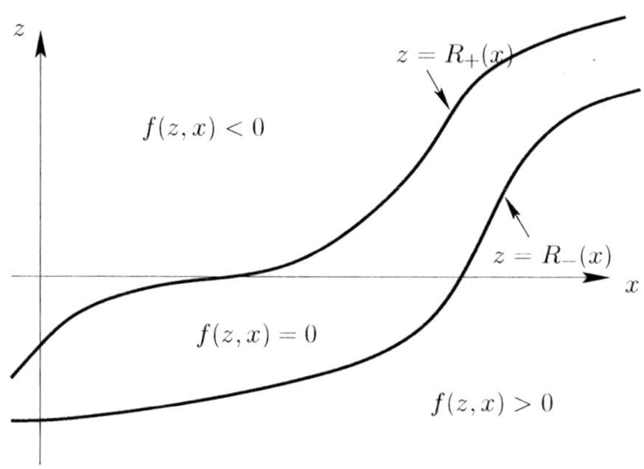

Figure 1.7. *Functions $R_-(x)$ and $R_+(x)$*

In this case for infinitesimally small $\varepsilon > 0$ the derivative $\dot{z}$ is equal to $+\infty$ for $z < R_-(x(t))$, is equal to $-\infty$ for $z > R_+(x(t))$, and is equal to 0 for $R_-(x(t)) \le z \le R_-(x(t))$. It indicates that the singular limit of the phase diagram for equation (1.6) can be described by Fig. 1.8.

In the case when

$$R_-(x) = x - h, \quad R_+(x) = x,$$

this diagram coincides with Fig. 1.3. That is, the singular limit $z(t, z_0)$ of the

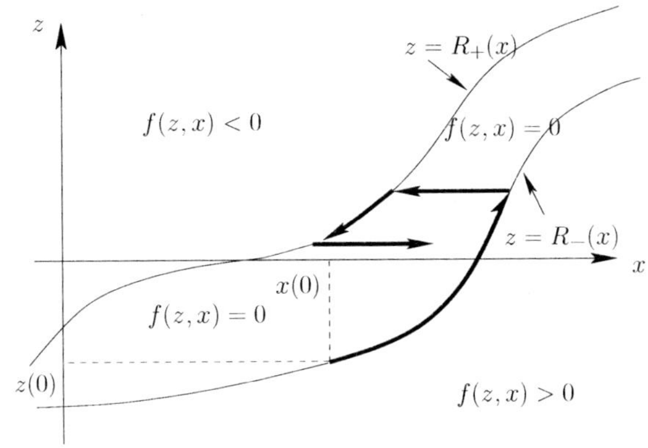

Figure 1.8. *A typical phase diagram for the singular limit nonlinearity*

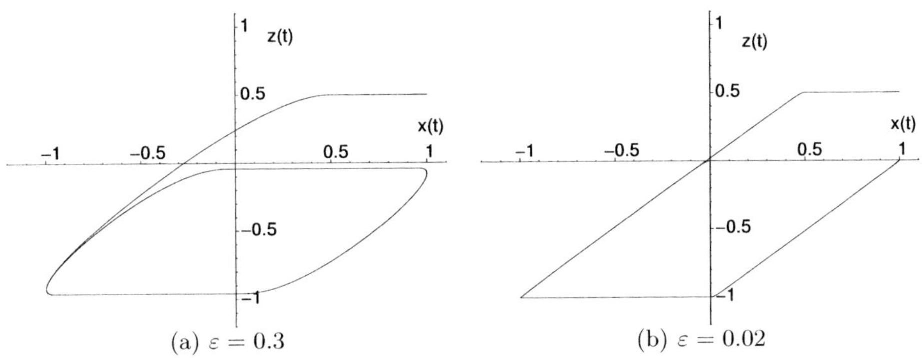

Figure 1.9. $h = 1, x(t) = \cos t$

equation

$$\varepsilon \dot{z} = G(z - x(t)), \tag{1.8}$$

describes the Play nonlinearity: $z(t, z_0) = P[z_0]x(t)$. Thus, the equality (1.1) may be rewritten as $S[y_0]x(t) = x(t) - z(t, z_0)$ with $y_0 + z_0 = x(0)$. This coincides with the Netushil's observation. To conclude, we present some graphs illustrating the convergence of the the phase diagrams of the singularly perturbed equation (1.8) to the corresponding input-output diagrams of the Play nonlinearity, see Fig. 1.9, 1.10 and 1.11.

1.1.3 General principle

Netushil's observation is a manifestation of the following guiding principle:

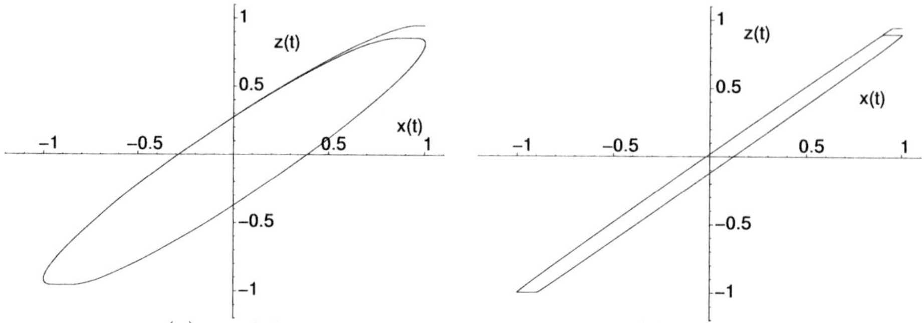

(a) $\varepsilon = 0.3$ (b) $\varepsilon = 0.02$

Figure 1.10. $h = 0.1, x(t) = \cos t$

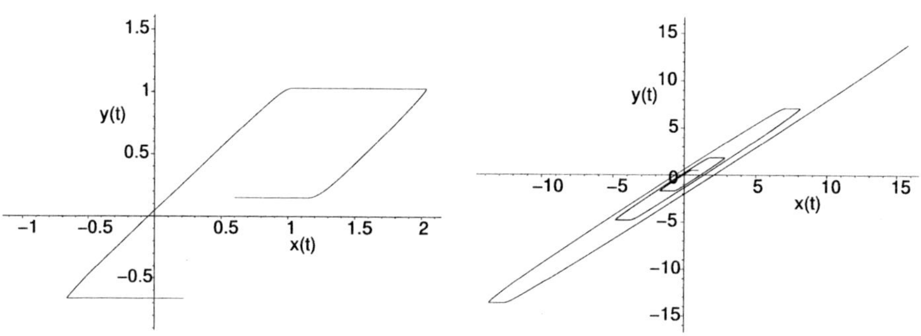

(a) $x(t) = e^{-0.15t}(0.6\cos 0.4t$
$+3.2\sin 0.4t)$

(b) $x(t) = e^{0.1t}\cos 0.6t$

Figure 1.11. $h = 1,\ \varepsilon = 0.1$

Hysteresis operators can be often represented/approximated by

$$y(t) = H(x(t), z(t)),$$

where $z(t)$ is a singular limit of an appropriate singularly perturbed equation.

We list a few observations which support this principle.

Firstly, we note that the Netushil representation implies immediately that the guiding principle holds for some classical models of elasto-plasticity, such as the Besseling Model [6] and the Ishlinskii Model [25]. We recall that the Besseling Model is a weighted sum of a finite number of Stop nonlinearities with different widths, whereas the Ishlinskii model uses appropriate integrals instead of finite sums. Moreover, a natural modification of our constructions shows that this principle is intact for classical von Mises and Tresca models of vector plasticity phenomena, see [28] and further references therein. This is again straightforward, since these models can be interpreted as multi-dimensional Stop nonlinearities.

Secondly, we observe that Figure 1.8 represents so-called *Generalized Play* [28], p.8. The role of this nonlinearity is based on the Identification Theorem, [28], p.59, stating that any hysteresis nonlinearity Γ with a memory which is stable with respect to noises of small amplitude may be reduced to a form

$$\Gamma x(t) = H(x(t), \mathcal{P}[z_0]x(t)),$$

where $H(x, z)$ is continuous and strictly monotone in z, and $\mathcal{P}$ is a Generalized Play. In other words, any canonical hysteron [28], p.31, has a singular limit representation satisfying the guiding principle given above.

Thirdly, later in this Chapter and in Chapter 3 it will be demonstrated that some other types of hysteresis nonlinearities, e.g. non-ideal relays, can also be represented as singular limits of equation (1.7) (with functions $f(x, z)$ for which the set of solutions of $f(x, z) = 0$ has a structure different from that considered above).

Finally, one more partial explanation for the validity of this principle is the following. *The dependence between the input $x(t)$ and the singular limit of the solutions $z(t, t_0, z_0)$ of the corresponding initial value problem for equation (1.7) is always rate independent*, although it is not true for any given strictly positive value of the parameter ε. The italicized statement is easy to prove. Indeed, let, for a given $\varepsilon > 0$ and given $x(t)$, the function $z(t, \varepsilon)$ be a solution of equation (1.7), and let us consider for some $\alpha > 0$ the functions $w(t, \varepsilon) = z(\alpha t, \varepsilon)$. Then, clearly, we can write

$$\delta \dot{w} = f(w, x(\alpha t)),$$

where $\delta = \varepsilon/\alpha$. That is,

$$w(t, \varepsilon/\alpha) = z(\alpha t, \varepsilon).$$

Let now $z_*(t)$ be a singular limit of the equation (1.7). Then the limit transition as $\varepsilon \to 0$ proves that the the function $w_*(t) = z_*(\alpha t)$ is a singular limit for the equation (1.7). This proves our assertion.

1.1.4 Closed loop systems

The hysteresis nonlinearities considered above are 'open-loop' systems with an input $x(t)$ and an output $y(t)$. In many cases they are, however, parts of a closed loop system where other elements are described by differential equations.

Consider, for instance, forced oscillations of a pendulum on an elastic-plastic element with Young's modulus equal to one both in the elastic and plastic components and with the plasticity limits $0, h$. This can be described by the differential-operator equation

$$\begin{aligned} \ddot{x} + x + y(t) &= \sin(\omega t), \\ y(t) &= S[y_0]x(t). \end{aligned} \tag{1.9}$$

This is a differential-operator equation, where the initial state $y(0) = y_0 \in [0, h]$ of the Stop nonlinearity is a parameter. It can be shown that for any initial condition

$$x(0) = x_0, \quad \dot{x}(0) = \dot{x}_0, \tag{1.10}$$

this equation has a unique solution $(x(t), y(t))$.

In this situation we can link the solutions of the equation with hysteresis to limit solutions of the singularly perturbed system (using an additional auxiliary variable z),

$$\begin{aligned} \ddot{x} + x + y &= \sin(\omega t), \\ y &= x - z, \\ \varepsilon \dot{z} &= G(z - x). \end{aligned} \tag{1.11}$$

Recall that the singular limit of the initial value problem

$$x(0) = x_0, \quad \dot{x}(0) = \dot{x}_0, \quad z(0) = z_0$$

for the system (1.11) is the limit of the corresponding solutions $(x_\varepsilon(t), z_\varepsilon(t))$ as $\varepsilon \to 0$. Namely, by Netushil's observation, the following is true.

The solution of the initial value problem (1.9), (1.10) coincides with the pair $(x(t), x(t) - z(t))$, where $(x(t), z(t))$ denotes the singular limit of the solutions of equation (1.11) with the initial conditions

$$x(0) = x_0, \quad \dot{x}(0) = \dot{x}_0, \quad z(0) = x(0) - y_0.$$

We present some graphs illustrating the convergence of phase diagrams, in the (x, y)-plane of the singularly perturbed equation (1.11), to the corresponding phase diagrams of the equation with the Play nonlinearity ((1.9), see Fig. 1.12 and 1.13 (in both cases $\omega = 2$).

The italicized observation above is again a manifestation of the general principle.

Equations with hysteresis nonlinearities are singular limits of appropriate singularly perturbed equations. Vice versa: Singular limits of equations with singular perturbations are solutions of equations with appropriate hysteresis nonlinearities.

This principle is very useful, and different aspects of its application will be discussed later in this chapter. Here we mention only that it highlights the role of equations with hysteresis in the asymptotic analysis of some multi-rate systems. Consider, as an example, a system of the form

$$\begin{aligned} \dot{x} &= \varepsilon f(x, y, z), \\ \dot{y} &= g(x, y, z), \\ \delta \dot{z} &= h(x, y, z), \end{aligned}$$

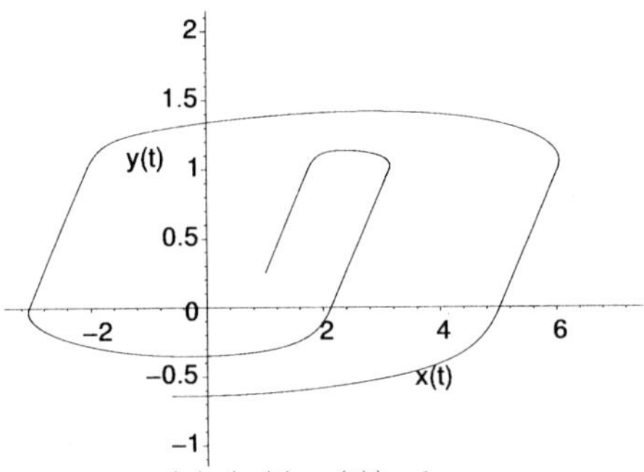

(a) $(x(t), y(t))$-plane

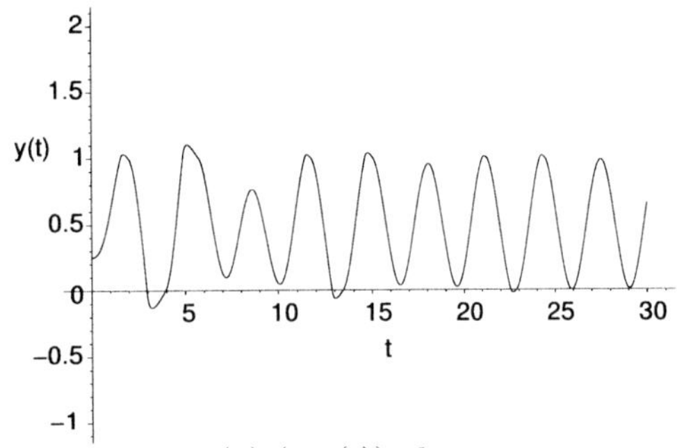

(b) $(t, y(t))$-plane

Figure 1.12. $h = 1, \varepsilon = 0.1$

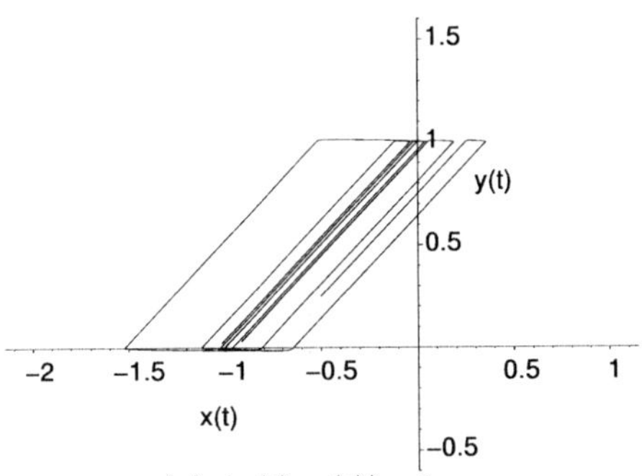

(a) $(x(t), y(t))$–plane

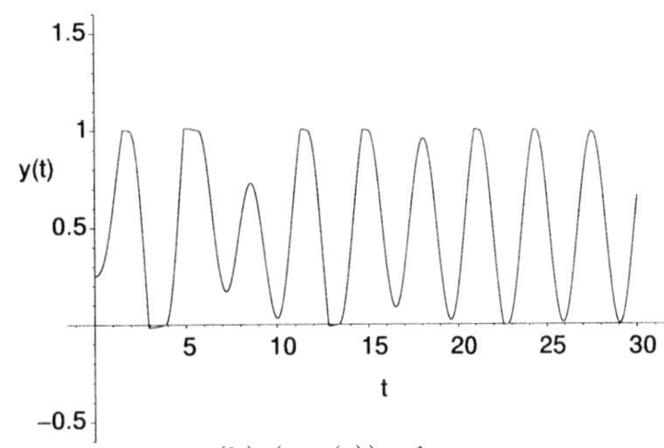

(b) $(t, y(t))$–plane

Figure 1.13. $h = 1, \varepsilon = 0.001$

where $\varepsilon, \delta > 0$ are small. Then the first stage of the asymptotic analysis of the system is replacing the last equation with its singular limit, with the subsequent use of an appropriate averaging procedure [52]. As we discussed in the case when the function h is not monotone in z, this singular limit is a hysteretic nonlinearity. Thus, the asymptotic analysis of the whole system reduces to the analysis of the differential operator equations

$$\begin{aligned}\dot{x} &= \varepsilon f(x, y, z),\\ \dot{y} &= g(x, y, z),\\ z(t) &= \Gamma_x y(t),\end{aligned}$$

where Γ is a hysteresis-type operator which depends on the parameter x.

To finalize this long introduction, we would like to express our view that bearing in mind the simple guiding principles mentioned above could be of great benefit to both the 'hysteretic' and the 'singularly perturbed' scientific communities.

The remainder of this chapter is dedicated to some further 'promotion' of this view. Firstly, in Section 1.2, we will discuss an approach to mathematical models of hysteresis as suggested by Krasnosel'skii, and especially to a modern interpretation of the Preisach model developed by Krasnosel'skii (with co-authors) and I. Mayergoyz. The Preisach Model nowadays plays a major role in different subject areas, especially in micromagnetism; see, for instance, the Special Issue of Physica B; Condensed Matter, 306, December 2001, No 1-4. This Issue devoted the main bulk of its 270 pages to applications of the Preisach model of hysteresis to different physical problems. The basis of many contributions is the Krasnosel'skii-Mayergoyz concept of the Preisach operator, which is the topic of the first part of the introductory chapter. We mention also the 5th International Symposium on Hysteresis and Micromagnetic Modeling (May 30 - June 1, 2005, Budapest, Hungary,

http://www.HMM2005.bme.hu),

which is devoted to the 100th anniversary of the birth of F. Preisach. In the second part of this chapter (Section 1.3) we discuss important phenomena related to singularly perturbed equations via simple piece-wise linear examples. This approach has allowed us illustrate such phenomena as initial layer, jump points, canard trajectories etc by explicit formulas. We note also that the piece-wise linear singularly perturbed equations play an key role in engineering applications. Afterwards, in Section 1.4, we will come back to the relationship between hysteresis and singular perturbation phenomena.

We mention that in writing this text we were inspired by the famous paper [37].

1.2 Hysteresis Phenomena

1.2.1 Why bother with equations with hysteresis?

We consider an iron pendulum oscillating in an external magnetic field, see Fig. 1.14.

Such oscillators are common in nature. Without a magnetic field, the pendulum dynamics are described (in a linear approximation) by the equation

$$\ddot{x} + a\dot{x} + x = \sin(\omega t),$$

which can be integrated explicitly. The situation when the magnetic field is present is, however, much more complex. The iron of the pendulum will be magnetized and demagnetized. This process is called *ferromagnetic hysteresis.* The magnetization will interact with the external magnetic field, and thus the equation will have the form

$$\ddot{x} + a\dot{x} + x = \sin(\omega t) + y(t), \quad y(t) = \Gamma x(t), \tag{1.12}$$

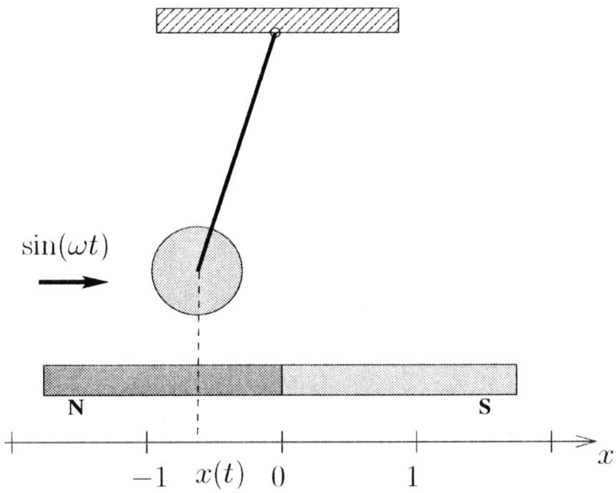

Figure 1.14. *An iron pendulum in a magnetic field*

where the nonlinearity $\Gamma x(t)$, which describes the interaction between the external magnetic field and the magnetized pendulum itself, will be described more fully later. This term, $\Gamma x(t)$, depends on the pendulum's position as well as on its magnetization. The pendulum's current magnetization depends in turn on the whole previous history of the pendulum's motion. Thus the actual equation is of a differential-operator type.

To proceed with a quantitative and qualitative analysis of equation (1.12), we need a convenient description of the operator Γ that acts on the function $x(t)$. Some important requirements for such a description are:

1. It should be amenable to physical observation;
2. It should admit a convenient numerical implementation (to allow computer modelling of the equations);
3. It should have "nice" properties as an operator in some function space (to develop a qualitative theory of the equation with hysteresis).

It is not clear, a priori, whether such a description of ferromagnetic hysteresis exists. However, we shall discuss this question within a broader setting below.

The outline of the remainder of this section is as follows:

In Subsection 1.2.2 we discuss the modern understanding of the phrase *hysteresis phenomenon.* In Subsections 1.2.3-1.2.6 we discuss one of the important mathematical models of hysteresis: the *Preisach nonlinearity.* In Subsection 1.2.6 we explain how this particular model satisfies requirement 2 above. In Subsections 1.2.7-1.2.8 we discuss the important *Identification Principle* that justifies requirement 1. Finally in Subsection 1.2.9 we discuss the numerical and qualitative analysis of closed loop systems using the Preisach model (requirement 3).

1.2.2 Generalities

The word hysteresis is of Greek origin and means etymologically 'to lag behind'. It was introduced into the scientific vocabulary about 120 years ago by the Scottish physicist Alfred Ewing as follows *"when there are two quantities M and N, such that cyclic variations of N cause cyclic variations of M, then if the changes of M lag behind those of N, we may say that there is hysteresis in the relation of M and N"* [18]. Further interesting historical details are presented in [10].

Nowadays, "hysteresis" is one of the most 'multi-valued' of terms (like, say, 'entropy'). For example, it is widely used in mechanics (plastic hysteresis) [30], physics (ferromagnetic hysteresis) [31], phase transitions [9], hydrology (soil-moisture hysteresis) [38], and economics (shock analysis) [11]. In many scientific and general encyclopedias, the meaning of the word is illustrated by a picture like Fig. 1.15:

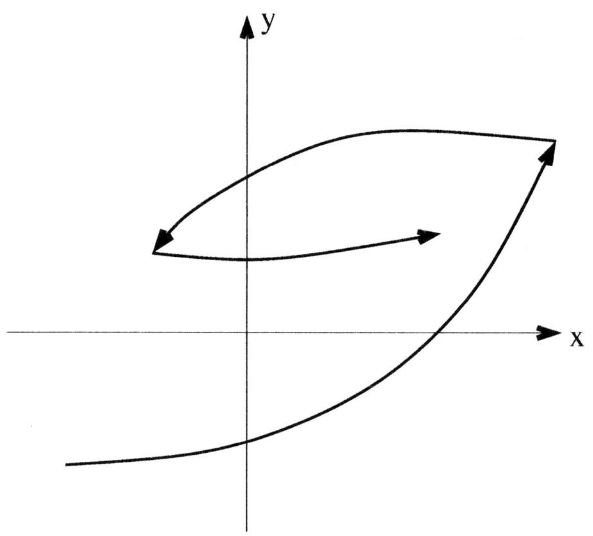

Figure 1.15. *A typical hysteresis input-output diagram*

The quantities x, y, marked as the horizontal and the vertical coordinates, can have different physical meanings, such as deformation versus stress (plastic hysteresis) or external magnetic field versus magnetization (ferromagnetic hysteresis).

In fact, this picture alone contains the basic points for an operational definition of hysteresis. We can suggest immediately that, in engineering language, a hysteresis nonlinearity is a kind of transducer Γ which relates the variable output $y(t)$ to a variable input $x(t)$, see Fig. 1.16.

This transducer Γ is not a function: for the same current value $x(t_*)$ of the input, different output values $y(t_*)$ can be observed (geometrically some vertical lines have multiple intersections with the input-output curve in Fig. 1.15). In other words, an output $y(t)$, after a certain reference time t_*, depends not only the input

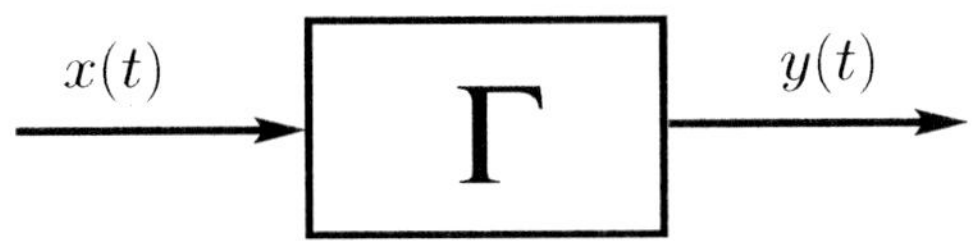

Figure 1.16. *A hysteresis transducer*

$x(t)$, $t \geq t_*$, but also on an internal/initial state $\omega = \omega(t_*)$ of the transducer Γ. Sometimes, this internal state can be characterized in suitable physical terms; in other cases, it is a kind of condensed memory of the history for $t < t_*$; sometimes it is just an abstract parameter. Moreover, since there is no mention of a time scale in Fig. 1.15, we can suggest that the input-output relationships are (at least in the first approximation) *rate-independent*, or, equivalently, *invariant with respect to time scaling*. That is, if the pair $(x(t), y(t))$ is an admissible input-output correspondence, then the correspondence $(x(at+b), y(at+b))$ is also admissible for any real b and positive a, see Fig. 1.17 below.

In formal mathematical language the 'input-output' relationships are to be understood as operators in a suitable function space. Rate independence means that these operators are invariant with respect to the action of the group G of affine transformations of the time-scale.

We have now arrived at one of the modern formal definitions of a hysteresis nonlinearity. Omitting a few technicalities, hysteresis nonlinearities are defined as *deterministic, rate independent operators*[3] [56]). Surprisingly, this general definition is sufficient for developing interesting formal concepts with various applications. This fact was understood by a group of Russian mathematicians in the early 1970s, see [28] and the bibliography therein. The general structure of these formal concepts is as follows:

1. Choose elementary hysteresis nonlinearities, so called *hysterons* (such as a non-ideal relay, Generalized Play [28], Prandtl or Duhem models, etc).

2. Treat complex hysteresis nonlinearities as *block-diagrams* of hysterons.

3. Establish *identification principles*.

Nowadays, this approach to hysteresis is standard [27] and it contains a wide variety of 'branches', depending on the choice of hysterons in item 1 and/or the basic type of the block-diagrams in item 2. Below we shall discuss only one such branch: the *Preisach model*. In this model the role of hysterons is that of *non-ideal relay nonlinearities*, or, as they are also called, *thermostat nonlinearities*. Block-diagrams are essentially the standard parallel connection of a number of hysterons.

[3]We note, in passing, a result of F. Holland (a personal communication): the only linear integral operator satisfying the rate independence property is the Hilbert transform.

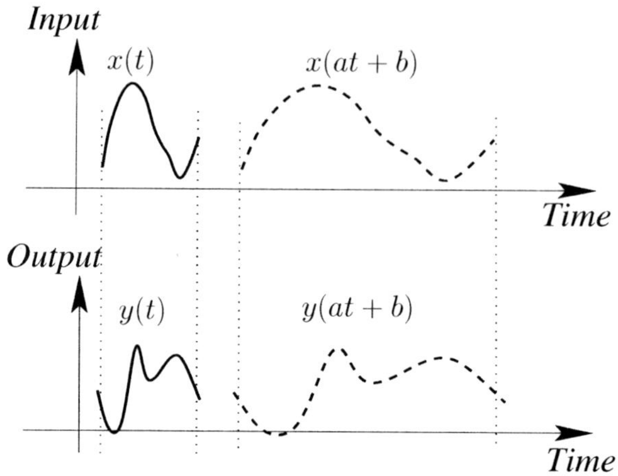

Figure 1.17. *Rate independence*

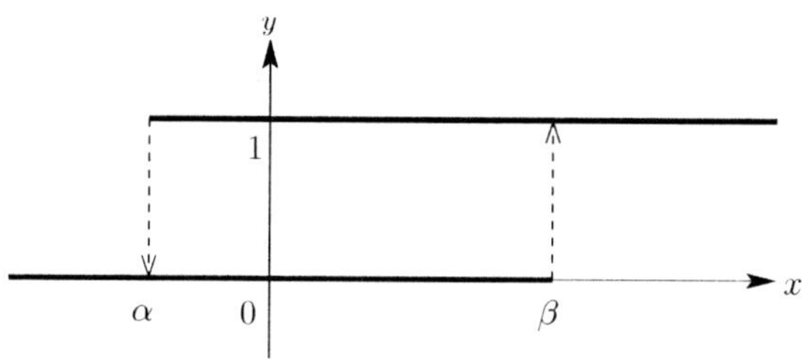

Figure 1.18. *Non-ideal relay* $R_{\alpha,\beta}$

1.2.3 Non-ideal relay

The non-ideal relay (with threshold values $\alpha < \beta$) is the simplest hysteretic transducer. This transducer is denoted by $R_{\alpha,\beta}$; its output $y(t)$ can take one of two values 0 or 1: at any moment the relay is either 'switched off' or 'switched on'. The dynamics of the relay are usually described as in Fig. 1.18.

The variable output $y(t)$

$$y(t) = R_{\alpha,\beta}[t_0, \eta_0]x(t), \qquad t \geq t_0,$$

depends on the variable input $x(t)$ ($t \geq t_0$) and on the initial state η_0. Here the input is an arbitrary continuous scalar function; η_0 is either 0 or 1. The scalar function $y(t)$ has at most a finite number of jumps on any finite interval $t_0 \leq t \leq t_1$.

The output behaves rather 'lazily': it prefers to be unchanged, as long as the phase pair $(x(t), y(t))$ belongs to one of two bold lines in Fig. 1.18. The value of

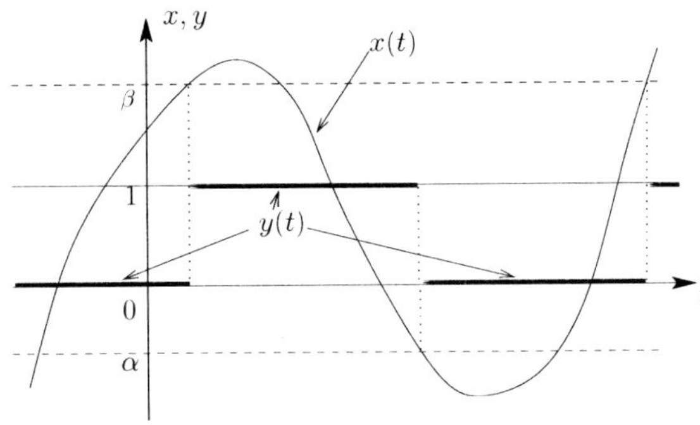

Figure 1.19. *Input and output or non-ideal relay $R_{\alpha,\beta}$*

the function y at a moment t is defined by the following explicit formula:

$$y(t) = R_{\alpha,\beta}[t_0, \eta_0]x(t) = \begin{cases} \eta_0, & \text{if } \alpha < x(\tau) < \beta \text{ for all } \tau \in [t_0, t]; \\ 1, & \text{if there exists } t_1 \in [t_0, t] \text{ such that} \\ & x(t_1) \geq \beta,\ x(\tau) > \alpha \text{ for all } \tau \in [t_1, t]; \\ 0, & \text{if there exists } t_1 \in [t_0, t] \text{ such that} \\ & x(t_1) \leq \alpha,\ x(\tau) < \beta \text{ for all } \tau \in [t_1, t]. \end{cases}$$

The equalities $y(t) = 1$ for $x(t) \geq \beta$ and $y(t) = 0$ for $x(t) \leq \alpha$ always hold for $t \geq t_0$.

Fig. 1.19 gives another geometrical illustration of the definition of a non-ideal relay. Here we combine on the same graph a typical input and the corresponding output of the relay.

From the physical point of view, the definition of the output is verifiable only for those inputs that haven't got local minima equal to α or local maxima equal to β. The question whether the corresponding switching would happen or not for such inputs is affected by small uncontrollable noises, which are present in any physical system. The physically meaningful definition of (the set) of possible outputs for such 'critical' inputs requires a special construction.

We note, finally, that such relay operators are discontinuous in any reasonable sense. A nice and important property of the non-ideal relay is its monotonicity with respect to both the input and the initial state: *if $\eta_{0,1} \leq \eta_{0,2}$ and $x_1(t) \leq x_2(t)$ for all $t \in [t_0, t_1]$, then $y_1(t) \leq y_2(t)$ for all $t \in [t_0, t_1]$.*

There is another way to represent the non-ideal relay which will be of great use later on. Any non-ideal relay R_{α_0,β_0}, $\alpha_0 < \beta_0$, can be represented as a point (α_0, β_0) in the half plane $\alpha < \beta$, and any point above the diagonal represents a non-ideal relay. This half plane will be known hence forth as the *'Preisach Half Plane'* or for convenience the *'Preisach Plane'*. The reader may have noticed that this new representation involves no reference to input or output. It is in the calculation of output that the true elegance of this representation comes to the fore. In this

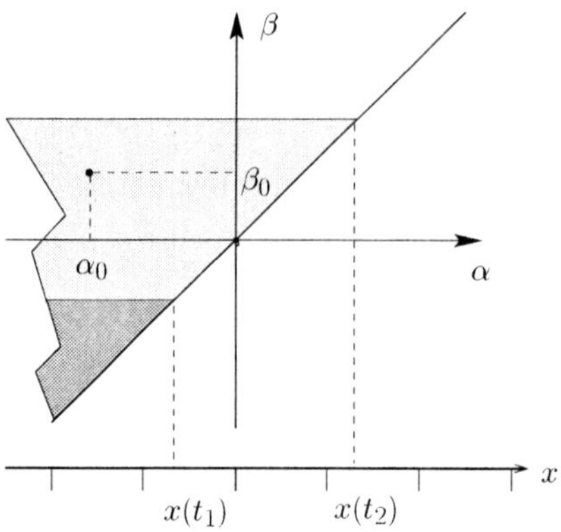

(a) Increasing input

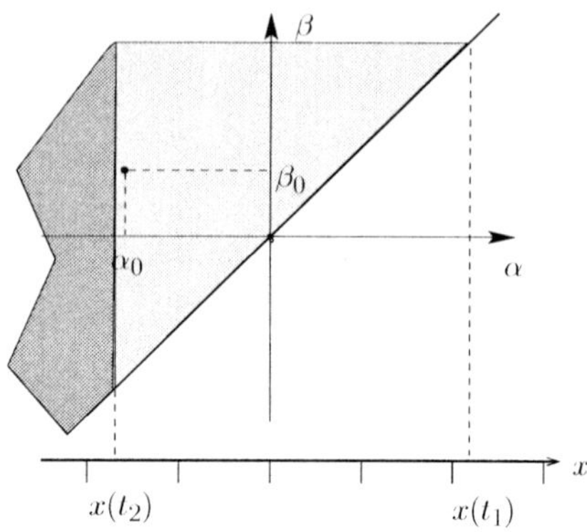

(b) Decreasing input

Figure 1.20. *Non-ideal relay R_{α_1,β_1}, on Preisach plane with variable input $x(t)$*

representation the input, $x(t)$, moves along the horizontal and controls the point on the diagonal $\alpha = \beta$ directly above itself. When moving toward the upper right corner, the point on the diagonal drags the horizontal line with itself, and 'shades' the domain below this line and above the diagonal. When moving towards the bottom left corner, the point on the diagonal drags the vertical line with itself, and 'un-shades' everything to the right of this line and above the diagonal. The output, $y(t)$, is then determined by whether the point (α_0, β_0) is within the shaded' region or not. Those relays with coordinates (α_0, β_0) which are within the shaded region are 'turned on' and thus their output is 1. Those relays that are not within the shaded region are 'turned off', so yield an output of 0.

In 1.20(a), with initial input $x(t_1)$, the relay R_{α_0,β_0} is not within the shaded region at the moment t_1 (this region is dark shaded in the picture) and so has an output of 0. As the input is increased to $x(t_2)$ the shaded area increases as described above and the extra lighter shaded area is incorporated. Thus, the region shaded at the moment t_2 is the union of the darker and lighter shaded areas. Therefore, at the input $x(t_2)$, the relay R_{α_0,β_0} is within the shaded portion of the Preisach plane and thus its output is changed to 1. Let us move now to Fig. 1.20(b) which represents the situation when an input $x(t)$ decreases from $x(t_1)$ to $x(t_2)$. The region shaded at the moment t_1 is here the union of the darker and lighter shaded areas; the relay R_{α_0,β_0} is contained in the shaded region and contributes 1 to the output. As the input is decreased to $x(t_2)$ the vertical line moves through the black dot, and the lightly shaded portion is removed from the gray area, leaving only the darker shaded piece. That is, the region shaded at the moment t_2 coincides with the darker shaded area. In particular, at the moment t_2 the relay R_{α_0,β_0} is no longer contained in the shaded region and thus the output for that relay is then 0.

While these pictorial representations hopefully aid the understanding of the process, those that may not immediately grasp the concept are encouraged to try the applets (interactive web pages) at http://phys.ucc.ie/~oll/hysteresis/node1.htm

1.2.4 The parallel connection of non-ideal relays

Take, now, a number of relays: $R^j = R_{\alpha_j,\beta_j}$, $1 \leq j \leq n$. Consider a parallel connection of the relays R^j, with the weights $\mu_j = \mu(j) > 0$. Fig. 1.21 illustrates this situation.

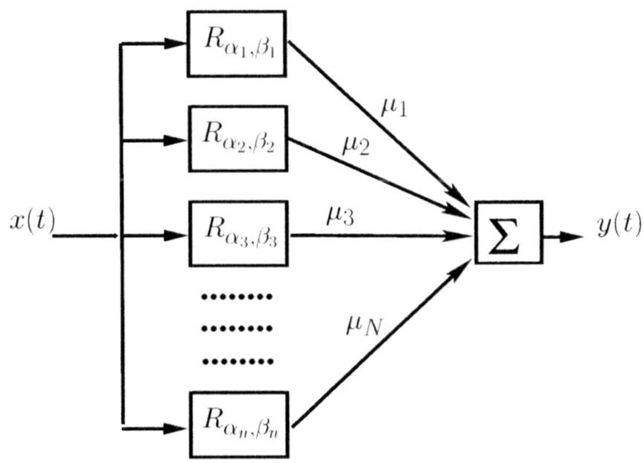

Figure 1.21. *Weighted parallel connection of a finite number of non-ideal relays*

As shown in Fig. 1.21 the same input, $x(t)$, is fed into each of the non-ideal relays (transducers) that are connected in parallel. Each output is then multiplied by its corresponding weight μ_j as defined above. The output of the entire system then, is the sum as shown in the equation

$$y(t) = y[t_0, \eta_0](t) = \sum_{j=1}^{n} \mu_j R^j[t_0, \eta_0(j)]x(t), \qquad t \geq t_0.$$

To get a visualization of how this parallel connection of non-ideal relays works, we shall represent the situation graphically in two ways. The first is as in Fig. 1.21. In this situation the relays are stacked one on top of another. This situation is shown in Fig. 1.22(a) for 3 random relays in parallel. As mentioned the same input is fed into each relay. The reader can imagine each relay being switched on in turn and visualize the output increasing as each relay is switched on. Fig. 1.22(b) then shows the 3 relays as points on the Preisach plane, the dynamics of which were introduced above. The result of identical inputs, $x(t)$, to either 'block diagram' (the relays or the Preisach plane) with identical initial states will be identical outputs, $y(t)$, for all time $t > t_0$.

Though it may seem like overkill to present both representations for the case of 3 relays in parallel, it is simply an illustration to show that each relay can indeed be represented on the Preisach plane. The power of the Preisach plane representation will become apparent in sections 1.2.6.

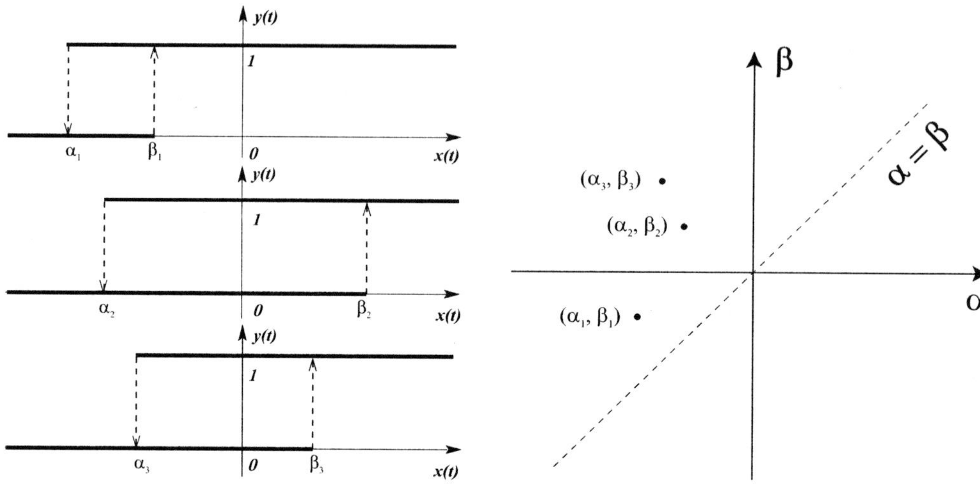

(a) Input-output representation (b) Displayed on the Preisach plane

Figure 1.22. *Parallel connection of 3 non-ideal relays*

1.2.5 Preisach model

Definition

Consider a family $\mathcal{R}$ of relays $R^\omega = R_{\alpha_\omega,\beta_\omega}$ with threshold values α_w, β_w, $w \in \Omega$. The index set Ω may be finite or infinite, and we shall call such a family a *bundle* of relays. Suppose that the set Ω is endowed with a probability measure μ. We further suppose that both functions $\alpha_\omega, \beta_\omega$ are measurable with respect to μ.

We call any measurable function $\eta(\omega) : \Omega \to \{0, 1\}$ the *initial state* of the bundle $\mathcal{R}$.

For any initial state $\eta_0(w)$ and any continuous input $x(t)$, $t \geq t_0$, we define the function

$$y(t) = y[t_0, \eta_0](t) = \int_\Omega R^\omega[t_0, \eta_0(\omega)]x(t)\, d\mu, \qquad t \geq t_0. \tag{1.13}$$

We refer to this model as a *Preisach nonlinearity* (or when convenient, a Preisach model or Preisach operator). Here $y(t)$ is the output of the Preisach model from the initial state η_0 and the input $x(t)$.

If the measure μ has a finite support $w_1, \ldots, w_n$, then the definition (1.13) may be rewritten as a parallel connection of the relays $R^{\omega_j} = R_{\alpha_{\omega_j},\beta_{\omega_j}}$ with the weights $\mu_j = \mu(\omega_j) > 0$:

$$y(t) = y[t_0, \eta_0](t) = \sum_{j=1}^{n} \mu_j R^{\omega_j}[t_0, \eta_0(\omega_j)]x(t), \qquad t \geq t_0.$$

Thus, from the point of view of system theory, the Preisach nonlinearity cor-

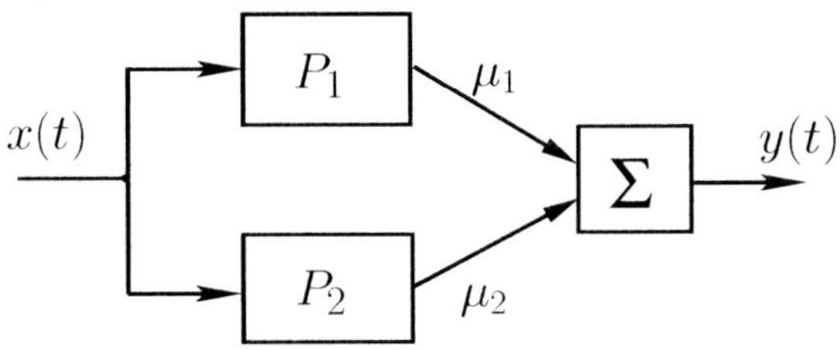

Figure 1.23. *Parallel connection of two Preisach transducers*

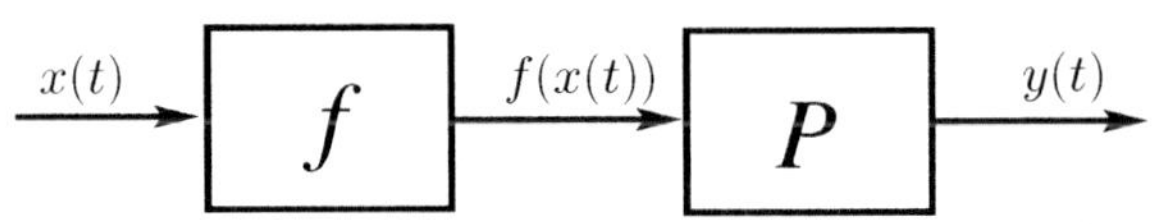

Figure 1.24. *Cascade connection of a function and a Preisach transducer*

responds to a parallel connection of a continuous bundle of non-ideal relays.

For $t > t_0$, it is also convenient to interpret the function

$$\eta(\omega) : \omega \to R^{\omega}[t_0, \eta_0(\omega)]x(t)$$

as the *state* of the Preisach nonlinearity at the moment t.

Although the individual relays are discontinuous in any reasonable sense, the Preisach operators are often continuous with respect to the uniform norm and have other nice properties, see [28].

The class of Preisach transducers is closed with respect to some natural operations. For instance, the weighted parallel connections of a number of Preisach transducers P_i, with positive weights, is a Preisach transducer, see Fig. 1.23. This fact follows immediately from the definitions.

We also note the following useful fact.

Proposition 1.2.1. *Let the function $f(x)$, be strictly increasing for $x_- \leq x \leq x_+$, and $g(x)$ be strictly positive on the same interval. Then the cascade connection Pf of a functional link $f : x \to f(x)$ with a Preisach transducer P, as shown in Fig. 1.24, is a Preisach transducer (with another measure μ). Also the transducer $g * P$, which multiplies the output $y(t)$ of the Preisach transducer by factor $g(x(t))$, is a Preisach transducer.*

This fact is nontrivial. It follows from the so-called Identification Principle , see below. We note that generally the connection fP is not a Preisach transducer.

The Preisach model was first suggested almost two thirds of a century ago [47] to describe ferromagnetism. The same model was independently invented and extensively studied in the 1950's in relation to adsorption hysteresis by D. H. Everet, see [14, 15, 16, 17]. Hysteresis in phase transitions is another area where Preisach type models have been used successfully. The monograph [9] is devoted to this topic. In the remainder of this section we mention briefly more areas where there is a convincing naive explanation of the applicability of the Preisach model.

First Impressions Last: Preisach Model in a Social Context

It has been noted that hysteresis helps to describe the dynamics of many physical and technical systems. Another area in which hysteresis may be involved is in a social context.

The saying goes that: 'first impressions last' and this is exactly what happens with a non-ideal relay. Once a person's mind is made up, one way or the other, it is more difficult to change it to the opposite than it is to convince him in the first place. Obviously it is a discontinuous situation when a person changes his mind.

So, take as an example the business concept of *goodwill* – the idea that there can be a monetary value attached to the customers that a business presently has. The idea goes far beyond that fact that the customers buy goods from the agent, to the respect that the customer has for that particular agent. It is easy to imagine that the first time a consumer becomes a customer of a particular firm and is satisfied with the goods or services provided that they would prefer to bring their custom to that firm. So, that consumer has goodwill towards that firm.

When a consumer becomes dissatisfied with their present supplier they will take their custom elsewhere. The level of satisfaction at which a consumer abandons their supplier will not be the same as that at which they first became a customer. Thus the satisfaction-custom relation can be represented by a non-ideal relay.

$$y(t) = R_{\alpha,\beta}[t_0, \eta_0]x(t) = \begin{cases} \eta_0, & \text{if } \alpha < x(\tau) < \beta \text{ for all } \ \tau \in [t_0, t]; \\[2ex] 1, & \text{if there exists } t_1 \in [t_0, t] \text{ such that} \\ & x(t_1) \geq \beta,\ x(\tau) > \alpha \text{ for all } \tau \in [t_1, t]; \\[2ex] 0, & \text{if there exists } \ t_1 \in [t_0, t] \text{ such that} \\ & x(t_1) \leq \alpha,\ x(\tau) < \beta \text{ for all } \tau \in [t_1, t]. \end{cases}$$

Here η_0 represents the initial inclination of the consumer, he may have purchased goods from the firm but has not developed any loyalty to the firm. Enough trust has not yet been built up. 0 then means that that particular consumer is not a customer of the firm. At some point he was indeed a customer but became so dissatisfied that he ceased to do business with the firm. The output value of 1 obviously means that that consumer is a customer of the firm and does have goodwill toward the firm.

The amount of goodwill that a consumer has toward a particular firm varies from one customer to another. One consumer is easily pleased, say, while another is more difficult to satisfy. The width of the non-ideal relay $r = \beta - \alpha$ is representative of how difficult it is to please a given consumer.

One might expect that the switch up and switch down satisfaction rates would be symmetric about zero satisfaction, but the asymmetry can be justified by the interaction between consumers. Those consumers with $|\beta| > |\alpha|$ are influenced by customers that are happy with the firm and those for which $|\beta| < |\alpha|$ are influenced by consumers that are dissatisfied with the firm. In the same way, as the satisfaction of present customers increases they will influence those that are not presently customers of the firm. i.e. as satisfaction increases so does the customer base of the firm. As mentioned earlier the amount of goodwill that a customer has for that firm depends on the consumer.

Summing over all consumers then leads to the total value of goodwill. It is easy to see that this summation is identical to some Preisach type model similar to those presented in the previous chapter. The probability distribution of the model would most likely however not be so simple as the grid points of the discrete model described earlier. The majority of consumers would have close to symmetric 'up' and 'down' satisfaction rates, thus the relays would be concentrated close to the line $\alpha = -\beta$. Neither would there be a vast difference in the width of the relays.

Consider also slander. A mere retraction of the slur against a person or group is of course not a satisfactory resolution. The damage has been done, people will remember the slur. It is like trying to un-ring a bell. Many people have turned against the person and will not return their favor until a great deal of trust is again built up, again like the non-ideal relay.

Of course some time dependence is required to model these situations since few of us are blessed with perfect memories.

If a situation arose that the popularity of a person were crucial at a particular time, would there be an optimum time at which to do the 'good deed'? In a election for example, would there be a perfect time in the campaign to release a particular piece of news, such that its impact induces the electorate to vote for you. Say, for example, that oil prices suddenly dropped or that taxes were cut just before election day. Even though they might have increased dramatically during the tenure of the present government the meagre drop might be fresh in the mind of the electorate and outweigh the steady increase that went before. See Chapter 2 and the references therein for further discussion on hysteresis in some sociological situations.

Preisach model in hydrology

This section introduces the idea of hysteresis in hydrology.

So-called *soil-moisture hysteresis* is important in terrestrial hydrology. The essence of this phenomenon is that "Less mechanical work is required to insert water into unsaturated soil than to remove it - good for agriculture!" [23].

It has been shown experimentally, as far back as 1941 [51], that there is a hysteresis effect in the relation between water retention and soil-moisture tension. The hysteresis effect is evident in the Soil-Water Characteristic Curve (SWCC), in that it is different depending on whether it is for wetting or drying. The origin of the hysteretic effect may be attributed to several factors [23]:

1. Geometric nonuniformity of individual pores, resulting in the so called "ink bottle effect".
2. Different spatial connectivity of pores during drying and wetting.
3. Variations in liquid-solid contact angle.
4. Air entrapment.

Important and successful mathematical descriptions of soil-moisture hysteresis have been suggested. The fundamental work on the subject appears in [1, 24, 38, 39], with further references therein. More recently some new one parameter classes of Preisach operators were introduced which were used as models of soil-moisture hysteresis for particular soils [19].

To describe the SWCC an appropriate equation is required. The most popular one is the van Genuchten equation, which takes the following form:

$$\theta = F(h) = \theta_r + (\theta_s - \theta_r)\left[1 + \left(\frac{h}{h_g}\right)^n\right]^{-m},$$

where θ is the volumetric water content, θ_r is the residual water content (in most cases this is zero), θ_s is the volumetric water content at natural saturation which is chosen as the water content scale parameter, h is the soil water pressure head (also known as the matric potential), where this variable is taken to be negative and expressed in cm of water, and h_g is the van Genuchten pressure head scale parameter. The two dimensionless water retention shape parameters, m and n, are related by $m = 1 - 1/n$.

All of these parameters for many different soil types can be found in [22], where experimental and field data were successfully fitted with Van Genuchten type functions. The Van Genuchten type Main Drying curves can be used as the basis for a Preisach model, which may be examined quantitatively. Note that the Van Genuchen form for the curves discussed is not the only possible equation to describe them. Reference should be made to [22] and the bibliography therein for further details.

1.2.6 Controllable Preisach nonlinearities

The analysis of the Preisach nonlinearity P can be reduced to the case when the set Ω is the Preisach half-plane $\Pi = \{(\alpha, \beta) : \beta > \alpha\}$ (see [28]). Moreover, it is sufficient to consider the case when the initial and varying states η_0, $\eta(t)$ are characteristic functions of sets S, which look like the shaded set in Fig 1.26. That is, S is located to the left of the line $\alpha = \beta$ and below a continuous piece-wise linear curve L, whose links are parallel to one of the coordinate axes. (If the measure μ is not absolutely continuous the horizontal links of L, excluding their right endpoints, must belong to S, whereas the vertical links, excluding their bottom endpoints, do not belong to S.) Additionally, in this standard case, the coordinates of the intersection of L with the diagonal coincide with the current value of the input x. This case describes all Preisach nonlinearities which are controllable in a natural sense [28].

This case is convenient because the evolution of the varying states admits a simple geometrical interpretation, see Fig. 1.27, 1.28. Here the input $x(t)$, moves along the horizontal scroll-bar, and controls the point on the diagonal $\alpha = \beta$ immediately above itself. When moving toward the upper right corner, this point on the diagonal drags the horizontal line, and shades the domain below this line and above the diagonal. (For instance, if, in Fig. 1.27, x increases from the value u

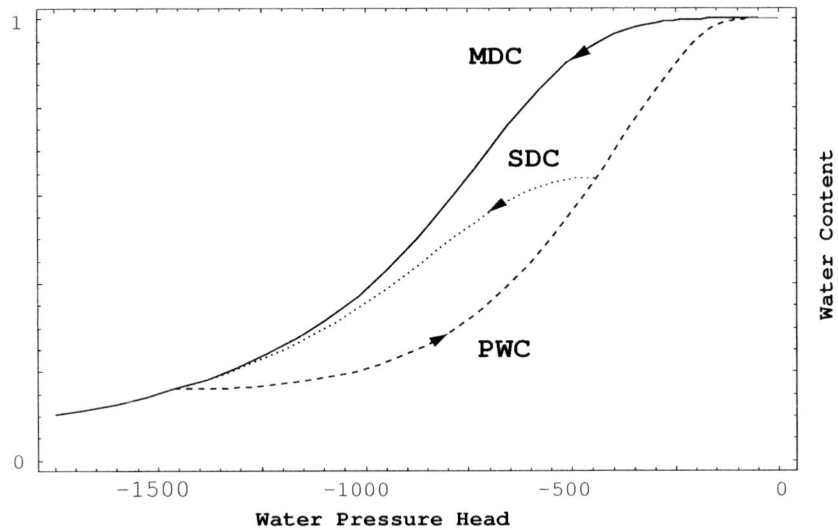

Figure 1.25. *Schematic diagram of the hysteresis model with the main drying curve (MDC), primary wetting curve (PWC) and a secondary drying curve (SDC) branching off the PWC*

to the value v, the gray area is increased by the lighter shaded triangle.) When moving towards the bottom left corner, the diagonal point drags the vertical line, and 'clears' everything to the right of this line and above the diagonal as in Fig. 1.28. The output $y(t)$ is the area of the shaded domain with respect to the measure μ.

The rules to calculate outputs of a controllable Preisach nonlinearity follow immediately from the description of a non-ideal relay presented at the end of Section 1.2.3.

We also mention the important *wiping out principle* as formulated in [31]. Each local input maximum wipes out those vertices of $L(t)$ whose α-coordinates are less than this maximum, and each local input minimum wipes out the vertices whose β-coordinates are above this minimum. That is, only the alternating series of dominant input extrema are stored by the Preisach model. All other input extrema are wiped out. See further details in [31].

Numerical implementation

For a given measure μ, it is easy to write a computer program to calculate the output of the Preisach model for an input function $x(t)$. We encourage readers to experiment with such a program via the interactive homepage

http://physics.ucc.ie/~oll/hysteresis/node8.htm

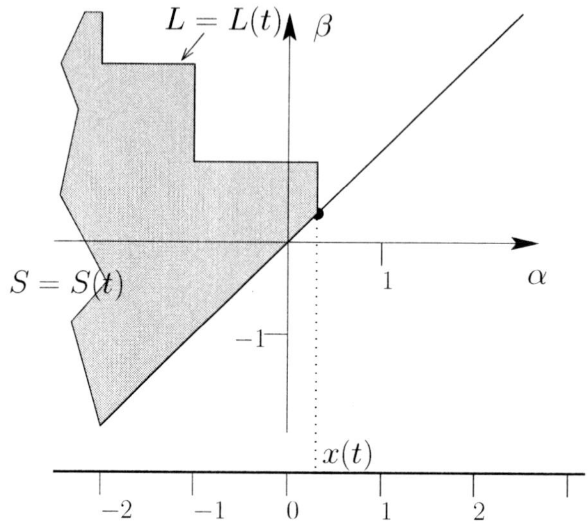

Figure 1.26. *Typical state of a Preisach model*

Fig. 1.29 gives some outputs of this program for an irregular input, and for the standard Lebesgue measure as μ.

1.2.7 Minor loops

Definition

The phrase *hysteresis loop* is commonly used in the natural sciences to indicate closed curves in the (x, y) plane associated with outputs of hysteresis nonlinearities corresponding to periodic inputs. In this section we will discuss important properties of hysteresis loops for Preisach nonlinearities.

We consider a harmonic input of the Preisach nonlinearity

$$x(t) = x(t; a, b) = \frac{a+b}{2} + \frac{a-b}{2}\cos(t), \quad t \geq 0.$$

For any initial state η, the corresponding output

$$y(t) = y(t, a, b; \eta)$$

is 2π-periodic for $t \geq \pi$.

The corresponding phase diagram $(x(t), y(t))$, $t > \pi$, is a closed curve called a *minor hysteresis loop* (with the limits a, b); denoted by $L(a, b; \eta)$. In contrast, the curve with the parametric representation $(x(t), y(t))$, $0 \leq t < \pi$, corresponds to the *transient process*, see Fig. 1.30.

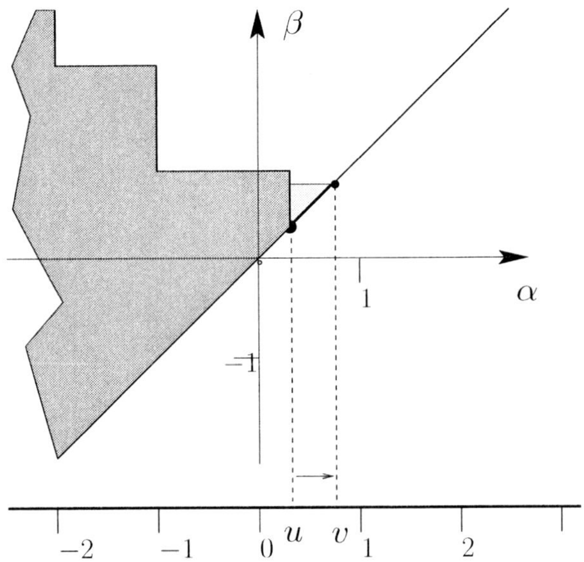

Figure 1.27. *Dynamics of the state of a Preisach model (an increasing input)*

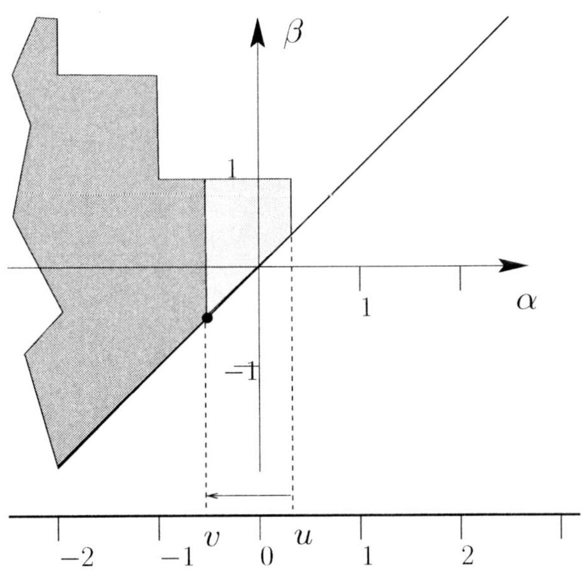

Figure 1.28. *Dynamics of the state of a Preisach model (a decreasing input)*

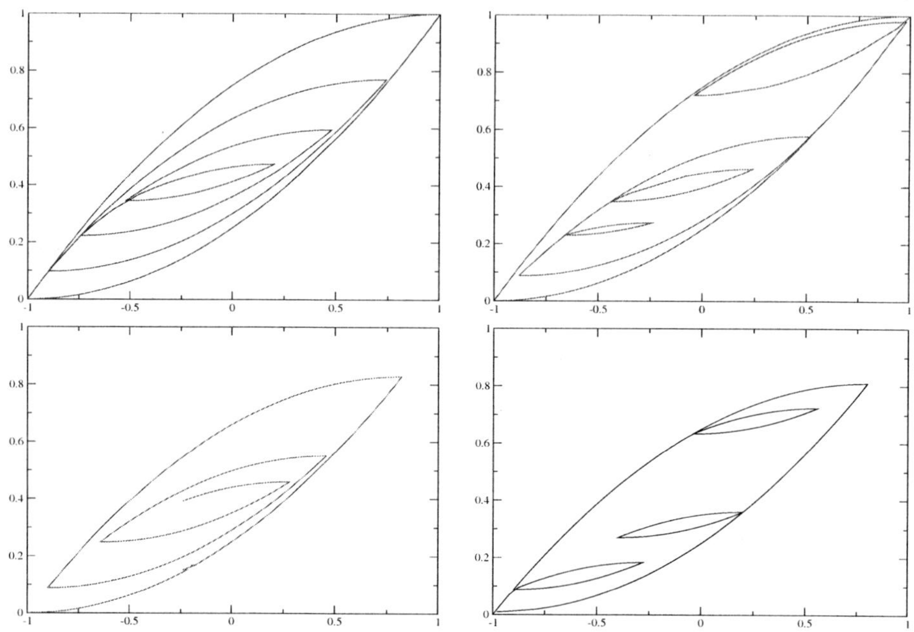

Figure 1.29. *Some typical snapshots of input-output phase diagrams for a Preisach model*

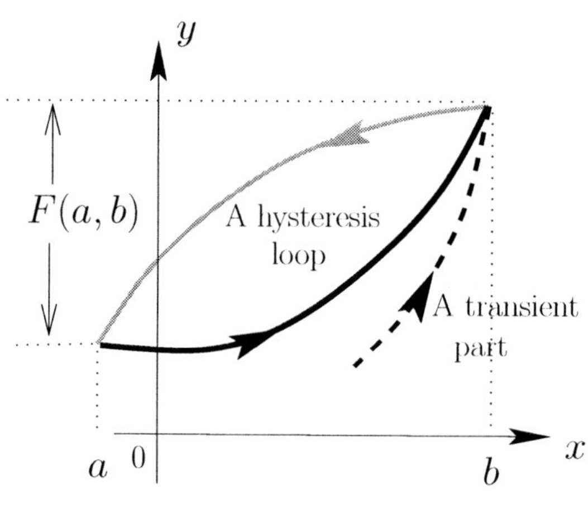

Figure 1.30. *A phase portrait of a hysteresis loop and a transient part*

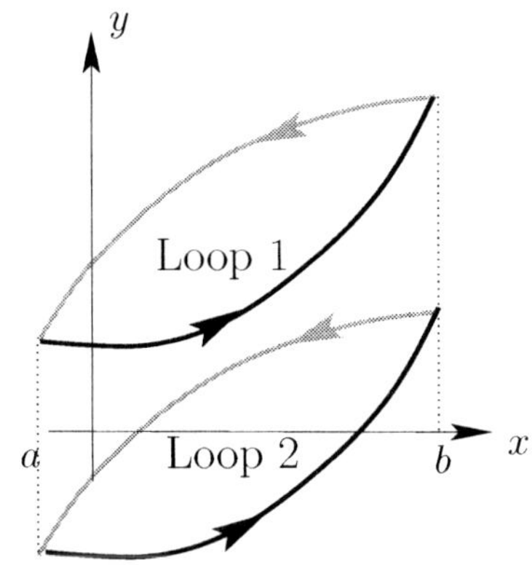

Figure 1.31. *Congruence of hysteresis loops*

While a hysteresis loop depends on the initial state $\eta_0 = \eta(0)$, the loops corresponding to different η_0 are congruent. This is the *Congruence Property* of the Preisach nonlinearity, see Fig. 1.31.

From Fig. 1.27 and 1.28 we see that the lower branch of a minor loop $L(a, b; \eta)$ is congruent to the graph of the function $y_-(x) = \mu(\Delta(a, x))$, where $\Delta(a, x)$ is the triangle with the vertices $(a, a), (a, x), (x, x)$:

$$\Delta(a, x) = \{(\alpha, \beta) : a \leq \alpha < \beta \leq x\},$$

see Fig. 1.32. In the same way the upper branch is congruent to the graph of the function $y_+(x) = -\mu(\Delta(x, b))$.

In particular, the value

$$F(a, b) = y(2\pi, a, b; \eta) \quad y(\pi, a, b; \eta) \tag{1.14}$$

is not dependent on η by the Congruence Property. It is called the *loop-magnitude function.* By definition, $F(a, b)$, $a < b$, coincides with the μ-measure of the triangle $\Delta(a, b)$:

$$F(a, b) = \mu(\Delta(a, b)) \geq 0,$$

and

$$F(a, b) = -F(b, a).$$

Moreover,

$$F(a_1, b_1) + F(a_2, b_2) - (F(a_2, b_1) + F(a_1, b_2)) \geq 0 \tag{1.15}$$

for any $a_1 \leq a_2 \leq b_2 \leq b_1$, since the expression in the left hand side coincides with the μ-measure of the rectangle $Q(a_1, b_1, a_2, b_2)$ with vertices

$$(a_1, b_1), (a_1, b_2), (a_2, b_1), (a_2, b_2),$$

see Fig. 1.32.

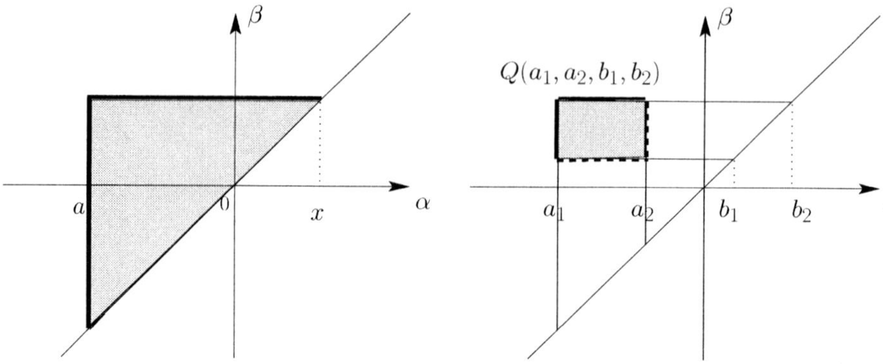

Figure 1.32. *Left: triangle* $\Delta(a, x)$ *(shaded). Right: rectangle* $Q(a_1, a_2, b_1, b_2)$ *(shaded)*

Fast settling property and extended congruence property

We consider in more detail the minor loops for an arbitrary continuous 2π-periodic input $x(t), t \geq 0$. Let τ_+ be the first time that $x(t)$ achieves its global maximum, and τ_- be the first time that $x(t)$ achieves its global minimum, and let $\tau = \max\{\tau_-, \tau_+\}$.

Lemma 1.2.1. *The output* $y(t, \eta_0)$ *is* 2π*-periodic for* $t \geq \tau$.

This assertion is called the *Fast Settling Property* of the Preisach model.

We also consider the curves, in the phase plane, given by

$$y_+(x; \eta_0) = \sup_{t: x(t) = x} y(t; \eta_0), \qquad y_-(x; \eta_0) = \inf_{t: x(t) = x} y(t; \eta_0). \tag{1.16}$$

That is, for a given value $x_* \in [\min_t x(t), \max_t x(t)]$, the number $y_+(x; \eta_0)$ is the ordinate of the highest intersection of the vertical line $x = x_*$ with the phase portrait of the trajectory $(x(t), y(t))$. Similarly, the number $y_-(x; \eta_0)$ is the ordinate of the lowest intersection of the vertical line $x = x_*$ with the phase portrait of the trajectory $(x(t), y(t))$.

The union of the curves (1.16) (or the area between them) is called the *principal minor loop*; it depends on the initial condition η_0, as well as on the function $x(t)$. The following observation holds

Lemma 1.2.2. *Any principal minor loop is congruent to a minor loop with the limits* $\min x(t)$, $\max x(t)$.

We will call this observation the *Extended Congruence Property* of the Preisach model. It is illustrated in Fig. 1.33.

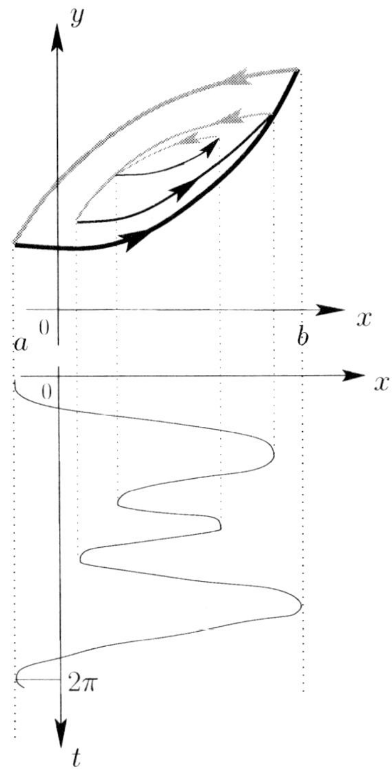

Figure 1.33. *Extended Congruence Property of Preisach model, with a principal minor loop shown by a directed bold line*

1.2.8 Identification Principle

Formulation

A Preisach model was first suggested almost two thirds of a century ago [47] to describe ferromagnetism. According to the Weiss theory, ferromagnets are composed of a large number of elementary magnets (domains). Under an applied external magnetic field $H(t)$ the state of each elementary magnet depends on the external field as well as an internal interaction with other domains which is a function of the magnetization state, $M(t)$. The resulting magnetization state contains a positive feedback mechanism which leads to hysteresis, and the Preisach model was suggested to capture some features of this process. Individual relays reflect the role of individual domains.

The approach described in the previous paragraph was based on ideas about the microstructure of ferromagnets. Formally, it does not take into account the interaction between domains and thus might seem inadequate even for this subject. However, the Preisach model appears to be extremely successful, and is now recognized as a fundamental tool in describing a wide range of hysteresis phenomena in quite different subject areas (physics, mechanics, superconductivity, tectonics, economics, etc.) The interested reader might consult the recent special issue of Physica B [41], and the bibliography therein. The modern explanation of the pivotal role of the Preisach nonlinearity is based on an operational, phenomenological approach and on special *identification theorems.* It will be briefly discussed in this subsection.

We begin with the following observation: careful physical experiments show the following for ferromagnetic hysteresis in a wide range of experimental conditions. *The input-output correspondence satisfies the Extended Congruence Property, and the Fast Settling Property. Moreover, the corresponding loop-magnitude function $F(a, b)$ satisfies the inequality (1.15).*

Thus, Preisach nonlinearities can be used to describe some ferromagnetic phenomena. However, the Preisach model has much wider and fundamental implications. The following fundamental Principle is (in a different form) due to I. Mayergoyz, see [31].

Identification Principle. *Let a rate-independent transducer Γ satisfy the Extended Congruence Property and the Fast Settling Property. Let, further, the corresponding function (1.14) satisfy the inequality (1.15). Then Γ can be identified with a parallel connection of a suitable Preisach nonlinearity and a functional transducer $x \to \varphi(x)$.*

Thus, if a rate independent phenomenon satisfies the conditions of the Identification Principle, within the accuracy of experimental observations, then that phenomenon can be identified with an appropriate Preisach model. For example, this guarantees that the Preisach model describes a range of plastic hysteresis phenomena, embracing the classical Prandtl, Besseling, and Ishlinskii models.

Discussing the Identification Principle

The Identification Principle , as formulated above, is not yet a rigorous mathematical theorem. However, we will give a "physical argument" for why it should hold, and also formulate the simplest rigorous statement related to this principle.

The informal justification of the principle can be divided into four steps.

Step 1. As explained above, *for the Preisach model* the combination

$$F(a_1, b_1) + F(a_2, b_2) - (F(a_2, b_1) + F(a_1, b_2))$$

coincides with the μ-measure of the rectangle $Q(a_1, b_1, a_2, b_2)$ with vertices

$$(a_1, b_1), (a_1, b_2), (a_2, b_1), (a_2, b_2).$$

On this basis, we *define* the measure μ by the equality

$$\mu\,(Q(a_1, b_1, a_2, b_2)) = F(a_1, b_1) + F(a_2, b_2) - (F(a_2, b_1) + F(a_1, b_2)).$$

Step 2. Now we define the function φ, up to an additive constant, as the function $\psi(x)$ given by

$$\psi(x) = \sum_{n=1}^{N} F\left(\frac{(n-1)x}{N}, \frac{nx}{N}\right)$$

for large N (or, more rigorously, as the corresponding limit as $N \to \infty$).

Step 3. Using the extended congruence property, it is easy to demonstrate that the settled phase portrait of any hysteresis loop for the transducer Γ is congruent to the hysteresis loop of the parallel connection $P_\mu + \varphi$.

Step 4. In a physical paradigm only repeatable input-output relationships are of interest; that is, we can concentrate on controllable transducers. But this means that each fragment of the input-output pair $(x(t), y(t))$ can be realized as part of a settled periodic correspondence.

It is important that the first two steps provide a basis for the approximate identification of the measure μ and the function φ.

The task of converting this plausible justification of the Identification Principle into a rigorous mathematical theorem would seem to be a nontrivial task. We now give the simplest rigorous statement related to the Identification Principle.

Let B be a black box. Further, let the corresponding function (1.14) satisfy the inequality (1.15) with continuous inputs. In mathematical language this black box is the totality $\mathcal{B}$ of possible input-output pairs $(x(t), y(t))$. Let $\mathcal{X}(M)$ be the set of admissible continuous inputs $x(t)$, $t \geq 0$, satisfying the estimate $|x(t)| \leq M$, and where the possible corresponding outputs $y(t)$, $t \geq 0$ are continuous. Suppose that B has the Fast Settling Property and the Extended Congruence Property for these inputs. Then we can introduce the loop-magnitude function $F(a, b)$, $|a|, |b| \leq M$, by (1.14). Let us suppose that this function satisfies the inequality (1.15), and that for some $\lambda > 0$ the Lipschitz estimate $F(a, b) \leq \lambda |a - b|$ holds.

Consider the subset $\mathcal{X}_0$ of $\mathcal{X}(M)$ consisting of the functions which satisfy the equalities

$$x(0) = -M, \quad x(1) = M.$$

Theorem 1. *There exist a measure μ on the triangle $-M \leq \alpha < \beta \leq M$ and a continuous function $\varphi(x)$ which have the following property. If $x(t) \in \mathcal{X}_0$, and $y(t)$ is a possible output of the black box for this input, then, for $t \geq 1$, the function $y(t)$ coincides with the function*

$$\varphi(x(t)) + P_\mu[1, \eta]x(t),$$

where the initial state satisfies $\eta(\alpha, \beta) \equiv 1$, $-M \leq \alpha < \beta \leq M$.

1.2.9 Closed loop systems with hysteresis

Basic example

The strength of the Preisach model resides in its combination of physical generality and mathematical simplicity. These features are especially important when analyzing closed loop systems with hysteresis. As an illustration we return to the forced oscillation of an iron pendulum in a magnetic field, Fig. 1.34.

For small oscillations, this is described by the equation

$$\ddot{x} + a\dot{x} + x = \sin(\omega t) + y(t), \quad y(t) = \Gamma x(t), \tag{1.17}$$

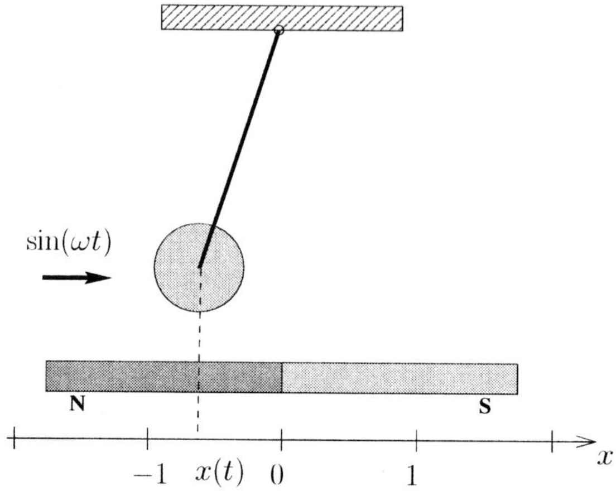

Figure 1.34. *An iron pendulum in magnetic field*

where $x(t)$ is the displacement from its equilibrium position at $x = 0$, and the nonlinearity $\Gamma x(t)$ describes the interaction between the external magnetic field and the magnetized pendulum itself. Since the ferromagnetic substance is an aggregate of dipoles, this equation for $t \geq t_0$ can be rewritten as

$$\ddot{x} + a\dot{x} + x = \sin(\omega t) + H'(x)M(t).$$

Here $H(x)$ is magnetic field at the point x, and $M(t)$ is the resulting magnetization of the pendulum. As we mentioned above, $M(t)$ can be often represented as

$$M(t) = P_\mu[t_0, \eta_0]H(x(t)),$$

where P_μ is the Preisach operator with an appropriate measure μ. Let us suppose further that the function $H(x)$ is monotone. Then we can use Proposition 1.2.1, and rewrite our equation in a simpler form

$$\ddot{x} + a\dot{x} + x = \sin(\omega t) + P_{\tilde{\mu}}[t_0, \eta_0]x(t),$$

with a new measure $\tilde{\mu}$.

We emphasize once more that equation (1.17) is a differential-operator equation, not a purely differential one.

Numerical experiments

We have an algorithm and a computer program to calculate the Preisach operator. Thus we can immediately calculate trajectories of equation (1.17) numerically, using modifications of Euler or Runge-Kutta type methods. For instance, Fig. 1.35, 1.36 give a picture of a numerical solution of equation (1.17) with $a = 0$ i.e., no damping, and we note the presence of beats when hysteresis is absent. Here we have chosen $\omega = \sqrt{2}$, and μ proportional to the Lebesgue measure.

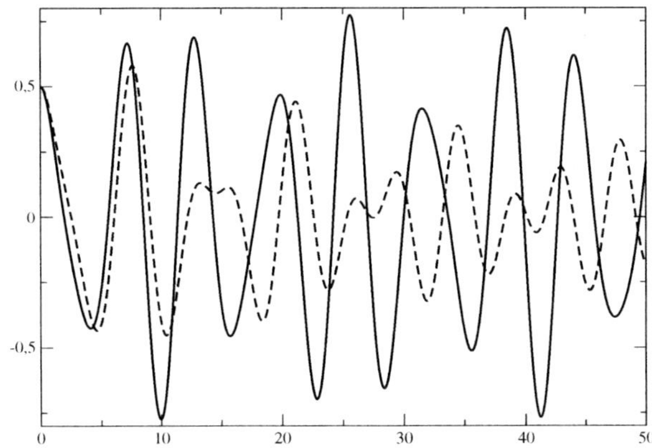

Figure 1.35. *Pendulum with (dashed), and without (solid), hysteresis; early stage of oscillations*

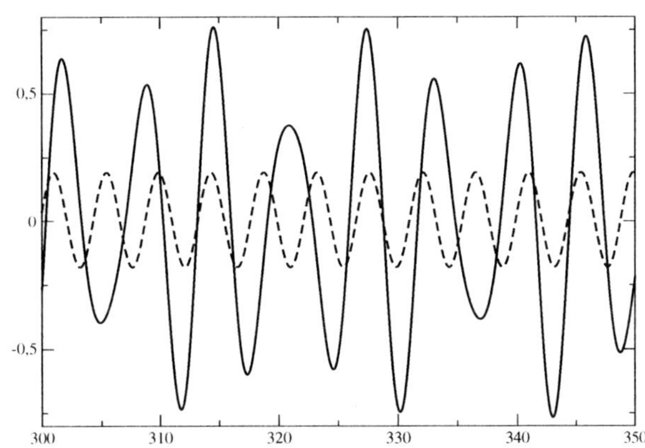

Figure 1.36. *Pendulum with (dashed), and without (solid), hysteresis; late stage of oscillations*

We observe that the solution of the equation with hysteresis approaches a periodic function with the frequency ω. Readers can proceed with further experiments at the interactive homepage

http://physics.ucc.ie/~oll/hysteresis/node22.htm

Asymptotic analysis: non-resonant case

Since the qualitative properties of the Preisach model are nowadays well understood, we can investigate rigorously the qualitative features of equations with hysteresis.

We consider the equation

$$\ddot{x} + x = \sin(\omega t) + \varepsilon y(t), \quad y(t) = \Gamma x(t), \tag{1.18}$$

where the forcing frequency ω is irrational, and ε is a small positive parameter. This equation describes non-resonant forced oscillations of a ferromagnetic pendulum in a weak magnetic field with negligible viscous damping. It is well known that solutions of the unperturbed equation are beats i.e., the sums of two periodic functions with different frequencies, and there exists only one $2\pi/\omega$-periodic solution given by

$$x_0(t) = \frac{1}{1-\omega^2}\sin(\omega t). \tag{1.19}$$

The principal question here is whether the role of hysteresis is similar to that of the standard damping given by a term like $a\dot{x}$. Both physical reasons and the results of numerical experiments, see Fig. 1.35, 1.36, suggest the answer is in the affirmative. The theorem below has been proved rigorously.

Theorem 2. *Let ω be irrational and $\varepsilon > 0$ be small. Then any solution of the equation (1.18) is attracted to a small neighbourhood of the function (1.19).*

We emphasize that this result requires an understanding of subtle properties of the Preisach nonlinearity in the class of irregular, beat-like inputs. Some of the ideas can be found in [20].

We will return to the resonant case in Subsection 1.4.3.

Asymptotically stable oscillations

The next example deals with the equation

$$\ddot{x} + a\dot{x} + x = \varphi(t) + y(t) + f(x), \quad y(t) = \Gamma x(t).$$

Theorem 3. *Let $a \geq 2$, $\varphi(t)$ be T-periodic, and f be monotone and bounded. Then there exists a stable T-periodic solution $x(t)$. If additionally, f is a real analytic function, and Γ is a parallel connection of a finite number of relays, then there exists an* asymptotically *stable T-periodic solution.*

This is a particular case of results established in [43]. We note that the solution $x(t)$ in Theorem 3 is well defined (see the last paragraph in Subsection 1.2.3): it does not have local minima equal to α_k or local maxima equal to β_k, where α_k, β_k are the thresholds of the corresponding relays.

The proof of Theorem 3 relies upon the monotonicity property of the non-ideal relay, as described at the end of Subsection 1.2.3, and upon special theorems about solutions of equations with monotone discontinuous nonlinearities [56]. These theorems were originally designed for systems with non-ideal relays. However they seem to be of much broader applicability. For example, the assertion of Theorem 3 appears to be new and unexpected even in the case when no relays are present;

that is, the theorem applies to

$$\ddot{x} + a\dot{x} + x = \varphi(t) + f(x).$$

In this case, we can state

Corollary 1.2.1. *Let $a \geq 2$, $\varphi(t)$ be T-periodic and f be monotone and bounded. Then there exists a stable T-periodic solution $x(t)$. If additionally, f is a real analytic function, then there exists an* asymptotically *stable T-periodic solution.*

Under the conditions of Corollary 1.2.1 it is possible that there exist many T-periodic solutions, some of which may be unstable.

1.3 Singular Perturbation Phenomena

In this section, via extremely simple examples, some phenomena of singular perturbation theory are demonstrated. We have focussed our attention on piecewise linear systems for a number of reasons. From the methodological standpoint, systems of this kind are convenient because they are integrable. Such systems, on the other hand, are used extensively in the modelling of a wide range of physical processes and, more importantly, such systems are presented as continuous models of hysteresis-like behaviour.

1.3.1 Initial layer

Consider the planar initial value problem

$$\begin{aligned} \dot{x} &= y, x(0) = x_0; \\ \varepsilon\dot{y} &= -ax - by, y(0) = y_0, \end{aligned} \tag{1.20}$$

on $t \geq 0$, where $a, b(b \neq 0)$ are constants, and for some small positive parameter ε.

Setting $\varepsilon = 0$, we obtain the *degenerate problem*:

$$\begin{aligned} \dot{x} &= y, x(0) = x_0; \\ 0 &= -ax - by. \end{aligned} \tag{1.21}$$

The *slow curve* is described by the equation

$$0 = -ax - by.$$

The unique solution to (1.20) is

$$x(t) = C_1 e^{\lambda_1 t} + C_2 e^{\lambda_2 t},$$

$$y(t) = C_1\lambda_1 e^{\lambda_1 t} + C_2\lambda_2 e^{\lambda_2 t},$$

where

$$\lambda_{1,2} = \frac{-b \pm \sqrt{b^2 - 4\varepsilon a}}{2\varepsilon},$$

$$C_1 = \frac{\lambda_2 x_0 - y_0}{\lambda_2 - \lambda_1}, \quad C_2 = \frac{y_0 - \lambda_1 x_0}{\lambda_2 - \lambda_1}.$$

An important role is played by two trajectories, which are the straight lines

$$y = \lambda_1 x$$

and

$$y = \lambda_2 x$$

Using the asymptotic representations

$$\lambda_1 = -a/b - \varepsilon a^2/b^3 + O(\varepsilon^2),$$

$$\lambda_2 = \varepsilon^{-1}[-b + \varepsilon a/b + O(\varepsilon^2)]$$

it is easy to see that the trajectory $y = \lambda_1 x$ is *attractive* when $b > 0$ and *repulsive* when $b < 0$. Note that the solution of the degenerate problem (1.21) can be considered as a *limiting solution* (as $\varepsilon \to 0$) with respect to the solution of original problem (1.20) when $b > 0$; see [35, 36, 55].

It is possible to say that $y = \lambda_1 x$ is the *slow integral manifold* and $y = \lambda_2 x$ is the *fast integral manifold* [53] (see, for details, Chapters 7 and 8 in this volume and references therein). Any trajectory of (1.20) can be represented as a trajectory on the attractive slow integral manifold plus an *asymptotically negligible* term when $b > 0$ (see Fig. 1.37 which demonstrates that trajectories go through the slow curve and approach the slow integral manifold).

1.3.2 Jump point

Consider the following piecewise linear differential system

$$\dot{x} = 1,$$
$$\varepsilon \dot{y} = x + |y|,$$

corresponding to the phase plane equation

$$\epsilon \frac{dy}{dx} = x + |y|.$$

The slow curve is described by the equation

$$x + |y| = 0,$$

and consists of an attractive part ($y < 0$) and a repulsive one ($y > 0$), which are separated by the *jump point* $x = y = 0$.

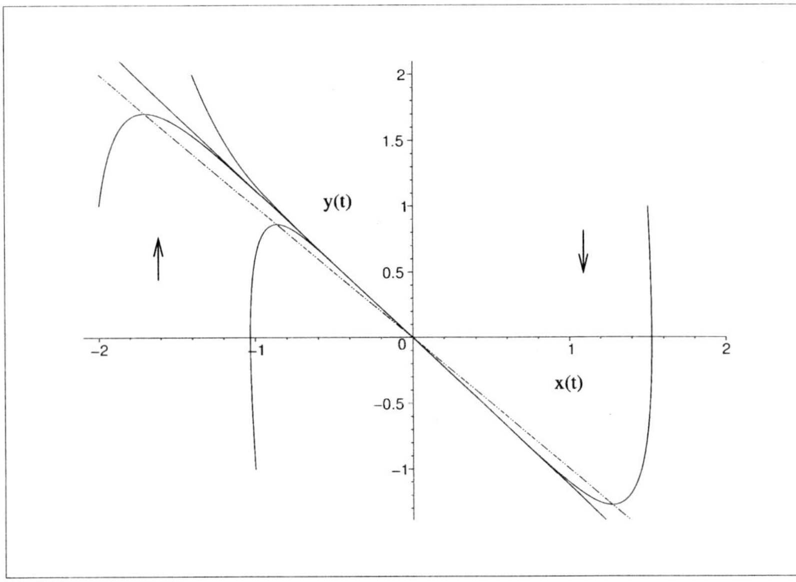

Figure 1.37. *Trajectories (solid lines) going through the slow curve (dotted line) and approach the slow invariant manifold (solid straight line)($a = b = 1, \varepsilon = 0.1$)*

As usual, an important role is played by the attractive slow invariant manifold

$$y = x - \varepsilon, y < 0$$

and the repulsive slow invariant manifold

$$y = -x - \varepsilon, y > 0,$$

hand in hand with their extensions

$$y = \begin{cases} x - \varepsilon, & x < \varepsilon \\ 2\varepsilon e^{(x-\varepsilon)/\varepsilon} - x - \varepsilon, & x \geq \varepsilon, \end{cases}$$

and

$$y = \begin{cases} -x - \varepsilon, & x < -\varepsilon, \\ 2\varepsilon e^{-(x+\varepsilon)/\varepsilon} + x - \varepsilon, & -\varepsilon < x < \varepsilon\nu, \\ \varepsilon(1+\nu)e^{(x-\nu\varepsilon)/\varepsilon} - x - \varepsilon, & \varepsilon\nu < x, \end{cases}$$

where ν is the root of $2e^{-1-\nu} + \nu - 1 = 0$ (see Fig. 1.38).

The phenomenon of a jump point is common in the theory of relaxation oscillations [32].

1.3.3 Canard trajectory

In some specific cases it is possible to glue together the attractive and the repulsive slow integral manifolds with the result that an attractive/repulsive trajectory

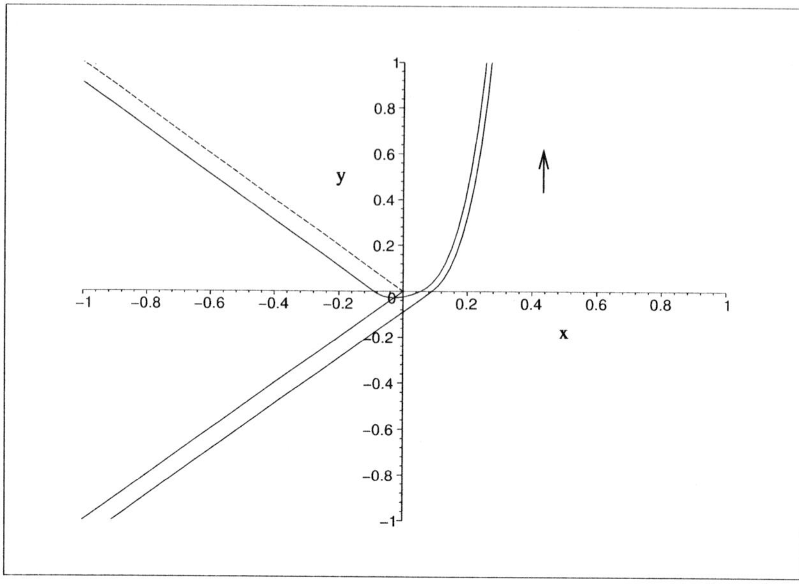

Figure 1.38. *The slow curve (dotted line), slow invariant manifolds (solid straight lines) and their extensions (solid lines)*($\varepsilon = 0.09$)

appears. Such behaviour can be illustrated with the system

$$\dot{x} = -y,$$
$$\varepsilon \dot{y} = ax + f(y),$$

where $a \geq 0$ and

$$f(y) = \begin{cases} \varepsilon p, & |y| < \varepsilon p \\ |y|, & |y| \geq \varepsilon p. \end{cases}$$

The slow curve is described by the equation $ax + f(y) = 0$. It is a straightforward exercise to verify that the curve

$$F(x, y, \varepsilon) = 0,$$

where F is the function

$$F(x, y, \varepsilon) = \begin{cases} (ax + \varepsilon p)^2/a + \varepsilon y^2 - \varepsilon^2 p^2 a(1/a - 1/\nu)^2 - \varepsilon^3 p^2, & |y| < \varepsilon p \\ |y| + \nu x, & |y| \geq \varepsilon p, \end{cases}$$

with

$$\nu = (1 - \sqrt{1 - 4a\varepsilon})/2\varepsilon,$$

is a trajectory of the differential system under consideration. This trajectory consists of three parts: attractive ($y < 0$) and repulsive ($y > 0$) slow integral manifolds given by

$$|y| + \nu x = 0, \ |y| \geq \varepsilon p,$$

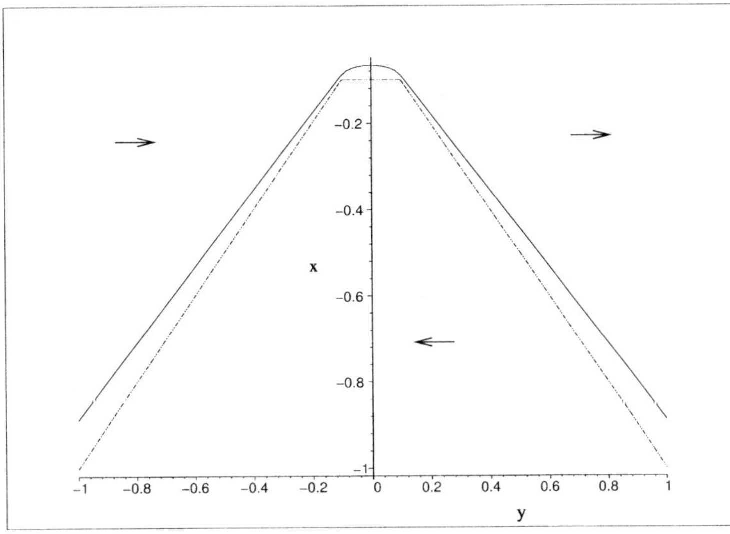

Figure 1.39. *The slow curve (dotted line) and the trajectory (solid line) ($a = p = 1$, $\varepsilon = 0.1$)*

and the curve

$$(ax + \varepsilon p)^2/a + \varepsilon y^2 = \varepsilon^2 p^2 a(1/a - 1/\nu)^2 + \varepsilon^3 p^2, \; |y| < \varepsilon p$$

which vanishes as $\varepsilon \to 0$ (see Fig. 1.39, the polygonal path corresponds to the slow curve of the system under consideration). Moreover, linearity makes the system solvable.

Such behaviour is an example of a fascinating phenomenon called a *canard* (see [3, 4, 5], and Chapters 4 and 8 of this volume with the references therein). More precisely, trajectories which at first pass along the stable integral manifold and then continue for a while along the unstable integral manifold are called *canards*. Trajectories of other types are displayed in Fig. 1.40.

The slow curve together with the three different canard trajectories are displayed in Fig. 1.41.

The situation considered above is not typical of canard theory, since we did not need an additional parameter to demonstrate the existence of canards. Usually a canard corresponds to a specific (bifurcation) value of an additional parameter. Such a situation will be considered in subsection 2.5 of this chapter. A more detailed analysis of systems without additional parameters will be presented in Chapter 8 (see Examples 2–5, therein).

1.3.4 Relaxation oscillations and hysteresis-like behaviour

The purpose of this subsection is to demonstrate a relaxation oscillation phenomenon using a piecewise linear model. The name relaxation oscillation was used by

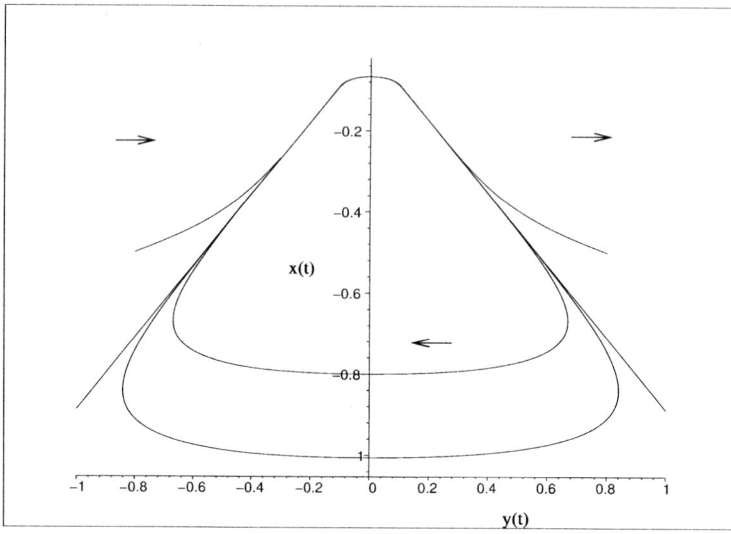

Figure 1.40. *Typical trajectories* $(a = p = 1,\ \varepsilon = 0.1)$

Balthasar van der Pol [48], who investigated the differential equation, now known as the van der Pol equation, as an application of the modelling of triode valves and heartbeats. A relaxation oscillation is characterized by an abrupt motion in a periodic orbit which exhibits two distinct and characteristic phases: one during which energy is stored up slowly and the other in which the energy is discharged nearly instantaneously when a certain critical threshold potential is attained. The mathematical theory of relaxation oscillations is set forth in Mishchenko and Rozov [32].

Consider the following piecewise linear differential system:

$$\dot{x} = y,$$

$$\varepsilon\dot{y} = -x + f(y),$$

where

$$f(y) = \begin{cases} -my - 1 - m, & y < -1, \\ y, & y \in [-1, 1], \\ -my + 1 + m, & y > 1, \end{cases}$$

ε is a small positive parameter, and m is a positive parameter This is a system with typical hysteresis behaviour (see Fig. 1.42 in which the behaviour of trajectories for different values of the parameters is exhibited).

Canard cycle

The occurrence of canards in systems with relaxation oscillations was demonstrated in [13] with the van der Pol equation. We demonstrate this with a piecewise linear

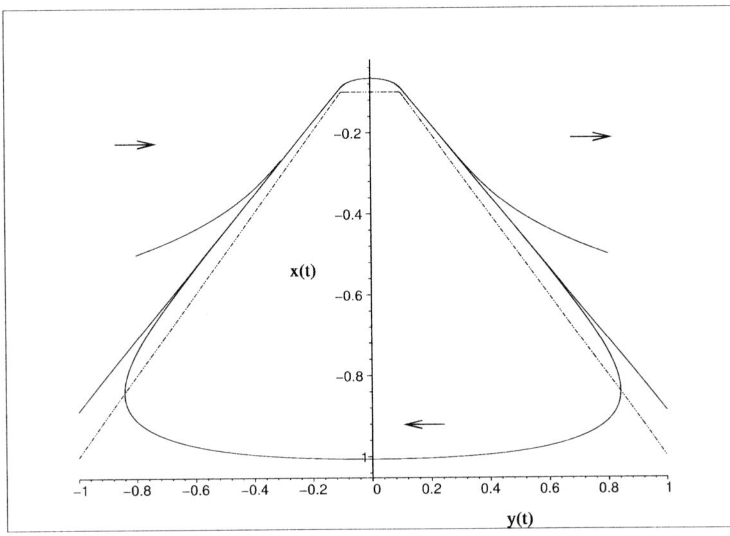

Figure 1.41. *Slow curve (dotted line) together with the three canards (solid lines) ($a = p = 1$, $\varepsilon = 0.1$)*

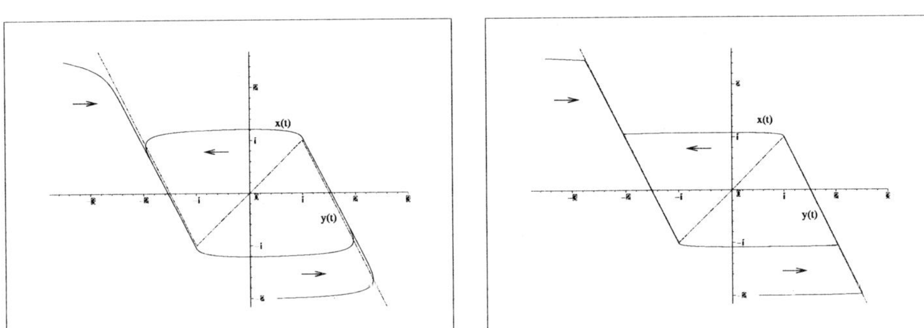

(a) $\varepsilon = 0.1$. (b) $\varepsilon = 0.02$.

Figure 1.42. *Hysteresis-type behaviour ($m = 2$). The slow curve (dotted line) and trajectories (solid lines)*

model.

Consider now the modified system with hysteresis

$$\dot{x} = y - a,$$

$$\varepsilon \dot{y} = -x + f(y);$$

with

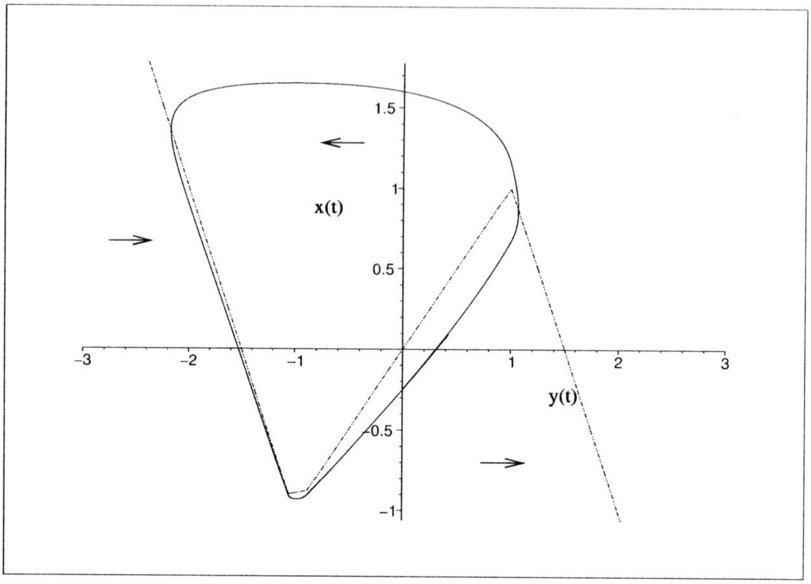

Figure 1.43. *The slow curve (dotted line) and the periodic canard trajectory (solid line) ($k = -0.226, m = 2, n = -0.12, p = 0.05, q = 0.120454545, \varepsilon = 0.2, a = -0.9765$)*

$$f(y) = \begin{cases} -my - 1 - m, & y < -1 - p, \\ -ny - 1 - k, & -1 - p \le y < -1 + q, \\ y, & -1 + q \le y \le 1, \\ -my + 1 + m, & y > 1, \end{cases}$$

where the choice $k := (n - m)(1 + p) + m; q := (n - k)/(n + 1)$ guarantees the continuity of f. It is possible to choose a value of the additional parameter a in such a way that the modified system possesses a periodic canard trajectory (see Fig. 1.43). This bifurcation value of a is called a *canard value* of the parameter. Such values of control parameters play a very important role in the construction of *separating solutions* corresponding to the critical regimes of chemical reactions (see Chapter 8).

1.4 Hysteresis vs Time Relaxation

1.4.1 Generalities

An instructive singularly perturbed or slow-fast system is the following

$$\dot{x} = f(t, x, y),$$
$$\varepsilon\dot{y} = g(x, y),$$

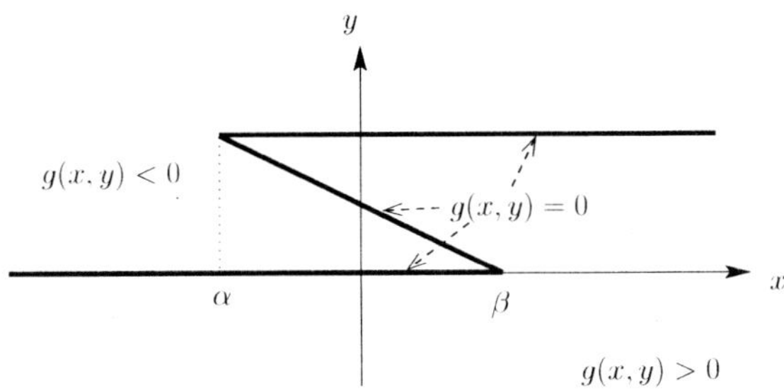

Figure 1.44. *Relay as a singularly perturbed ODE*

where $\varepsilon > 0$ is a small parameter. We concentrate first on the auxiliary equation

$$\varepsilon \dot{y} = g(x(t), y),$$

where $x(t)$ is an arbitrary function. We denote the solution of this equation, with a given initial condition $y(t_0) = y_0$, by

$$y(t) = V_\varepsilon[t_0, y_0]x(t).$$

Under suitable technical conditions it can be proved that for a generic function $x(t)$ there exists a limit

$$y_*(t) = V[t_0, y_0]x(t) = \lim_{\varepsilon \to 0} V_\varepsilon[t_0, y_0]x(t).$$

The solutions of the original system are then approximated as $\varepsilon \to 0$ by the solutions of the differential-operator equation

$$\dot{x} = f(t, x, y(t)), \quad y(t) = V[t_0, y_0]x(t). \tag{1.22}$$

The following observation is straightforward:

Lemma 1.4.1. *The limit operator V is rate-independent.*

Thus, *the global analysis of singularly perturbed equations becomes, in the first approximation, an investigation of equations with hysteresis nonlinearities.* This is the point of contact between equations with hysteresis and singularly perturbed equations. We will demonstrate how this observation works in Subsection 1.4.2

Example 1.4.1. *Let the function g be such that it can be described by the picture in Fig. 1.44.*

Suppose $x(t)$ does not have local minima equal to α or local maxima equal to β. Then the operator V is well defined on $x(t)$ and

$$V[t_0, y_0]x(t) = R_{\alpha,\beta}[t_0, y_0]x(t)$$

for all $t > t_0$ *such that*

$$x(t) \neq \alpha \quad \text{and} \quad x(t) \neq \beta.$$

(See Section 1.2.3 for the definition of $R_{\alpha,\beta}$*.)*

This example shows that *some singularly perturbed equations can approximate equations with hysteresis.* This is the second point of contact between equations with hysteresis and singularly perturbed equations. This approach is especially important if there are some physical reasons to believe that the relay (or any other discontinuous hysteresis nonlinearity) is an idealization of a limit operator similar to V. Sometimes, however, discontinuous hysteresis operators are idealizations of quite different continuous operators. The application of methods used in systems with singular perturbations to these new types of singular approximations is the third point of contact between equations with hysteresis and singularly perturbed equations. We will discuss this briefly in Subsection 1.4.3.

As the next step in the same direction we mention that the Preisach model with an absolutely continuous measure can be uniformly approximated by a finite number of relays. Thus we conclude that, for example, equation (1.22) can be approximated by a finite system of singularly perturbed equations of the following form:

$$\begin{aligned} \dot{x} &= f(t, x, y_1, \ldots, y_N), \\ \varepsilon \dot{y}_i &= g_i(x, y), \quad i = 1, \ldots, N. \end{aligned}$$

We note in conclusion that numerous mathematical models of physical, mechanical and other systems can be described by equations of the form

$$\begin{aligned} \dot{x} &= f(t, x, y, z), \\ y &= \Gamma x(t), \\ \varepsilon \dot{z} &= g(x, y, z), \end{aligned}$$

where Γ is a continuous hysteresis nonlinearity acting on a function $x(t)$ (for, instance, a Preisach nonlinearity P_μ with an absolutely continuous measure μ). The analysis of such systems will require the combination of ideas in both the areas of systems with hysteresis and singularly perturbed systems. The main technical problem lies in the fact that hysteresis operators are not smooth in a standard sense: they only satisfy a Lipschitz condition as operators in function spaces. Thus even the notions of stable and unstable manifolds must be adjusted accordingly.

The work begins here. We mention the method of split-hyperbolicity [44], and, especially, the results of O. Rasskazov [49] on the existence and properties of stable and unstable manifolds for systems with hysteresis nonlinearities.

1.4.2 Hysteresis technique in the analysis of singularly perturbed equations

General ideas

Today, powerful methods of analysis for equations with hysteresis are known. These methods can be used to understand the global dynamics of equation (1.22), and thus yield some progress on the global analysis of systems featuring singular perturbations.

We will discuss just one example:

$$\begin{aligned} L(d/dt)x &= \varphi(t) + \psi(x) + y, \\ \varepsilon\dot{y} &= M(y) - x. \end{aligned} \tag{1.23}$$

Here $L(p)$ and $M(y)$ are polynomials with $\deg(L) \geq 1$ and $\deg(M) > 1$. The forcing function φ is continuous and 2π-periodic, and the function ψ is real and analytic.

Lemma 1.4.2. *Let all roots of L be real and negative. Further suppose that ψ is bounded and monotone, that $\deg(M)$ is odd, and that the leading coefficient of M is positive.*

Then there exists at least one 2π-periodic Lyapunov stable solution $x_*(t)$ of the limit equation

$$\begin{aligned} L(d/dt)x &= \varphi(t) + \psi(x) + y, \\ y &= V[t_0, y_0]x(t). \end{aligned}$$

We note the availability of a constructive algorithm to find $x_*(t)$.

This lemma and some standard constructions imply the following assertion.

Theorem 4. *Under the conditions of Lemma 1.4.2, for sufficiently small $\varepsilon > 0$, the system (1.23) admits at least one 2π-periodic Lyapunov stable solution.*

The proofs of Lemma 1.4.2 and Theorem 4 follow the general scheme suggested in [43]. This scheme also suggests the existence of *asymptotically stable and isolated* 2π-periodic solutions; we hope to prove this soon in collaboration with P. Krejci.

1.4.3 Hysteresis phenomena as a source of new singularly perturbed problems

Discussion

We consider again the pendulum in a magnetic field, described by the equation

$$\ddot{x} + x = b\sin(t) + y(t), \quad y(t) = 2R_{-\beta,\beta}x(t) - 1. \tag{1.24}$$

From the physical point of view, this is a case of a resonance and the magnet is a ferromagnetic monocrystal. It can be shown that for $b < 4/\pi$ the system

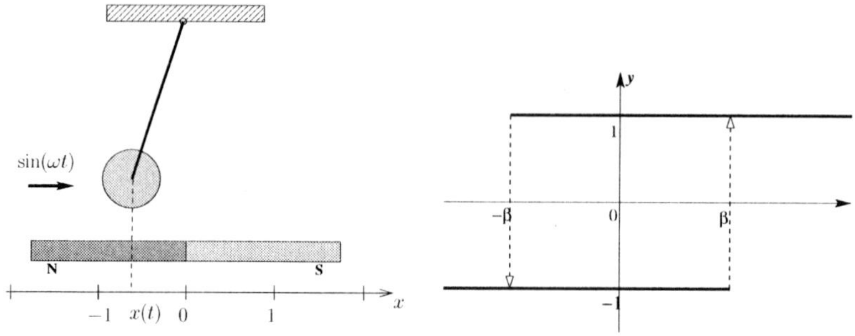

Figure 1.45. *Pendulum (left), and the transducer* $2R_{-\beta,\beta}x(t) - 1$ *(right)*

is dissipative. Thus we can expect periodic solutions with period 2π, and these solutions are important to understand the global behaviour of the system. However, equation (1.24) does not have periodic solutions in the standard sense. The root of the problem is, of course, in the discontinuity of the relay nonlinearity.

Now we will apply ideas of singular perturbations to the more delicate analysis of this equation. The idea is to approximate the equation (1.24) by a continuous system and to investigate the limit behaviour of 2π-periodic solutions of this continuous system.

We first suggest that (1.24) is an idealization of the singularly perturbed system

$$\begin{aligned} \ddot{x} + x &= b\sin(\omega t) + y, \\ \varepsilon\dot{y} &= g(x, y). \end{aligned}$$

Here g is as depicted in Fig. 1.44. It can be shown that for small ε this system has a canard-type solution with phase diagrams represented by Fig. 1.46.

This is fine from a mathematical point of view, but not quite satisfactory from the physical one. The trouble is that ferromagnetic monocrystals should be considered as the limit case of a different procedure. This procedure will be introduced in the next subsection; afterwards we will come to the corresponding canard-type solutions.

Generalized Play

Ferromagnetic monocrystals can be considered as an idealization of an interesting hysteresis nonlinearity called *Generalized Play* [28]. This will be denoted by L_ε, and is illustrated in Fig. 1.47.

Formally, this nonlinearity can be described as follows. We denote by Ω the set of pairs (x, y) satisfying

$$\gamma_-(x) \leq y \leq \gamma_+(x).$$

For a continuous input $x(t)$ and for an initial state η_0 satisfying $(x(t_0)), \eta_0) \in \Omega$,

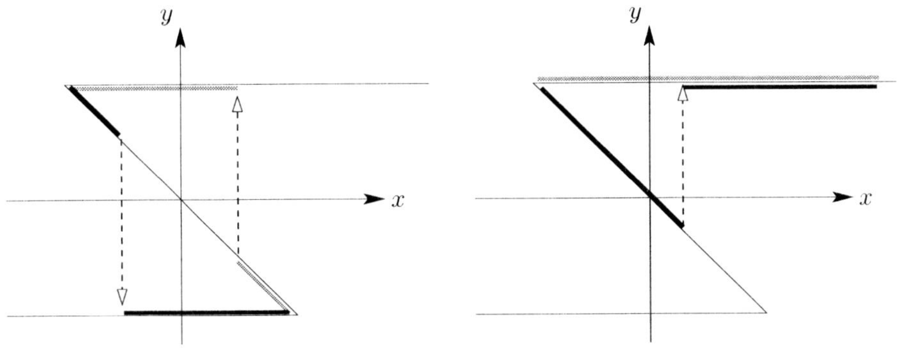

(a) A symmetric canard (b) An asymmetric canard

Figure 1.46. *Parametric plots of canards*

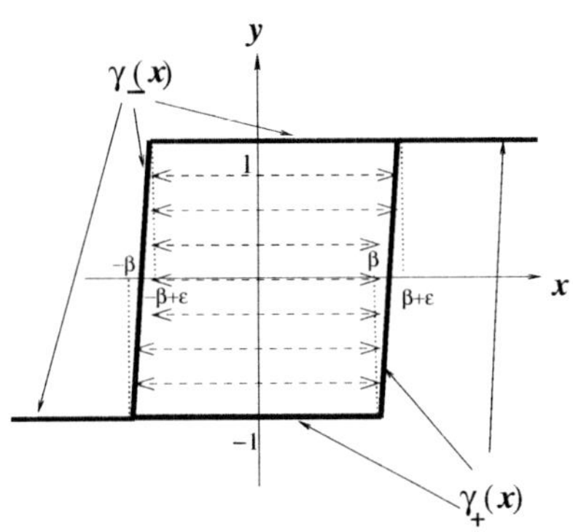

Figure 1.47. *Generalized Play L_ε*

the output

$$y(t) = L_\varepsilon[t_0, \eta_0]x(t)$$

is the unique continuous function such that, firstly, $y(t_0) = \eta_0$; secondly,

$$(x(t), y(t)) \in \Omega, \quad t \geq t_0;$$

thirdly, the inclusion

$$(x(t), y(\tau)) \in \Omega, \quad \tau \leq t \leq \sigma,$$

implies that $y(t) \equiv y(\tau)$ for $\tau \leq t \leq \sigma$. Thus, the output behaves 'lazily': it prefers to remain unchanged when the phase pair $(x(t), y(t))$ belongs to Ω. Fig. 1.48 provides the graph of a typical phase trajectory.

L_ε can be also represented by the Preisach model with an appropriate measure.

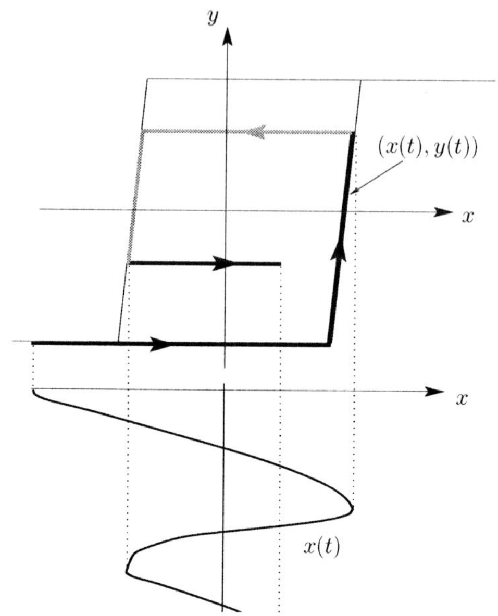

Figure 1.48. *Phase portrait of a typical trajectory for the Generalized Play* L_ε

Lemma 1.4.3. *Suppose* $x(t)$ *has no local minima equal to* α *or local maxima equal to* β*, and let* y_0 *be either 0 or 1. Then the limit*

$$y_*(t) = \lim_{\varepsilon \to 0} L_\varepsilon[t_0, y_0]x(t)$$

is well defined, and

$$y_*(t) = R_{-\beta,\beta}[t_0, y_0]x(t)$$

for all $t > t_0$ *such that*

$$x(t) \neq \alpha \quad \text{and} \quad x(t) \neq \beta.$$

In other words, for generic input functions $x(t)$ the discontinuous nonlinearity $2R_{-\beta,\beta}x(t) - 1$ is a limit of continuous nonlinearities $L_\varepsilon[t_0, y_0]$ as $\varepsilon \to 0$.

Hysteretic canards

We return to the resonance equation (1.24):

$$\ddot{x} + x = b\sin(t) + y(t), \quad y(t) = 2R_{-\beta,\beta}x(t) - 1.$$

By Lemma 1.4.3 this is to be considered as an idealization of the equation

$$\ddot{x} + x = b\sin(\omega t) + y(t), \quad y(t) = L_\varepsilon x(t), \tag{1.25}$$

where L_ε is a Generalized Play from the previous subsection, and $\varepsilon > 0$ is a small parameter.

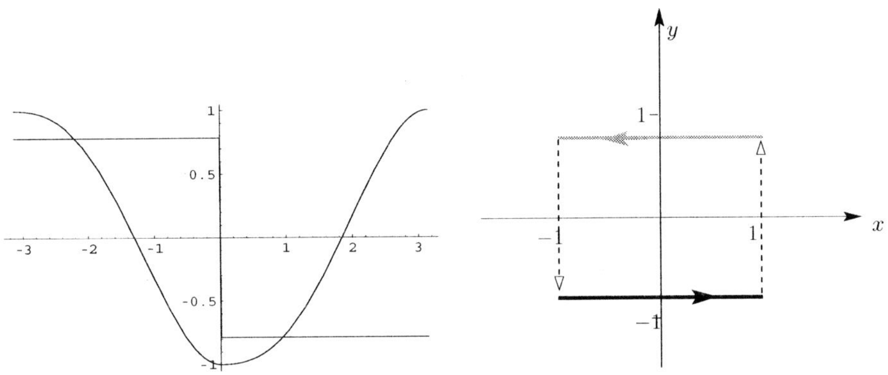

(a) Functions $x_0(t)$ (sine like function) and $y_0(t)$ (step like function)

(b) Parametric plot of $(x_0(t), y_0(t))$

Figure 1.49. *Symmetric hysteretic canard*

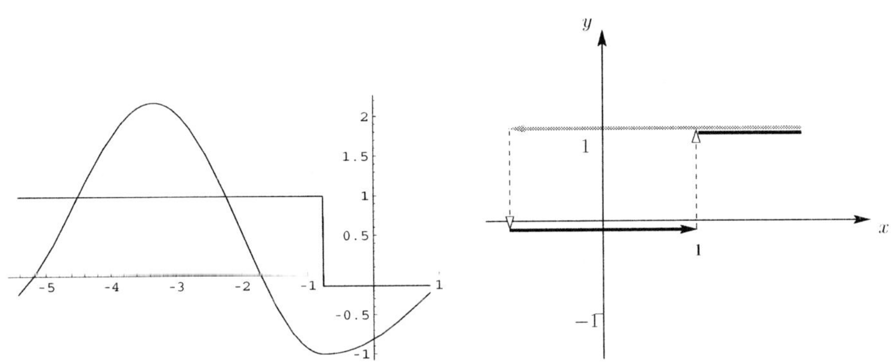

(a) Functions $x_1(t)$ (sine like function) and $y_1(t)$ (step like function)

(b) Parametric plot of $(x_1(t), y_1(t))$

Figure 1.50. *Asymmetric hysteretic canard*

We will say that a pair $(x_*(t), y_*(t))$ is a *limiting periodic solution of the equation (1.25)*, if it can be approximated by some solutions $(x_\varepsilon(t), y_\varepsilon(t))$ of the system (1.25) with arbitrary small ε. That is, if for any $\sigma > 0$, there exists $\varepsilon_0 > 0$ such that for $0 < \varepsilon < \varepsilon_0$ the equation (1.25) has a periodic solution $(x_\varepsilon(t),\ y_\varepsilon(t))$ satisfying

$$\max |x_\varepsilon(t) - x_*(t)| < \sigma, \qquad \int_0^{2\pi} |x_\varepsilon(t) - x_*(t)|\, dt < \sigma.$$

Theorem 5. *Let $0 < b < 4/\pi$, then equation (1.25) has exactly three limiting periodic solutions:*

$$(x_0(t), y_0(t)), \quad (x_1(t), y_1(t)), \quad (x_2(t), y_2(t)).$$

We now give explicit formulas for these solutions. We introduce

$$x_0(t,b) = -\beta\cos t + \frac{b}{2}(\sin t - t\cos t) + \frac{\pi b}{4}(\cos t - 1)\operatorname{sign} t, \ -\pi \le t < \pi,$$

$$x_1\left(t + \frac{\pi-\tau}{2}; b\right) = -(1+\beta)\cos t + \frac{b}{2}\left(\sin\frac{\tau}{2}\sin t - t\cos\left(t + \frac{\pi-\tau}{2}\right)\right)$$
$$+1 + \frac{\pi b}{4\sin(\tau/2)}(\cos t - 1)(\operatorname{sign} t + 1), \ \tau - 2\pi \le t < \tau,$$

$$x_2(t) = -x_1(t-\pi).$$

Here $\tau = \tau(b)$, for $0 < b < 4/\pi$, is defined by the relationships

$$\arccos\frac{1-\beta}{1+\beta} < \tau < \pi, \quad b = 2\frac{\beta - 1 + (\beta+1)\cos\tau}{\sin(\tau/2)(\sin\tau + \tau - 2\pi)}.$$

We also introduce 2π-periodic functions $y_0(t)$, $y_1(t)$, $y_2(t)$ by the equalities

$$y_0(t;b) = -\frac{\pi b}{4}\operatorname{sign} t, \quad -\pi \le t < \pi,$$

$$y_1\left(t + \frac{\pi-\tau}{2}; b\right) = 1 - \frac{\pi b}{4}\sin\frac{\tau}{2}\left(1 + \operatorname{sign} t\right), \quad \tau - 2\pi \le t < \tau,$$

$$y_2(t) = -y_1(t-\pi).$$

Proposition 1.4.1. *These functions satisfy Theorem 5 .*

Theorem 5 and Proposition 1.4.1 are similar to those proved in [7].

Figures 1.49 graph $x_0(t)$, $y_0(t)$ for $\beta = 1$. Similarly, Figures 1.50 graph the functions $x_1(t)$, $y_1(t)$.

We claim that these limit solutions are analogues of canard solutions from Subsection 1.4.3. In particular, $(x_0(t), y_0(t))$ is an analogue of the symmetric canard from Fig. 1.46(a), and $(x_1(t), y_1(t))$ is an analogue of the asymmetric canard from Fig. 1.46(b).

- The first reason is that classical canards and these solutions are mathematical explanations of the same physical phenomena: the limit behaviour of periodic solutions of a family of differential equations approximating a given equation with a discontinuous, relay-type nonlinearity.

- Secondly, in both cases the limit solutions demonstrate strange behaviour: in the classical canard situation they follow unstable manifolds for a time; in the 'hysteretic canard' situation, the solutions can stop in the course of a jump, and then hover for a while at an intermediate level.

- Thirdly, similar methods can probably be applied to analyze both situations. For example, the topological degree approach used to prove the last theorem could be useful for the rough location of classical canards. On the other hand, the typical 'canard community' methods of investigation will be useful in a more subtle asymptotic analysis of hysteretic canards.

1.5 Concluding Remarks

Using simple elementary examples we have highlighted basic ideas of the theory of singularly perturbed equations and the theory of hysteresis. We have also shown that these two mathematical theories are intertwined. In particular:

- Hysteresis operators can often be represented, or approximated, by singular limits of an appropriate singularly perturbed equation. Therefore, singularly perturbed equations can approximate equations with hysteresis.

- Equations with hysteresis nonlinearities are singular limits of appropriate singularly perturbed equations. Vice versa: Singular limits of equations with singular perturbations are solutions of equations with appropriate hysteresis nonlinearities.

- The global analysis of singularly perturbed equations becomes, in the first approximation, an investigation of equations with hysteresis nonlinearities.

- Specific methods for the analysis of equations with hysteresis can yield some progress on the global analysis of systems with singular perturbations.

- Hysteresis nonlinearities can be considered a source of new modifications of exciting singularly perturbed phenomena, such as canards.

1.6 Acknowledgements

Part of this chapter (namely subsections 2.3–2.5) were written jointly with Mark Donegan. The authors are grateful to Robert O'Malley and Michael Mortell for long and very helpful discussions. The first author was partially supported by the Enterprise Ireland Grant SC/2003/376. The second author was supported in part by the Boole Centre for Research in Informatics, University College Cork, by the Russian Foundation for Basic Research under grant no. 04–01–96515, and by Presidium of Russian Academy of Science Program no. 19.

Bibliography

[1] R. Angulo-Jaramillo, D. Elrick, J. Y. Parlange, P. G. Marchant, and R. Haverkamp, *Analysis of short-time single-ring infiltration under falling-head conditions with gravitational effects*, AGU Hydrology Days 2003 (2003), pp. 16–23.

[2] B. Amable, J. Henry, F. Lordon, and R. Topol, *Hysteresis: What It Is and What It Is Not*, OFCE Working Paper 9216, Paris, 1992.

[3] V. I. Arnold, V. S. Afraimovich, Yu. S. Il'yashenko, and L. P. Shil'nikov, *Theory of Bifurcations,* in Dynamical Systems, 5, Encyclopedia of Mathematical Sciences, V. Arnold, ed., Springer Verlag, New York, 1994.

[4] E. Benoit, J. L. Callot, F. Diener, and M. Diener, *Chasse au canard*, Collect. Math., 31–32(1–3) (1981–1982), pp. 37–119.

[5] E. Benoit, ed., *Dynamic Bifurcations*, Lecture Notes in Math., 1493, Springer-Verlag, Berlin, 1991.

[6] J. F. Besseling, *A theory of elastic, plastic and creep deformations of an initially isotropic materials showing anisotropic strain-hardening*, J. Appl. Mech. 25 (1958), pp. 529–536.

[7] N. A. Bobylev, V. V. Boltyanskii, S. Yu. Vsechsvyatskii, V. V. Kalashnikov, V. B. Kolmanovskii, V. S. Kozyakin, A. A. Kravchenko, A. M. Krasnosel'skii,. and A. V. Pokrovskii, *Mathematical Systems Theory*, Nauka, Moscow, 1986 (in Russian, MR 88a:93001).

[8] M. Brokate and A. V. Pokrovskii, *Asymptotically stable oscillations in systems with hysteresis nonlinearities*, J. Differential Equations, 150 (1998), pp. 98–123.

[9] M. Brokate and J. Sprekels, *Hysteresis and Phase Transitions*, Springer-Verlag, Berlin, 1996.

[10] R. Cross and A. Allan, *On the history of hysteresis*, in Unemployment, Hysteresis & Natural Rate Hypothesis, R. Cross, ed., Blackwell, pp. 26–38, 1988.

[11] R. Cross, *Hysteresis, The Handbook of Economic Methodology*, Edward Edgar, 1995.

[12] R. Cross, J. Darby, J. Ireland, and L. Piscitelli, *Hysteresis and Unemployment: a Preliminary Investigation*, Society for Computational Economics, Computing in Economics and Finance, 1999. Available at http://ideas.repec.org/p/sce/scecf9/721.html .

[13] W. Eckhaus, *Relaxation oscillations including a standard chase on French ducks*, Lecture Notes Math., 925 (1983), pp. 449–494.

[14] D. H. Everett and W. I. Whitton, *A general approach to hysteresis*, Trans. of the Faraday Society, 48 (1952), pp. 749–757

[15] D. H. Everett and F. W. Smith, *A general approach to hysteresis. Part 2: Development of the domain theory*, Trans. of the Faraday Society, 50 (1954), pp. 187–197.

[16] D. H. Everett, *A general approach to hysteresis. Part 3: A formal treatment of the independent domain model of hysteresis*, Trans. of the Faraday Society, 50 (1954), pp. 1077–1096.

[17] ——, *A general approach to hysteresis. Part 4: An alternative formulation of the domain model*, Trans. of the Faraday Society, 51 (1955), pp. 1551–1557.

[18] J. A. Ewing, *Experimental research in magnetism*, Phil. Trans. of the Royal Society of London, 176(II) (1895), pp. 523–640.

[19] D. Flynn, H. McNamara, P. O'Kane, and A. Pokrovskii, *Application of the Preisach Model in Soil-moisture Hysteresis*, BCRI Preprint 15/03 Cork, Ireland, 2003.

[20] T. S. Gilman. and A. V. Pokrovskii, *Forced vibrations of an oscillator with hysteresis taken into account*, Soviet Math. Doklady, 25(2) (1982), pp. 424 – 427.

[21] M. Gocke, *Types of Hysteresis Applied in Economics, Westfalische Wilhelms-Universitat Munster, Volkswirtschaftliche Diskussionsbeitrage 292*, 1999. Prepared for the International Conference on Industrial and Applied Mathematics.

[22] R. Haverkamp, P. Reggiani, P. J. Ross, and J. Y.Parlange, *Soil water hysteresis prediction model based on theory and geometric scaling*, in Environmental Mechanics Water, Mass and Energy Transfer in the Biosphere, P. Raats, D. Smiles, and A. Warrick, eds., Geophysical Monograph Series, 129 (The Philip Volume), 2002.

[23] D. J. J. Hillel, *Introduction to Soil Physics*, Academic Press, New York, 1982.

[24] W. HOGARTH, J. HOPMANS, J. Y. PARLANGE, AND R. HAVERKAMP, *Application of a simple soil-water hysteresis model*, J. of Hydrology, 98 (1988), pp. 21–29.

[25] A. Y. ISHLINSKII, *Some applications of statistical methods to describing deformations of bodies*, Izv. Akad. Nauk SSSR, Techn. Ser., 9 (1944), pp. 580–590 (in Russian).

[26] M. P. KENNEDY AND L. O. CHUA, *Hysteresis in electronic circuits: A circuit theorist's perspective*, International J. of Circuit Theory and Applications, 19 (1991), pp. 471–515.

[27] *Kluwer Encyclopedia of Mathematics*, Supplement Volume 1, Kluwer Acad. Publ., pp. 310, 384, 1997.

[28] M. A. KRASNOSEL'SKII AND A. V. POKROVSKII, *Systems with Hysteresis*, Springer-Verlag, Berlin, 1989.

[29] P. KREJCI, *Hysteresis, Convexity and Dissipation in Hyperbolic Equations*, Gakkotosho, Tokyo, 1996.

[30] P. KREJCI AND J. SPREKELS, *Strong solutions to equations of visco-thermoplasticity with a temperature-dependent hysteretic strain-stress law*, in Variations of Domain and Free-boundary Problems in Solid Mechanics (Paris, 1997), pp. 237–244, Solid Mech. Appl., 66, Kluwer Acad. Publ., Dordrecht, 1999.

[31] I. D. MAYERGOYZ , *Mathematical Models of Hysteresis*, Springer–Verlag, New York, 1991.

[32] E. F. MISHCHENKO AND N. KH. ROZOV, *Differential Equations with Small Parameters and Relaxation Oscillations*, Plenum Press, New York, 1980.

[33] A. V. NETUSHIL, *Nonlinear element of the stop type*, Avtomat. Telemech., 7 (1968), pp. 175–179 (in Russian).

[34] ——, *Self-oscillation in systems with negative hysteresis*, in Proc of 5th International Conference on Nonlinear Ocsillations, 4, pp. 393–396, Izd. AN USSR, Kiev, 1970.

[35] J. A. NOHEL AND D. H. SATTINGER, eds., *Selected Works of Norman Levinson*, Birkhauser, Boston, 1998.

[36] R. E. O'MALLEY, *Singular Perturbation Methods for Ordinary Differential Equations*, Appl. Math. Sci., 89, Springer–Verlag, New-York, 1991.

[37] ——, *Naive singular perturbation theory*, Special issue in memory of Richard Weiss. Math. Models Methods Appl. Sci., 11(1) (2001), pp. 119–131.

[38] J.-Y. PARLANGE, *Water transport in soils*, Annu. Rev. Fluid Mech., 12 (1980), pp. 77-102.

[39] J. Y. Parlange, T. S. Steenhuis, R. Haverkamp, D. A. Barry, P. J. Culligan, W. L. Hogarth, M. B. Parlange, P. Ross, and F. Stagnitti, *Soil properties and water movement*, in Vadose Zone Hydrology - Cutting Across Disciplines, M. B. Parlange and J. W. Hopmans, eds., Oxford University Press, New York, pp. 99–129, 1999.

[40] L. Piscitelli, *Hysteresis in Economics*, University of Strathclyde, PhD thesis, Glasgow, Scotland, 1998,.

[41] *Proc. of the Third Int. Symposium on Hysteresis and Micromagnetic Modelling*, Physica B, Condensed Matter 306(1-4), 2001.

[42] L. Piscitelli, R. Cross, M. Grinfeld, and H. Lambar, *A test for strong hysteresis*, Computational Economics, 15(1) (2000), pp. 59-78, Aavailable at http://ideas.repec.org/a/kap/compec/v15y2000i1-2p59-78.html.

[43] A. V. Pokrovskii, *Shuttle algorithm in the analysis of systems with hysteresis nonlinearities*, in Models of Hysteresis, A. Visintin, ed., Pitman Research Notes in Mathematics Series, 286, Longman Sci. Tech., Harlow, pp. 124–142, 1993.

[44] ——, *Topological shadowing and split-hyperbolicity*, *J. for Difference and Differential Equations*, special issue dedicated to M. A. Krasnosel'skii, 4(3–4) (1997), pp. 335–360.

[45] L. Prandtl, *Spannungverteilung in plastischen Korpern*, *Proceedings of the First International Congress on Applied Mechanics*, pp. 43–54, 1924.

[46] ——, *Ein Gedankenmodell zur kinetischen Theorie der festen Korper*, Z. Angew. Math. Mech., 8 (1928), pp. 85–106.

[47] P. Preisach, *Über die magnetische Nachwirkung*, Zeitschrift für Physik, 94 (1938), pp. 277–302.

[48] van der Pol B., *Over relaxatie-trillingen*, Tijdschr. Ned. Radiogenoot. 3 (1926), pp. 25-40.

[49] O. Rasskazov, *Forward and backward stable sets of split-hyperbolic mappings*, Izvestiya of RAEN, Series MMMIU, 5(1–2) (2001), pp. 185–205.

[50] L. L. Rauch *Oscillation of a third order nonlinear autonomous system. Contributions to the Theory of Nonlinear Oscillations*, Ann. Maths. Studies, 20 (1950), pp. 39–88.

[51] L. A. Richards, *Uptake and retention of water by soil as determined by distance to a water table*, J. Amer. Soc. Agron., 33(1941), pp. 778–786.

[52] J. Sanders and F. Verhulst, *Averaging Methods in Nonlinear Dynamical Systems*, Applied Mathematical Sciences, 59, Springer-Verlag, New York, 1985.

[53] V. A. Sobolev, *Integral manifolds and decomposition of singularly perturbed systems*, System and Control Lett., 5 (1984), pp. 169–179.

[54] G. Tao and P. Kokotovic, *Adaptive Control of Systems with Actuator and Sensor Nonlinearities*, John Wiley & Sons, 1996.

[55] A. B. Vasil'eva, V. F. Butuzov, and L. V. Kalachev, *The Boundary Function Method for Singular Perturbation Problems*, SIAM, Philadelphia, 1995.

[56] A. Visintin, *Differential Models of Hysteresis*, Springer–Verlag, Berlin, 1994.

Chapter 2

Frustration Minimization, Hysteresis and the El Farol Problem

R. Cross, M. Grinfeld, H. Lamba, and A. Pittock

The parable, due to Arthur, of the El Farol bar provides an account of the aggregate dynamics resulting from individuals using heterogeneous inductive predictors when deciding whether or not to attend the bar. In the Arthur formulation individuals make their bar attendance decisions on the basis of observing how crowded the bar has been in the preceding weeks. We modify this approach to take account of the times an individual experienced the enjoyment of an uncrowded bar and the regret at either having been at a too crowded bar, or at not having come when the bar was uncrowded. In our formulation the attendance decision is driven by the dominant component of enjoyment or regret. We allow for hysteresis thresholds and our simulations show that this modification to our model leads to an increase of the periodicity of aggregate bar attendance.

2.1 Introduction

In this paper we introduce a simple conception of human motivation and investigate its consequences for the original formulation of the El Farol bar problem of W. Brian Arthur [2]. The decision facing the individual is whether or not to attend the El Farol bar in view of the possibility that it might be too crowded. In the Arthur representation, individuals observe the aggregate dynamics of bar attendance and choose from a set of predictors the ones that have most accurately predicted the actual recent bar attendance. The "active" or successful predictors are selected by a process of induction (see section 2.2 for more details).

This approach contrasts with the deductive approach to the "rationality" of decision-taking traditionally used in economics, such as in the expected utility theory: see [19] and [10] for critical survey assessments. In the traditional approach, "rationality" is defined in the instrumentalist sense [20] of consistency with a set of simple axioms such as those articulated by [17]. Many violations of such axioms have been observed (see [5]). Of the alternative explanations of decision-taking, prospect theory [11], regret theory [14] and cognitive dissonance [1] are amongst the most prominent.

The innovation in the present paper is to introduce some layers of psychology into the inductive learning involved in the formulation given in [2]: see [13]. Our procedure is to introduce the insights of the regret theory of [3, 8, 14] into the decision whether or not to attend the El Farol bar. This is done by taking into account the times the individual experienced the enjoyment of a non-crowded bar, the regret or disappointment at having attended the bar when it was too crowded, and the lost opportunities associated with not having attended when the bar was not crowded. The cognitive dissonance of [1] is introduced by allowing the individuals to have hysteresis thresholds with respect to enjoyment, disappointment, and lost opportunities. Thus attendance strategies can be retained even when the bar attendance dynamics that led to the strategies have been removed. While taking regret and cognitive dissonance into account, our analysis attempts to retain a strong flavour of the induction present in the original formulation in [2]. The behaviour described is rational only in the "bounded" sense of Herbert A. Simon [18]. We contend that rationality in the El Farol conditions of incomplete information requires not only inductive learning but also knowing one's psychological requirements, that is, knowing oneself (see [4]). We feel that Socrates would have agreed.

The resulting model is simple to state and relatively easy to simulate on a computer. Its analysis, due to the discontinuous strategy-switching which is a necessary consequence of our assumptions, is non-trivial. There are a myriad of possible modifications and elaborations of the basic setup that can be further explored.

The structure of the paper is as follows. In Section 2.2 we briefly review Arthur's formulation and approach to the El Farol problem. This is not so much a "problem" as a benchmark example which can be used to see how different approaches to the decision-making of individuals are reflected in the aggregate dynamics (occupation dynamics of the bar in the El Farol case). In Section 2.3 we put forward our approach to human motivation. In Section 2.4 we express cognitive dissonance in terms of our key variables and show that this gives rise to hysteretic behaviour at the level of the individual. Finally, in Section 2.5 we present results of some simulations of the model that indicate a huge richness of dynamics arising from our very simple assumptions; this, as work on even low-dimensional dynamical systems (e.g., [12]) shows, is not unexpected.

2.2 The Arthur Approach

In his classic paper Arthur [2] considers the following situation: there is only one bar in the town, the El Farol bar, and every citizen over licensing age, of which there are 100, has each week to make a decision whether or not to go to the bar on Thursday night, when they play Irish music. It is known that if the number of occupants in the bar on a particular night is larger than some critical number, say 60, no-one will have a good time (too crowded, service will be slow, etc.). The question is to understand how reasonable assumptions on the decision-making procedures used by the individuals are expressed in the occupancy dynamics of the bar.

Arthur's approach to the problem is as follows. He assumes that there is a set of n predictors $P = \{p_1, \ldots, p_n\}$, of which the j-th individual ($j = 1, \ldots, 100$) is "issued" with a subset P_j of cardinality $m < n$. Each of these predictors uses information on past occupancy to predict the occupancy next Thursday. Each individual picks, at each decision time (say, on Thursday afternoon), the prediction of the predictor that had done best the previous Thursday afternoon in predicting the occupancy on that Thursday night, and acts accordingly: if the prediction is that the number of people will be larger than 60, the individual in question will not go, and so on. Now, even though not taking into account the overall record of predictors and judging by last week's performance only, such a choice of a predictor on which to base a decision can be defended on the grounds that the individual is obliged to learn inductively in the absence of any deductive means of figuring out the aggregate bar attendance.

Before we suggest an arguably more realistic alternative, we would like to introduce a distinction between **data** and the **construction** placed on the data. Henceforth by data we will just mean the facts relevant to the situation and by the construction, data filtered by the individual and put into a form useful for decision-making. Thus, data in Arthur's approach consists of the history of the occupancy of the bar in previous weeks. For simplicity, we will assume infinite memory of the individuals; this is of course not realistic. Finite memory clutters the notation, but can be incorporated. If N_k is number of people in the bar in week k, after week i the data consists of the vector

$$\boldsymbol{N}_i = \{N_1 \ldots N_i\}.$$

(Thus, for any given i, a predictor $p \in P$ takes $\boldsymbol{N}_i$ as its input and predicts N_{i+1}.)

The construction for individual j after week i therefore consists of a number l defining the predictor $p_l \in P_j$ that best predicted the occupancy on the previous Thursday, i.e. the number l that minimizes $|p_l(\boldsymbol{N}_{i-1}) - N_i|$ and the number $p_l(\boldsymbol{N}_i)$.

Note that in this case data is shared, while the construction is private, since the sets P_j are different from individual to individual, and in fact, the sum total of individual differences is contained in the sets P_j.

2.3 An Alternative Approach to Decision Making

We would like to suggest a different picture of decision making. We introduce a layer of psychology and hereby attempt to populate our model with homo sapiens (see [21]). In our view, people are subject to multiple tensions and frustrations, which provide the motivation for action. One could organize the different tensions into one composite function, and try to minimize it, but we feel that the consequent mathematical simplicity has not much to commend it. Such tensions define the psychological state of an individual as a vector only in a very high-dimensional space. A person's actions, in our view, are mainly of the "fire-extinguishing" variety:

an individual will always attend to the dominant component of the vector, that is, will try to relieve the most prominent tension. Anyone who has sat on a needle while having tooth-ache will intuitively see the reason for this assumption.

We will now spell out the consequences of such assumptions for the El Farol situation. We are careful at each stage to distinguish between data and the psychological construction placed on it.

2.3.1 Data

Clearly, Arthur's model overlooks an obvious source of data: in addition to $\boldsymbol{N}_i$, which is data shared between all the individuals, each individual j also has access to her own record of bar attendance. We define the variable s_j^i to be 1 if individual j was in the bar in week i and 0 otherwise. Thus, individual j also knows $\boldsymbol{s}_j^i = \{s_j^1, \ldots, s_j^i\}$ and can use that when formulating a bar attendance strategy.

2.3.2 Construction

Consider the following numbers:

$$D_j^i = \sharp\{l \in \{1, \ldots, i\} |\ N_l \geq 60,\ s_j^l = 1\},$$

$$L_j^i = \sharp\{l \in \{1 \ldots, i\} |\ N_l < 60,\ s_j^l = 0\},$$

$$E_j^i = \sharp\{l \in \{1 \ldots, i\} |\ N_l < 60,\ s_j^l = 1\}.$$

Here $\sharp$ denotes the cardinality of a set. D_j^i counts the times up to and including the i-th week, when the individual did not have a good time in the bar (disappointment), L_j^i is the count of lost opportunities to have a good time, and E_j^i is the number of times when the individual enjoyed herself in the bar.

So far no account has been taken of human variability. Clearly, people differ in the way past experience impinges on their self-perception. One way to encode this is to give different weights to occurrences depending on their distance in time from the present. That is a possibility, but we shall pursue a different approach, by noting that some people attach little weight to lost opportunities while others attach more weight to disappointments than to good times. Hence we associate each of the numbers D_j^i, L_j^i and E_j^i with positive weights d_j, l_j, e_j (see [19]). The construction used to make a decision is encoded in the **state of the individual** j,

$$\boldsymbol{S}_j^i = (D_j^i d_j,\ L_j^i l_j,\ E_j^i e_j).$$

2.3.3 Decision making

One way to define "quasi-rational" (in the sense of [21]) behaviour for individual j would be to have her maximize $E_j^{i+1} e_j$, and simultaneously minimize $D_j^{i+1} d_j$ and $L_j^{i+1} l_j$; or, alternatively, to maximize an additive expression such as

$$E_j^{i+1} e_j - L_j^{i+1} l_j - D_j^{i+1} d_j,$$

given $\boldsymbol{S}_j^i$, that is, to make a choice of s_j^{i+1}. Were we to follow this approach, we would have mainly to explain the j-th individual's (quasi-rational) prediction for the number of visitors in week $i+1$, $\overline{N}_j^{i+1}$. Then, with slight modifications, this approach would collapse to that of Arthur, notwithstanding the layer of psychology that we have introduced.

However, if the "fire-extinguishing" approach suggested above is to be followed, it makes sense to define

$$C_j^i = \max\{D_j^i d_j,\ L_j^i l_j,\ E_j^i e_j\}.$$

Our representation of human motivation is then to say that the bar attendance decision is determined simply by C_j^i: if $C_j^i = E_j^i e_j$ or $C_j^i = L_j^i l_j$, then $s_j^{i+1} = 1$, and if $C_j^i = D_j^i d_j$, $s_j^{i+1} = 0$. Of course one could introduce probabilities to take into account inertia, but even without that the resulting well-defined (non-smooth) dynamical system seems interesting enough. Thus our model is iterated as follows: at time i compute C_j^i for all the individuals and update s_j^{i+1}. The parameters could in principle be calibrated by experimental economics techniques.

We are not saying that the individuals are not anticipatory systems in the sense of [16], in that they do not carry within themselves a predictive model of the environment. We just claim that our model only takes account of psychological constructions placed on different predictions. If what makes the individual suffer is the thought of lost opportunities, she will try to alleviate this suffering by making sure the weight of lost opportunities does not become heavier, even at the expense of disappointment from visiting an overcrowded bar.

It is not clear that, in relation to challenges occurring over a relatively short time-span, human behaviour should be expected to have been shaped by evolutionary forces (see the discussion of [7]). Attending to a pressing psychological need at the expense of a less pressing one can have survival value, though obviously does not encompass the whole gamut of behavioural strategies that would be "rational" in an evolutionary sense. It is possible to find parameter values for which the average bar occupancy is around 60%, in which case the behaviour of the cohort as a whole can be deemed "rational" (since on the average the bar is neither overcrowded nor underused). We do not restrict ourselves *a priori* to such parameter regimes, since our aim is to model actual, and not "desirable" behaviour.

Clearly, as more information becomes available, it can be incorporated in the decision-making process. Thus, if a person is driven by disappointment and at the same time knows for certain that the bar will be empty next Thursday night, there is no reason why she should not go. In the simple setup we are considering, just as in life, such infallible oracles do not exist.

2.4 Hysteresis

The possibility of hysteretic behaviour is mentioned by Arthur but without elaboration. For a discussion of hysteresis in economics see [6, 9, 15] for hysteresis set in a probabilistic context. In the present context, hysteresis involves the retention of a strategy once the stimulus that led to the adoption of the strategy has been removed. This is recognizably human: the expression in the economic arena of cognitive dissonance and habit formation.

We now describe one possible method of introducing hysteretic behaviour in terms of our variables: an individual is hysteretic with respect, say, to disappointment if

$$\begin{aligned}
&C_j^i = D_j^i d_j \quad \text{though} \quad D_j^i d_j \neq \max\{D_j^i d_j,\ L_j^i l_j,\ E_j^i e_j\} \\
&\text{whenever} \quad D_j^{i-1} d_j = \max\{D_j^{i-1} d_j,\ L_j^{i-1} l_j,\ E_j^{i-1} e_j\}, \\
&\text{and} \quad \max\{D_j^i d_j,\ L_j^i l_j,\ E_j^i e_j\} - D_j^{i-1} d_j < \epsilon_j,
\end{aligned}$$

where ϵ_j is the hysteresis, the strength of j-th individual's cognitive dissonance (with respect to disappointment in this case).

It should be clear that if people are hysteretic with respect to enjoyment or lost opportunities (i.e. envy) occupancy levels at the bar will be higher than if there is no hysteresis. This indicates that a successful bar management strategy would try to induce and keep high hysteresis levels with respect to these two tension sources.

2.5 Numerics

We now present some numerical simulations using the above model, both with and without hysteresis. The range of dynamical behaviour that can be observed is very large depending upon the parameters entered into the model. Of particular interest is the observation that the overall occupancy of the bar from week-to-week can display approximate periodicity even though a significant percentage of the participants do not show this periodicity in their individual behaviour. Chaotic (no discernible periodicity) and intermittent (switching between two apparently stable but distinct modes of behaviour at seemingly random times) parameter regimes can also be observed. We also examine the effects of introducing hysteresis on the behaviour of individuals and on the overall bar occupancy rate.

In what follows we simulate the behaviour of 100 individuals and set the optimum occupancy rate of the bar at 60. This leaves us with 300 parameters to be chosen, namely the weights d_j, l_j, e_j, $1 \leq j \leq 100$. We define these by drawing each d_j from a uniform probability distribution defined on a certain interval $[a_d, b_d]$. The same is done for each l_j and e_j using intervals $[a_l, b_l]$ and $[a_e, b_e]$ respectively. Thus, in a statistical sense, the model is reduced to one with only 6 parameters. While this method of assigning weights is certainly simplistic, it is sufficient to display a wide range of system dynamics and also allows for a controlled exploration of the parameter space by, for example, changing the average value of d_j and observing the effects.

The parameters used for the first simulations are $[a_d, b_d] = [3, 6]$, $[a_l, b_l] = [1, 3]$ and $[a_e, b_e] = [3, 4]$. At the start of the simulation the values of D_j, L_j and E_j are randomly assigned to be either 1,2 or 3 to define an initial attendance history for the jth individual. The model was then iterated for 100 weeks.

The first row of Fig. 1 shows two plots – the left plot is a plot of the attendance for the last 30 weeks displaying the asymptotic or long-time dynamics after any initial transients have decayed away. The right plot shows the attendance record of 20 randomly selected individuals over the same 30 week period. Circles denote attendance while the absence of a circle denotes abs(tin)ence for that particular week.

After 100 weeks, hysteresis was introduced into the model which was then iterated for a further 100 weeks. This hysteresis is with respect to disappointment, loss and enjoyment, that is to say

$$C_j^i = \max\{D_j^{i-1}d_j,\, L_j^{i-1}l_j,\, E_j^{i-1}e_j\}$$

$$\text{if} \quad \max\{D_j^i d_j,\, L_j^i l_j,\, E_j^i e_j\} - \max\{D_j^{i-1}d_j,\, L_j^{i-1}l_j,\, E_j^{i-1}e_j\} < \epsilon_j.$$

For simplicity $\epsilon_j = 2$ for all j. The lower plots of Fig. 1 were created in exactly the same way as the upper ones but show the last 30 weeks of the second 100-week period. It should be noted that the weights d_j, l_j and e_j are precisely the same for both sets of plots and the 20 randomly selected individuals are also the same. This permits a much more direct comparison of the effects of cognitive dissonance/hysteresis.

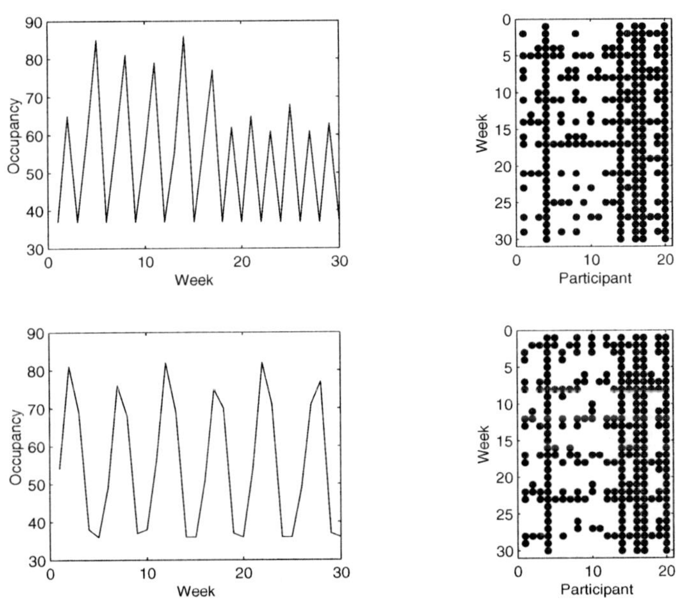

Figure 1. *The top row shows the behaviour of the system without hysteresis while the lower row shows the effects of including hysteresis.*

The most striking effect of including hysteresis is an increase in the approximate period of the aggregate behaviour. This was observed in a large majority of the simulations that were run, and is consistent over a very large parameter range. Also, as noted above, the behaviour of individuals is much less predictable than the overall bar-attendance. As can be seen from Fig. 1 a wide variety of bar-attending behaviour naturally emerges, ranging from people who are present almost every week to those who turn up very rarely.

Fig. 2 is produced in exactly the same way as Fig. 1 but with new parameters $[a_d, b_d] = [6, 7], [a_l, b_l] = [2, 3]$ and $[a_e, b_e] = [3, 10]$. The results without hysteresis are similar to those in Fig. 1 with an approximate periodic component of length 2-3 weeks. The addition of hysteresis once again appears to increase the length of any approximate periodicity that is present. However, the resulting aggregate attendance plot differs significantly from its counterpart in Fig. 1 and forcefully demonstrates that even very simple frustration minimization strategies at the individual level can result in highly complex and unpredictable group dynamics.

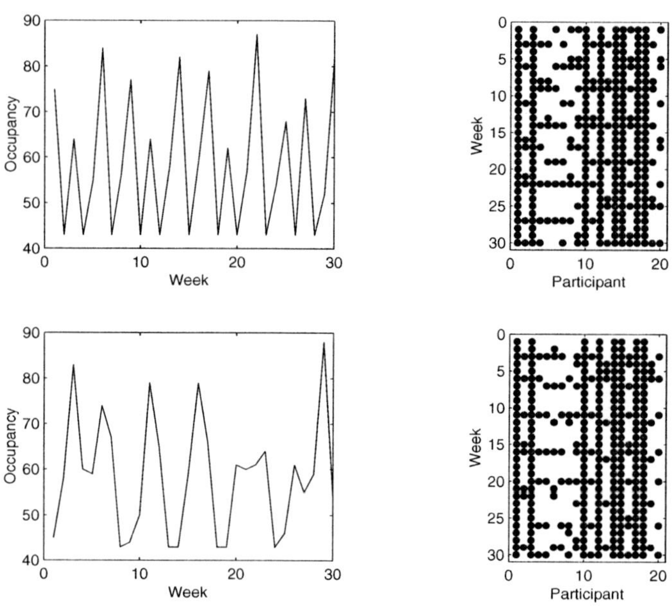

Figure 2. *As Fig. 1 but with data generated using different weighting parameters (see text).*

2.6 Concluding Remarks

The noticeable degree of periodicity present in the numerical simulations of the previous section – perhaps higher than would be expected in reality – can be explained by the fact that the El Farol bar problem is extremely simplified. Following Arthur, we have made no distinction between there being 61 patrons in the bar and 100. Clearly, in realistic situations the quality of the service, amount of noise, etc. depend "continuously" on the number of people, and one would expect that having been present in a bar with another 99 punters, one would give it a wide berth for a long time. It is reasonably obvious how to incorporate such dependencies into our construction-creation mechanisms by making d_j time-dependent through dependence on N_i, and so on. Other shortcomings, such as the absence of external random influences and the infinite memory of the participants can also be incorporated.

We would, however, argue that by endowing our El Farol punters with a capacity for disappointment and regret at lost opportunities, and by allowing for cognitive dissonance and stickiness in habits, the bar becomes a more recognizable place. It is also plausible to allow the punters to remember whether or not they attended the bar (assuming they were not too drunk for that).

We finally remark that Arthur's model can be regarded as a parable that applies to a wider class of problems. It could, for example, be applied to firms making market entry-exit decisions, in which case the different predictors stand for the opinions expressed around the different executive board tables. Our notions of disappointment, lost opportunity, enjoyment and frustration minimization extend quite naturally to such situations.

Bibliography

[1] G. A. Akerlof and W. T. Dickens, *The economic consequences of cognitive dissonance*, American Economic Review, 72(3) (1982), pp. 307–319.

[2] B. W. Arthur, *Inductive reasoning and bounded rationality*, American Economic Review, 84(2) (1994), pp. 406–411.

[3] D. Bell, *Regret in decision-making under uncertainty*, Operations Research, 30(5) (1982), pp. 961–981.

[4] R. Bénabou and J. Tirole, *Self-knowledge and self-regulation: an economic approach*, in: The Psychology of Economic Decisions, I. Brocas and J. D. Carrillo, eds., Centre for Economic Policy Research and Oxford University Press, New York, 2003, pp. 137–167.

[5] I. Brocas and J. D. Carillo, eds. *The Psychology of Economic Decisions*, Centre for Economic Policy Research and Oxford University Press, New York, 2003.

[6] R. Cross, *On the foundations of hysteresis in economic systems*, Economics and Philosophy, 9(1) (1993), pp. 53–74.

[7] J. H. Fetzer, *Evolution, rationality, and testability*, Synthese, 82(1990), pp. 423-439.

[8] P. C. Fishburn, *Nontransitive measurable utility*, Journal of Mathematical Psychology, 26(1982), pp. 31–67.

[9] M. Grinfeld, L. Piscitelli, and R. Cross, *A probabilistic framework for hysteresis*, Physica A, 287 (2000), pp. 577–586.

[10] D. Kahneman, *A psychological perspective on economics*, American Economic Review, 93(2) (2003), pp. 162–168.

[11] D. Kahneman and A. Tversky, *Prospect theory: an analysis of decision under risk*, Econometrica, 47(2) (1979), pp. 263–291.

[12] A. Katok and B. Hasselblatt, *Introduction to the Modern Theory of Dynamical Systems*, Cambridge, Cambridge University Press, 1999.

[13] G. LOEWENSTEIN, *The fall and rise of psychological explanations in the economics of intertemporal choice*, in: Choice over Time, G. Loewenstein and J. Elster, eds., Russell Sage Foundation, New York, 1992, pp. 3–34.

[14] G. LOOMES AND B. SUGDEN, *Regret theory: an alternative theory of rational choice under uncertainty*, Economic Journal, 92 (1982), pp. 805–824.

[15] L. PISCITELLI, *Hysteresis in Economics*, Ph.D. thesis, University of Strathclyde, Glasgow, Scotland, 1998.

[16] R. ROSEN, *Anticipatory Systems*, Oxford University Press, Oxford, 1985.

[17] L. J. SAVAGE, *The Foundations of Statistics*, Wiley and Sons, New York, 1954.

[18] H. A. SIMON, *A behavioural model of rational choice*, Quarterly Journal of Economics, 69(1) (1955), pp. 99–118.

[19] C. STARMER, *Developments in non-expected utility theory: the hunt for a descriptive theory of choice under risk*, Journal of Economic Literature, XXXVIII (2000), pp. 332–382.

[20] R. SUGDEN, *Rational choice: a survey of contributions from economics and philosophy*, Economic Journal, 101 (1991), pp. 751–785.

[21] R. H. THALER, *From homo economicus to homo sapiens*, Journal of Economic Perspectives, 14(1) (2000), pp. 133–141.

Chapter 3

Hysteresis in Singularly Perturbed Problems

P. Krejčí

A discontinuous hysteresis law is derived as a singular limit in differential equations with non-monotone nonlinearities arising, for example, in a model for instabilities of a fluid flow in a tube with pump and valve with uncertain parameters. The input-output relation is considered in the space of regulated functions, and it is shown that it preserves the property of uniformly bounded oscillation. Stability is obtained in a small neighbourhood of the equilibrium.

3.1 Introduction

The paper deals with a model of singular input-output $u \mapsto x$ behaviour described by the differential equation

$$\alpha\dot{x}(t) + g(x(t)) = u(t), \quad x(0) = x_0, \tag{3.1}$$

where u is a given function, $\alpha > 0$ is a small parameter, and $g : \mathbb{R} \to \mathbb{R}$ is a continuous function of the form

$$g(x) = \begin{cases} g_1(x) & \text{for } x \in J_1 =]-\infty, x_-], \\ g_2(x) & \text{for } x \in J_2 =]x_-, x_+[, \\ g_3(x) & \text{for } x \in J_3 = [x_+, +\infty[, \end{cases} \tag{3.2}$$

for some $x_- < x_+$, with g_2 decreasing and g_1, g_3 increasing in their respective domains J_i, $i = 1, 2, 3$, $g(\pm\infty) = \pm\infty$, see Fig. 3.1. We denote

$$g_1(x_-) := G_+ > G_- := g_3(x_+).$$

Following [5, 12], we show in detail in Section 3.2 how equations of the type (3.1) with sequences u_k of inputs and x_k of outputs arise in a model for spontaneous oscillations occurring in a pump-valve hydraulic system. Intuitively, it can be expected that the limit $u \mapsto x$ behaviour for α very small will have a hysteretic character. The mathematical problem consists in finding an appropriate functional framework in which the convergences $u_k \to u$ and $x_k \to x$ should take place if no lower bound for the uncertain values of α is available. Uniform convergence will certainly not be the right choice, as the hysteretic jumps are not a priori localized in time. On the other hand, weak-star convergence in L^∞ is too weak to guarantee convergence in the nonlinear term. If u has bounded variation and x stays in a left or right neighbourhood of a local maximum of g, then x may not have uniformly bounded variation independently of α.

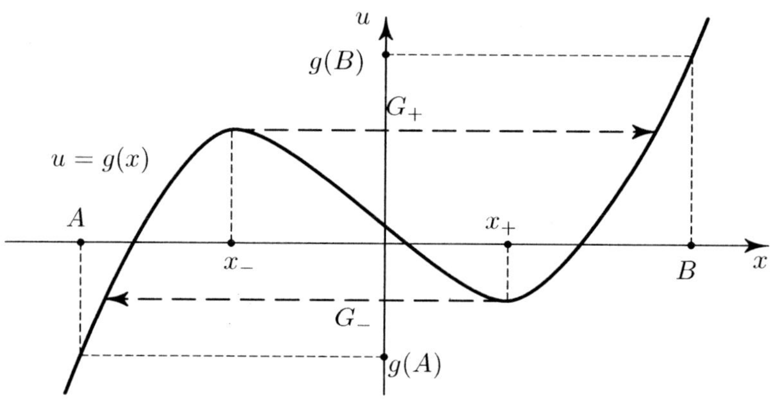

Figure 3.1. *The $u \mapsto x$ diagram*

A convergence concept (so-called *r-convergence*, see Definition 3.3 below) has been proposed in [11] in the space $G(0,T)$ of *regulated functions* defined in an interval $[0,T]$. Recall that a function $u : [a,b] \to \mathbb{R}$ is said to be regulated according to [1], if both one-sided limits $u(t+)$, $u(t-)$ exist at each point $t \in [a,b]$ with the convention $u(a-) = u(a)$, $u(b+) = u(b)$. It was shown in [11] that all solutions associated with an r-convergent sequence of inputs and with arbitrary $\alpha > 0$ constitute a set containing an r-convergent subsequence.

This result is not fully satisfactory because of the fact that the limit of an r-convergent sequence of regulated functions is not necessarily regulated, although it has at most countably many discontinuity points. Our goal here is, therefore, to derive a higher regularity result in this direction. We use the concept of *uniformly bounded ε-variation* introduced by Fraňková in [4], see Definition 3.2 below. The remarkable Theorem 3.8 of [4] states, as a generalization of the Helly Selection Principle, that every bounded sequence of regulated functions with uniformly bounded ε-variation contains a pointwise convergent subsequence and the limit is regulated. We use this result and prove even more, namely, that every bounded sequence in

$G(0,T)$ with uniformly bounded ε-variation contains an r-convergent subsequence. As the main results stated in Section 3.3, we show that all solutions x associated with a bounded set of inputs u with uniformly bounded ε-variation and with arbitrary $\alpha > 0$ constitute a set with uniformly bounded ε-variation (Theorem 3.6), and that the limit as $\alpha \to 0$ defines a deterministic pointwise $u \mapsto x$ hysteresis relation in $G(0,T)$ (Theorems 3.7, 3.8).

The argument relies substantially on the so-called *Play operator* defined as the solution operator of a variational inequality in integral form. We use only its scalar version with continuous inputs here; the general regulated Hilbert-valued case has been investigated in [2, 10]. Some basic properties of the play are listed in Section 3.4.

Sections 3.5 and 3.6 are reserved for proofs. In Section 3.5 we analyze the relationship between uniformly bounded ε-variation and various convergence concepts in the space of regulated functions, and in Section 3.6 we prove Theorems 3.6 – 3.8. In the concluding Section 3.7 we come back to the original problem presented in Section 3.2 and find sufficient conditions for the existence of a stable regime in which the undesirable oscillatory behaviour does not occur.

3.2 Physical Motivation

This research was originally motivated by the engineering problem of unstable spontaneous pressure oscillations having a hysteresis character and arising in a pump-valve system, see [5]. We briefly describe a simple mathematical model which reflects the main features. Consider a tube of length ℓ placed along the horizontal x-axis with a pump at $x = 0$ and valve at $x = \ell$, and assume that the fluid flow in the tube obeys the linear law

$$\left.\begin{aligned} p_t + c\,q_x &= 0 \\ q_t + c\,p_x &= 0 \end{aligned}\right. \qquad \text{for } (x,t) \in \,]0,\ell[\, \times \,]0,\infty[\,, \tag{3.3}$$

where p and q are the dimensionless pressure and discharge, respectively, and $c > 0$ is a constant sound speed. The system is coupled with initial and boundary conditions

$$\left.\begin{aligned} p(x,0) &= p^0(x), \\ q(x,0) &= q^0(x), \end{aligned}\right. \qquad \text{for } x \in [0,\ell]\,, \tag{3.4}$$

$$\left.\begin{aligned} \alpha q_t(0,t) + f(q(0,t)) + p(0,t) &= p_e + \bar{p}(t), \\ p(\ell,t) &= K\,q(\ell,t), \end{aligned}\right. \qquad \text{for } t > 0, \tag{3.5}$$

where p^0, q^0 are given continuous functions satisfying the compatibility condition

$$p^0(\ell) = K\,q^0(\ell)\,, \tag{3.6}$$

$\alpha > 0$ is a (small) constant called "pump inertance", f is a continuous (typically non-monotone) function describing the discharge-pressure characteristic of the pump, $K > 0$ is the valve parameter, p_e is a constant external pressure and $\bar{p}(t)$ is a given function which is regulated on each bounded interval and represents external pressure fluctuations. Using the dimensionless independent variables

$$t' = \frac{ct}{\ell}, \quad x' = \frac{x}{\ell}, \quad \alpha' = \frac{c\alpha}{\ell},$$

and omitting the primes, we can rewrite system (3.3) – (3.6) in the form

$$\begin{aligned} p_t + q_x &= 0, \\ q_t + p_x &= 0, \end{aligned} \qquad \text{for } (x,t) \in]0,1[\times]0,\infty[\,, \tag{3.7}$$

$$\begin{aligned} p(x,0) &= p^0(x), \\ q(x,0) &= q^0(x), \end{aligned} \qquad \text{for } x \in [0,1] \,, \tag{3.8}$$

$$\begin{aligned} \alpha q_t(0,t) + f(q(0,t)) + p(0,t) &= p_e + \bar{p}(t) \\ p(1,t) &= K\, q(1,t), \end{aligned} \qquad \text{for } t > 0 \,, \tag{3.9}$$

and

$$p^0(1) = K\, q^0(1) \,. \tag{3.10}$$

The solution to (3.7) – (3.10) will be constructed in the usual way by the method of characteristics; that is, we look for functions $\varphi : [-1,\infty[\to \mathbb{R}$, $\psi : [0,\infty[\to \mathbb{R}$ such that

$$\begin{aligned} (q+p)(x,t) &= \varphi(t-x), \\ (q-p)(x,t) &= \psi(t+x), \end{aligned} \qquad \text{for } (x,t) \in]0,1[\times]0,\infty[\,. \tag{3.11}$$

Conditions (3.8) imply that

$$\begin{aligned} \psi(x) &= q^0(x) - p^0(x) \,, \\ \varphi(-x) &= q^0(x) + p^0(x), \end{aligned} \qquad \text{for } x \in]0,1[\,.$$

Set

$$\lambda = \frac{K-1}{K+1} \in]-1,1[\,. \tag{3.12}$$

From the second condition in (3.9) and (3.11) it follows that

$$\psi(t+1) = -\lambda\, \varphi(t-1), \qquad \text{for } t > 0 \,.$$

This enables us to extend the domain of ψ by putting $\psi(t) = \psi_0(t)$ for $t \in [0,2]$, where

$$\psi_0(t) = \begin{cases} q^0(t) - p^0(t), & \text{for } t \in [0,1[\,, \\ -\lambda\,(q^0(2-t) + p^0(2-t)), & \text{for } t \in [1,2] \,, \end{cases} \tag{3.13}$$

and ψ_0 is continuous on $[0,2]$ by virtue of (3.10).

Let us consider now a sequence of equations with unknown functions y_k, $k = 0,1,2,\ldots$, in the form

$$\alpha \dot{y}_k(t) + g(y_k(t)) = \psi_k(t) + \bar{p}_k(t), \qquad \text{for } t \in [0,2], \tag{3.14}$$

analogous to (3.1), with initial conditions

$$y_0(0) = q^0(0)\,, \quad y_k(0) = y_{k-1}(2), \qquad \text{for } k = 1,2,\ldots, \tag{3.15}$$

where ψ_0 is given by (3.13) and

$$g(y) = y + f(y) - p_e, \qquad \text{for } y \in \mathbb{R}\,, \tag{3.16}$$
$$\bar{p}_k(t) = \bar{p}(t+2k), \qquad \text{for } k = 0,1,2,\ldots \text{ and } t \in [0,2]\,, \tag{3.17}$$
$$\psi_k(t) = \lambda\,(\psi_{k-1}(t) - 2\,y_{k-1}(t)), \qquad \text{for } k = 1,2,\ldots \text{ and } t \in [0,2]\,. \tag{3.18}$$

Note that by (3.13), (3.15), (3.18) we have $\psi_1(0) = \lambda\,(q^0(0) + p^0(0)) = \psi_0(2)$, and an easy induction yields

$$\psi_{k+1}(0) - \psi_k(2) = \lambda\,(\psi_k(0) - \psi_{k-1}(2) - 2\,(y_k(0) - y_{k-1}(2))) = 0,$$

for every $k = 1,2,\ldots$. Setting

$$\left.\begin{array}{rcl} \psi(t) & = & \psi_k(t-2k), \\ y(t) & = & y_k(t-2k), \end{array}\right. \qquad \text{for } t \in [2k, 2k+2[\,, \quad k = 0,1,2,\ldots, \tag{3.19}$$

and

$$\varphi(t) = \left\{ \begin{array}{rcll} 2y(t) & - & \psi(t), & \text{for } t \ge 0\,, \\ q^0(-t) & + & p^0(-t), & \text{for } t \in [-1,0[, \end{array}\right.$$

we see that ψ, φ, y are continuous, the functions

$$p(x,t) = \frac{1}{2}(\varphi(t-x) - \psi(t+x))$$

and

$$q(x,t) = \frac{1}{2}(\varphi(t-x) + \psi(t+x))$$

satisfy (3.8) – (3.10) pointwise, and equations (3.7) hold in the sense of distributions.

We have thus transformed the investigation of the long time behaviour of solutions to (3.7) – (3.10) into the problem of convergence for sequences $\{y_k\}$, $\{\psi_k\}$ defined by (3.14) – (3.18). The function g in (3.16) typically has the shape as in Fig. 3.1, and hence we are in the situation of equation (3.1) with a recurrent relation between the right-hand sides and previously computed solutions. The next four sections are devoted to a general analysis of the $u \mapsto x$ relation in (3.1), in the last section we derive a stability result for the particular problem (3.14) – (3.18).

3.3 Statement of the Problem and Main Results

As has been mentioned in the introduction, we work in the space $G(a,b)$ of regulated functions $[a,b] \to \mathbb{R}$ which we endow with the family of seminorms

$$\|u\|_{[c,d]} = \sup\{|u(t)|\,;\, a \le c \le t \le d \le b\}\,.$$

Indeed, $\|\cdot\|_{[a,b]}$ is a norm which transforms $G(a,b)$ into a Banach space, see [1]. More about regulated functions can be found e.g., in [4].

We denote by $G_L(a,b)$ and $G_R(a,b)$ the subspaces of $G(a,b)$ of left-continuous and right-continuous functions, respectively, and by $BV(a,b)$, $BV_L(a,b)$, $BV_R(a,b)$ the corresponding (dense) subsets of $G(a,b)$, $G_L(a,b)$, $G_R(a,b)$, respectively, of functions of bounded variation. The closed subspace of $G(a,b)$ with respect to the norm $\|\cdot\|_{[a,b]}$ consisting of all continuous real functions on $[a,b]$ will be denoted by $C(a,b)$, and we set $CBV(a,b) = C(a,b) \cap BV(a,b)$.

Here, and subsequently, we consider a fixed interval $[0,T]$ with some $T > 0$, and denote by $\mathbb{R}^+$ the open interval $]0,+\infty[$.

Definition 3.1. *A set $U \subset G(0,T)$ is said to have* uniformly bounded oscillation *if*

(i) *there exists a constant $R > 0$ such that*

$$|u(t) - u(s)| \;\le\; R\,, \qquad \forall u \in U \quad \forall s,t \in [0,T]\,,$$

(ii) *there exists a non-increasing function $N : \mathbb{R}^+ \to \mathbb{R}^+$ such that for every $r > 0$ and every system $\{\,]a_k,b_k[\;;\, k = 1,\dots,m\}$ of pairwise disjoint intervals $]a_k,b_k[\,\subset [0,T]$ the implication*

$$\Big(|u(b_k) - u(a_k)| \ge r \quad \forall k = 1,\dots,m\Big) \quad \Rightarrow \quad m \le N(r)$$

holds for every $u \in U$.

Definition 3.2. ([4]) *A set $U \subset G(0,T)$ is said to have* uniformly bounded ε-variation *if there exists a non-increasing function $L : \mathbb{R}^+ \to \mathbb{R}^+$ such that*

$$\forall \varepsilon > 0 \;\; \forall u \in U \;\; \exists \psi \in BV(0,T) : \quad \|u - \psi\|_{[0,T]} \le \varepsilon\,, \;\; \operatorname*{Var}_{[0,T]} \psi \le L(\varepsilon)\,.$$

Definition 3.3. ([11]) *Let $\{u_n\,;\, n \in \mathbb{N}\}$ be a sequence of functions $[0,T] \to \mathbb{R}$, and let $u : [0,T] \to \mathbb{R}$ be a given function. For $\delta > 0$ put*

$$M_\delta \;=\; \{t \in [0,T]\,;\; \exists s_n \to t \,:\; \limsup_{n\to\infty} |u_n(s_n) - u(t)| \ge \delta\}\,. \tag{3.20}$$

We say that u_n r-converges *to u, and write $u_n \stackrel{r}{\to} u$ as $n \to \infty$, if M_δ is finite for every $\delta > 0$.*

In order to characterize the relationship between the concepts introduced above, we first mention, without proof, the following special case of Theorem 2.2 in [2].

Proposition 3.4. *A set $U \subset G(0,T)$ has uniformly bounded oscillation if and only if it has uniformly bounded ε-variation.*

In Section 3.5, we prove the following hierarchy in convergence concepts in $G(a,b)$.

Proposition 3.5. *Let $\{u_n\,;\, n \in \mathbb{N}\}$ be a sequence of functions $[0,T] \to \mathbb{R}$.*

(i) *If $u_n \xrightarrow{r} u$, then there exists a countable set $M^* \subset [0,T]$ such that each $t \in [0,T]\backslash M^*$ is a continuity point of u, and $u_n(t) \to u(t)$ for every $t \in [0,T]\backslash M^*$ as $n \to \infty$.*

(ii) *If $\{u_n\,;\, n \in \mathbb{N}\}$ is a bounded sequence in $G(0,T)$ with uniformly bounded ε-variation, then there exists $u \in G(0,T)$ and a subsequence $\{u_{n_k}\}$ such that $u_{n_k} \xrightarrow{r} u$ and $u_{n_k}(t) \to u(t)$ for every $t \in [0,T]$ as $k \to \infty$.*

(iii) *If all u_n are in $G(0,T)$ and converge uniformly to a function $u \in G(0,T)$, then $\{u_n\,;\, n \in \mathbb{N}\}$ has uniformly bounded ε-variation.*

As a generalization of Helly's Selection Principle, it was shown in Theorem 3.8 in [4] that every bounded sequence of regulated functions with uniformly bounded ε-variation contains a pointwise convergent subsequence and the limit is regulated. Parts (i), (ii) of Proposition 3.5 thus improve this result. Part (iii) follows, for instance, from Proposition 5.6 in [10], but we give an elementary alternative proof here.

We now state the main results of this paper.

Theorem 3.6. *Let $U \subset G_L(0,T)$ be a bounded set with uniformly bounded oscillation, and let $c > 0$ be a constant. Then the set $X \subset W^{1,\infty}(0,T) \subset G_L(0,T)$ of all solutions x to (3.1) – (3.2), with $u \in U$, $x_0 \in [-c,c]$, and $\alpha > 0$, is bounded and has uniformly bounded oscillation.*

Let now $u \in G_L(0,T)$ and $x_0 \in \mathbb{R}$ be fixed. As we intend to let α tend to 0, we denote by x_α the solution of (3.1) for each value α of the singular parameter. We have the following two convergence results.

Theorem 3.7. *Let $u \in G_L(0,T)$ and $x_0 \in \mathbb{R}$ be given, and let $[t_1,t_2] \subset [0,T]$ be an arbitrary interval.*

(i) *If $u(t_1+) > G_+$ and $u(t) \geq G_-$ for every $t \in\,]t_1,t_2]$, then for every $t^* \in\,]t_1,t_2]$ there exists $\alpha_0 > 0$ such that*

$$x_\alpha(t) \in J_3\,, \qquad \forall t \in [t^*,t_2] \text{ and } \forall \alpha \in\,]0,\alpha_0]\,, \tag{3.21}$$

$$\lim_{\alpha \to 0+} x_\alpha(t) = g_3^{-1}(u(t))\,, \qquad \forall t \in\,]t_1,t_2]\,. \tag{3.22}$$

(ii) *If $u(t_1+) < G_-$ and $u(t) \leq G_+$ for every $t \in\,]t_1,t_2]$, then for every $t^* \in\,]t_1,t_2]$ there exists $\alpha_0 > 0$ such that*

$$x_\alpha(t) \in J_1\,, \qquad \forall t \in [t^*,t_2] \text{ and } \forall \alpha \in\,]0,\alpha_0]\,,$$

$$\lim_{\alpha \to 0+} x_\alpha(t) = g_1^{-1}(u(t))\,, \qquad \forall t \in\,]t_1,t_2]\,.$$

Theorem 3.8. *Let $u \in G_L(0,T)$ and $x_0 \in \mathbb{R}$ be given. Assume that one of the following two conditions holds:*

(i) $x_0 \notin J_2$,

(ii) $x_0 \in J_2$, $u(0+) \neq g(x_0)$.

Then there exists a function $x \in G_L(0,T)$ such that $x(t) \notin J_2$ and $g(x(t)) = u(t)$ for every $t \in]0,T]$, and $\lim_{\alpha\to 0+} x_\alpha(t) = x(t)$ for every $t \in [0,T]$.

Theorem 3.6 can be considered in the context of Propositions 3.4, 3.5 as a regularity result with respect to Theorem 3.2 in [11] which states that if $u_n \xrightarrow{r} u$, $x_0^n \to x_0$, and $\alpha_n \to 0$, then the corresponding sequence $\{x_n\}$ of solutions to (3.1) – (3.2) contains an r-convergent subsequence.

The example

$$u(t) = \begin{cases} 0, & \text{for} \quad t = 0\,, \\ \sin\frac{1}{t}, & \text{for} \quad t \in]0, 1/\pi]\,, \end{cases} \qquad u_n(t) = \begin{cases} 0, & \text{for} \quad t \in [0, 1/(n\pi)]\,, \\ \sin\frac{1}{t}, & \text{for} \quad t \in]1/(n\pi), 1/\pi]\,, \end{cases}$$

illustrates the main drawback of the r-convergence: we have $u_n \xrightarrow{r} u$ and $u_n(t) \to u(t)$ for every $t \in [0, 1/\pi]$, but the r-limit is not regulated although all functions u_n are continuous and the sequence is uniformly bounded. The advantage of Theorem 3.6 thus lies in the fact that it allows us to stay within the framework of regulated functions after passing to the limit.

Theorems 3.7 and 3.8 show the hysteresis character of the limit $u \mapsto x$ relation. The values of x leave the interval J_1 only if u exceeds G_+ from above, and the interval J_3 only if u exceeds G_- from below, cf. Fig. 3.1. The interval J_2 is the instability region. Note also that in the case $u(0+) = g(x_0)$, $x_0 \in J_2$ there is no convergence in general, see Remark below, cf. also Theorem 3.3 in [11].

We now devote Section 3.4 to a brief overview of results on the Play operator as the main tool in our analysis. The proof of Proposition 3.5 will be given in Section 3.5, and Theorems 3.6 – 3.8 will be proved in Section 3.6.

3.4 The Play Operator

In this section we consider the problem

Problem ($\mathcal{P}$). *For a given $r > 0$, $u \in C(0,T)$, and $z_0 \in [-r, r]$, find $\xi \in CBV(0,T)$ such that*

$$u(t) - \xi(t) \in [-r, r]\,, \qquad \forall t \in [0,T]\,, \tag{3.23}$$

$$u(0) - \xi(0) = z_0\,, \tag{3.24}$$

$$\int_0^T (u(\tau) - \xi(\tau) - y(\tau))\, d\xi(\tau) \geq 0\,, \quad \forall y \in C(0,T)\,,\ \|y\|_{[0,T]} \leq r. \tag{3.25}$$

The integral in (3.25) is the Riemann-Stieltjes integral. A vectorial counterpart of Problem ($\mathcal{P}$) was investigated in [8] and the extension to regulated inputs was done in [10]. The following results, Proposition 3.9 and Proposition 3.10, can be found in Proposition II.1.1, Remark II.1.3, and Exercise I.3.2 of [8].

Proposition 3.9. *For every $r > 0$ and $(z_0, u) \in [-r, r] \times C(0,T)$, there exists a unique $\xi \in CBV(0,T)$ satisfying (3.23) – (3.25). Moreover, if ξ_1, ξ_2 are solutions of (3.23) – (3.25) corresponding to $(z_0^i, u_i) \in [-r, r] \times C(0,T)$, $i = 1, 2$, respectively, then for every $t \in [0,T]$ we have*

$$\|\xi_1 - \xi_2\|_{[0,t]} \;\le\; \max\{|\xi_1(0) - \xi_2(0)|, \|u_1 - u_2\|_{[0,t]}\}\,.$$

Proposition 3.10. *Let $(z_0, u) \in [-r, r] \times C(0,T)$ be given, and let $\xi \in CBV(0,T)$ satisfy (3.23) – (3.25). Then for every $0 \le a < b \le T$ we have*

$$\int_a^b (u(\tau) - \xi(\tau) - y(\tau))\, d\xi(\tau) \;\ge 0\,, \qquad \forall y \in C(a,b)\,,\; \|y\|_{[a,b]} \le r\,.$$

As an easy consequence of Proposition 3.9, we have

Corollary 3.11. *Let $\xi = \mathfrak{p}_r[z_0, u]$ for some $(z_0, u) \in [-r, r] \times C(0,T)$. Then for every $0 \le s < t \le T$ we have*

$$|\xi(t) - \xi(s)| \;\le\; \|u(\cdot) - u(s)\|_{[s,t]}\,.$$

Proposition 3.9 enables us to define the one-parameter family of solution operators

$$\mathfrak{p}_r : [-r, r] \times C(0,T) \to CBV(0,T) \;:\; (z_0, u) \mapsto \xi = \mathfrak{p}_r[z_0, u]$$

of Problem $(\mathcal{P})$ called the *Play operators*. They were originally introduced in [7], and their various aspects have been systematically studied e. g. in [3, 8, 13, 14]. The extension to arbitrary measurable inputs has been done in [9] in a different setting. In Sections 3.5, 3.6 we make substantial use of the following characterization of the play which is typical for the scalar situation, see Fig. 3.2.

Proposition 3.12. *Let $r > 0$ and $(z_0, u) \in [-r, r] \times C(0,T)$ be given, and let $\xi = \mathfrak{p}_r[z_0, u]$. Then there exists a partition*

$$0 = s_0 \le t_0 < s_1 < t_1 < \ldots < s_m \le t_m = T$$

such that ξ is

(i) *monotone in $[t_{k-1}, s_k]$ for $k = 1, \ldots, m$,*

(ii) *constant in $[s_k, t_k]$ for $k = 0, \ldots, m$,*

(iii) *non-monotone in $[t_{k-1}, t_k + \delta]$ for any $\delta > 0$ and $k = 1, \ldots, m-1$,*

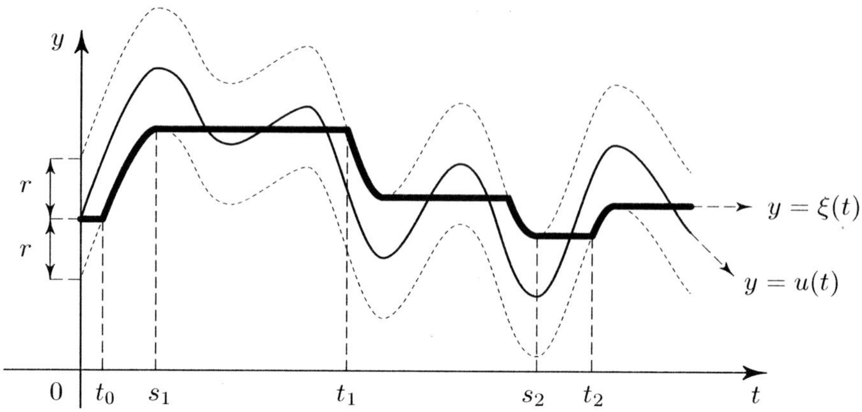

Figure 3.2. *A diagram of the Play operator* $\xi = \mathfrak{p}_r[0, u]$

and

$$(u(s_k) - \xi(s_k))(u(t_k) - \xi(t_k)) = -r^2, \qquad \text{for } k = 1, \dots, m-1, \tag{3.26}$$

$$|u(s_k) - u(s_{k-1})| \geq 2r, \qquad \text{for } k = 2, \dots, m-1. \tag{3.27}$$

Before giving the proof of Proposition 3.12, we start with the following auxiliary result.

Lemma 3.13. *Let* $(z_0, u) \in [-r, r] \times C(0, T)$ *be given, and let* $\xi = \mathfrak{p}_r[z_0, u]$*. Then for every* $t \in [0, T]$ *the following implications hold.*

(i) *If* $u(t) - \xi(t) > -r$*, then there exists* $\delta > 0$ *such that* ξ *is non-decreasing in* $[t - \delta, t + \delta] \cap [0, T]$*;*

(ii) *If* $u(t) - \xi(t) < r$*, then there exists* $\delta > 0$ *such that* ξ *is non-increasing in* $[t - \delta, t + \delta] \cap [0, T]$*;*

Proof. The argument is the same in each of the cases (i), (ii). In (i), for instance, we find $\delta > 0$ and $\varrho > 0$ such that $u(\tau) - \xi(\tau) \geq -r + \varrho$ for $\tau \in [t - \delta, t + \delta] \cap [0, T]$. For every $[a, b] \subset [t - \delta, t + \delta] \cap [0, T]$ we define $y(\tau) = u(\tau) - \xi(\tau) - \varrho$ for $\tau \in [a, b]$. Then $-r \leq y(\tau) \leq u(\tau) - \xi(\tau) \leq r$ for every $\tau \in [a, b]$, and Proposition 3.10 yields

$$\varrho(\xi(b) - \xi(a)) = \int_a^b (u(\tau) - \xi(\tau) - y(\tau))\, d\xi(\tau) \geq 0,$$

hence ξ is non-decreasing in $[t - \delta, t + \delta] \cap [0, T]$. The proof of case (ii) is similar. □

Proof of Proposition 3.12.

If $|u(\tau) - \xi(\tau)| < r$ for every $\tau \in [0, T[$, then ξ is constant in $[0, T]$ and the assertion is trivial. If this is not the case, we put

$$t_0 = \min\{t \in [0, T[\ ;\ |u(t) - \xi(t)| = r\}.$$

By Lemma 3.13, we have $\xi(\tau) = \xi(0)$ for all $\tau \in [0, t_0]$. Assume, for instance, that $u(t_0) - \xi(t_0) = r$; the other case is obtained by symmetry. By Lemma 3.13, there exists $\delta > 0$ such that ξ is non-decreasing in $[t_0, t_0 + \delta]$, and we may put

$$t_1 = \sup\{t \in]t_0, T]\ ;\ \xi \text{ is non-decreasing in } [t_0, t]\}.$$

We stop the algorithm if $t_1 = T$. Otherwise, we have by Lemma 3.13 that $u(t_1) - \xi(t_1) = -r$. We continue by induction and construct a sequence $0 \le t_0 < t_1 < t_2 < \dots$ passing from t_k to t_{k+1} provided $t_k < T$, with the properties

$$u(t_k) - \xi(t_k) = (-1)^k r, \tag{3.28}$$

$$(-1)^{k-1}\xi \quad \text{is non-decreasing in} \quad [t_{k-1}, t_k], \tag{3.29}$$

$$\xi \quad \text{is non-monotone in} \quad [t_{k-1}, t_k + \delta] \quad \text{for any} \quad \delta > 0. \tag{3.30}$$

We fix any $\ell \in \mathbb{N}$ such that $t_\ell < T$, and for $k = 1, 2, \dots, \ell$ we define the points

$$s_k = \min\{t \in [t_{k-1}, t_k]\ ;\ \xi(t) = \xi(t_k)\}.$$

Assume that $(-1)^{k-1}(u(s_k) - \xi(s_k)) < r$. We may use Lemma 3.13 to obtain that $(-1)^{k-1}\xi$ is non-increasing in $[s_k - \delta, s_k]$ for some $\delta > 0$, and hence, by (3.29), ξ is constant in $[s_k - \delta, s_k]$, which contradicts the definition of s_k. Consequently,

$$(-1)^{k-1}(u(s_k) - \xi(s_k)) = r, \quad t_{k-1} < s_k < t_k \quad \text{for } k = 1, 2, \dots, \ell. \tag{3.31}$$

For $k = 2, \dots, \ell$, we have by (3.29), (3.31) that

$$\begin{aligned}(-1)^{k-1}(u(s_k) - u(s_{k-1})) &= (-1)^{k-1}(u(s_k) - \xi(s_k) - u(s_{k-1}) \\ &\quad + \xi(s_{k-1})) + (-1)^{k-1}(\xi(t_k) - \xi(t_{k-1})) \\ &\ge 2r.\end{aligned} \tag{3.32}$$

Since u is continuous, the number ℓ cannot be arbitrarily large, and there exists necessarily $m \in \mathbb{N}$ such that $t_m = T$, $\ell \le m - 1$. Consequently, (3.26) follows from (3.28) and (3.31), (3.27) follows from (3.32), and Proposition 3.12 is proved. □

3.5 Uniformly Bounded Oscillation

This section is devoted to the proof of Proposition 3.5.

Proof of Proposition 3.5.

(i) Let M_δ be the sets (3.20). Then $M^* := \bigcup_{n=1}^{\infty} M_{1/n}$ is countable, and for every $t \in [0, T] \setminus M^*$ and every sequence $s_n \to t$ we have

$$\lim_{n\to\infty} u_n(s_n) = u(t).$$

In particular, $u_n(t) \to u(t)$ for every $t \in [0,T] \setminus M^*$. We now check that

$$u(t-) \;=\; u(t) \;=\; u(t+)\,, \quad \forall t \in [0,T] \setminus M^*\,.$$

Let $t \in]0,T] \setminus M^*$ and $\delta > 0$ be given, and let $t_k \nearrow t$ be an arbitrary sequence in $[0,T]$. There exists $k_0 \in \mathbb{N}$ such that $t_k \notin M_{\delta/2}$ for $k \geq k_0$. For each such k and for n sufficiently large put $s_k^n = t_k - 1/n$. We have $\limsup_{n\to\infty} |u_n(s_k^n) - u(t_k)| < \delta/2$, hence for every $k \geq k_0$ we can find $n_k > k$ such that $|u_{n_k}(s_k^{n_k}) - u(t_k)| \leq \delta/2$. The points $\sigma_k := s_k^{n_k}$ satisfy $t_k - 1/k \leq \sigma_k \leq t_k$, hence

$$\lim_{k\to\infty} \sigma_k = t\,, \qquad \lim_{k\to\infty} |u_{n_k}(\sigma_k) - u(t)| = 0\,.$$

We find $k_1 \geq k_0$ such that for $k \geq k_1$ we have $|u_{n_k}(\sigma_k) - u(t)| < \delta/2$, hence

$$|u(t_k) - u(t)| \;<\; \delta\,, \qquad \text{for} \quad k \geq k_1\,.$$

We conclude that $u(t) = u(t-)$ for all $t \in [0,T] \setminus M^*$. The argument for $u(t+)$ is similar.

(ii) According to Theorem 3.8 in [4], we may assume that

$$\lim_{n\to\infty} u_n(t) = u(t)\,, \qquad \forall t \in [0,T]\,.$$

The argument will be based on a gradual selection of subsequences and diagonalization. For this purpose, we denote by $\mathcal{E}(\mathbb{N})$ the system of all infinite subsets of $\mathbb{N}$.

Assume for the purpose of contradiction the hypothesis that

$$\{u_n\} \text{ does not contain any } r\text{-convergent subsequence.} \tag{3.33}$$

Then the number

$$\delta_1 \;:=\; \inf\{\delta > 0\,;\, M_\delta \text{ is finite}\}\,,$$

where M_δ is the set from (3.20), is positive. We fix $t_1 \in M_{\delta_1/2}$, a set $K_1 \in \mathcal{E}(\mathbb{N})$, and a sequence $\{s_n^1\,;\, n \in K_1\}$ in $[0,T]$ such that

$$\lim_{n\to\infty,\, n\in K_1} s_n^1 \;=\; t_1\,,$$

$$|u_n(s_n^1) - u(t_1)| \;\geq\; \delta_1/4\,, \qquad \forall n \in K_1\,.$$

By hypothesis (3.33), $\{u_n\,;\, n \in K_1\}$ is not r-convergent, hence we may put

$$\delta_2 \;:=\; \inf\{\delta > 0\,;\, M_\delta(K_1) \text{ is finite}\} \;>\; 0\,,$$

where $M_\delta(K)$ for some $K \in \mathcal{E}(\mathbb{N})$ denotes the set (3.20) related to the subsequence $\{u_n\,;\, n \in K\}$. Note that for $K, K' \in \mathcal{E}(\mathbb{N})$, $K' \subset K$, we have

$$M_\delta(K') \subset M_\delta(K)\,, \qquad \forall \delta > 0\,. \tag{3.34}$$

This implies in particular that $\delta_2 \le \delta_1$. The set $M_{\delta_2/2}(K_1)$ contains infinitely many points, hence we may fix $t_2 \in M_{\delta_2/2}(K_1)$, $t_2 \neq t_1$, a set $K_2 \subset K_1$, $K_2 \in \mathcal{E}(\mathbb{N})$, and a sequence $\{s_n^2 ; n \in K_2\}$ in $[0,T]$ such that

$$\lim_{n\to\infty,\, n\in K_2} s_n^2 = t_2 \,,$$

$$|u_n(s_n^2) - u(t_2)| \ \ge \ \delta_2/4 \,, \qquad \forall n \in K_2 \,.$$

We continue by induction and construct a sequence $\{K_j ; j \in \mathbb{N}\}$ of sets $K_j \in \mathcal{E}(\mathbb{N})$, $\mathbb{N} \supset K_1 \supset K_2 \supset \ldots$, positive numbers $\delta_1 \ge \delta_2 \ge \ldots$, and sequences $\{s_n^j ; n \in K_j\}$ of elements of $[0,T]$ such that for all $j \in \mathbb{N}$ we have

$$\delta_j := \inf\{\delta > 0 ;\, M_\delta(K_{j-1}) \text{ is finite}\} \,,$$

$$t_j \in M_{\delta_j/2}(K_{j-1}) \,, \quad t_j \neq t_i \quad \text{for } i \neq j \,,$$

$$\lim_{n\to\infty,\, n\in K_j} s_n^j = t_j \,,$$

$$|u_n(s_n^j) - u(t_j)| \ \ge \ \delta_j/4 \,, \qquad \forall n \in K_j \,.$$

We proceed by diagonalization, and construct a set $K^* = \{n_j ; j \in \mathbb{N}\} \in \mathcal{E}(\mathbb{N})$ of integers such that $n_j \in K_j$, $n_{j+1} > n_j$ for every $j \in \mathbb{N}$. By hypothesis (3.33), the sequence $\{u_{n_j} ; j \in \mathbb{N}\}$ does not r-converge, hence

$$\delta^* := \inf\{\delta > 0 ;\, M_\delta(K^*) \text{ is finite}\} > 0 \,.$$

The argument of (3.34) yields

$$M_\delta(K^*) \subset M_\delta(K_j) \,, \qquad \forall \delta > 0 \ \text{ and } \forall j \in \mathbb{N} \,,$$

hence $\delta_j \ge \delta^*$ for every $j \in \mathbb{N}$. We further have

$$\lim_{n\to\infty,\, n\in K^*} s_n^j = t_j \,, \quad \forall j \in \mathbb{N} \,,$$

$$|u_n(s_n^j) - u(t_j)| \ \ge \ \delta_j/4 \ \ge \ \delta^*/4 \,, \qquad \forall n \in K^* \,,\ n \ge n_j \,.$$

We now obtain the contradiction by proving that the sequence $\{u_n ; n \in K^*\}$ does not have uniformly bounded oscillation, and using Proposition 3.4. This will be done in the following way. For each $m \in \mathbb{N}$ we find $n(m) \in K^*$ such that for all $j = 1, \ldots, m$ we have

$$|s_{n(m)}^j - t_j| \ < \ \frac{1}{2} \min\{|t_j - t_i| ;\, i = 1, \ldots, m \,,\ i \neq j\} \,,$$

$$|u_{n(m)}(s_{n(m)}^j) - u(t_j)| \ \ge \ \frac{\delta^*}{4} \,,$$

$$|u_{n(m)}(t_j) - u(t_j)| \ \le \ \frac{\delta^*}{8} \,.$$

This yields that $|u_{n(m)}(s_{n(m)}^j) - u_{n(m)}(t_j)| \ge \delta^*/8$ for each $j = 1, \ldots, m$, hence the sequence $\{u_n ; n \in K^*\}$ does not have uniformly bounded oscillation, and the proof is complete.

(iii) For $r > 0$ and $n \in \mathbb{N}$ put

$$\begin{aligned}
N_n(r) &= \max\{m \in \mathbb{N}\,;\ \exists\{\,]a_k, b_k[\, \subset [0,T]\,;\ k = 1, \dots, m\} \text{ pairwise disjoint,} \\
&\qquad |u_n(b_k) - u_n(a_k)| \geq r\}\,, \\
N_\infty(r) &= \max\{m \in \mathbb{N}\,;\ \exists\{\,]a_k, b_k[\, \subset [0,T]\,;\ k = 1, \dots, m\} \text{ pairwise disjoint,} \\
&\qquad |u(b_k) - u(a_k)| \geq r\}
\end{aligned}$$

All these numbers are well defined, as u_n, u are regulated. Let $n(r) \in \mathbb{N}$ be such that $\|u_n - u\|_{[0,T]} < r/4$ for $n \geq n(r)$. For such n we then have $N_n(r) \leq N_\infty(r/2)$, hence

$$\sup_{n \in \mathbb{N}} N_n(r) \ \leq\ \max\{N_\infty(r/2),\ \max\{N_n(r)\,;\ n = 1, \dots, n(r)\}\}\,,$$

hence the system $\{u_n\,;\ n \in \mathbb{N}\}$ has uniformly bounded oscillation, and the assertion follows from Proposition 3.4. □

3.6 Singularly Perturbed Equations

We start this section with an easy Lemma.

Lemma 3.14. *Let $[a,b] \subset \mathbb{R}$ be an interval such that $g(a) \leq g(b)$. Let $\alpha > 0$, $u \in G_L(0,T)$, and $x_0 \in [-c,c]$ be arbitrary, and let x be the corresponding solution of (3.1). Assume that there exists an interval $[s,t] \subset [0,T]$ such that*

(i) *$u(\tau) \in [g(a), g(b)]$ for every $\tau \in\,]s,t]$,*

(ii) *$x(s) \in [a,b]$.*

Then $x(\tau) \in [a,b]$ for all $\tau \in [s,t]$.

Proof. Set

$$\bar{g}(y) = \begin{cases} g(a), & \text{for } \ y < a\,, \\ g(y), & \text{for } \ y \in [a,b]\,, \\ g(b), & \text{for } \ y > b\,, \end{cases}$$

and consider the differential equation in $[s,t]$

$$\alpha \dot{z}(\tau) + \bar{g}(z(\tau)) = u(\tau)\,, \qquad z(s) = x(s)\,.$$

We test the identity

$$\alpha \dot{z}(\tau) + \bar{g}(z(\tau)) - g(b) \ = \ u(\tau) - g(b)$$

by $(z(\tau) - b)^+ := \max\{0, z(\tau) - b\}$. The implication $z(\tau) > b \Rightarrow \bar{g}(z(\tau)) = g(b)$ yields

$$\frac{1}{2}\frac{d}{d\tau}\left((z(\tau) - b)^+\right)^2 \ \leq\ 0 \quad \text{a.e. in } \ ]s,t[\,,$$

and hence $(z(\tau)-b)^+ \leq (x(s)-b)^+ = 0$ for every $\tau \in [s,t]$. Similarly, testing the identity

$$\alpha\dot{z}(\tau) + \bar{g}(z(t)) - g(a) \;=\; u(t) - g(a)$$

by $(z(\tau)-a)^- := \max\{0, -z(\tau)+a\}$, we obtain $(z(\tau)-a)^- \leq (x(s)-a)^- = 0$. Hence $x(\tau) = z(\tau) \in [a,b]$ for every $\tau \in [s,t]$ and $\alpha > 0$, and the proof is complete. □

The proof of Theorem 3.6 consists of several steps.

Proof of Theorem 3.6.

Step 1. *Boundedness of X.*

We fix real numbers $A < B$ such that

$$A \;\leq\; \min\{-c, g_1^{-1}(G_-)\}\,, \qquad B \;\geq\; \max\{c, g_3^{-1}(G_+)\}\,,$$

and $g(A) \leq u(t) \leq g(B)$ for each $u \in U$ and $t \in [0,T]$, see Fig. 3.1. From Lemma 3.14 we conclude that

$$x(t) \in [A,B] \quad \text{for every} \quad t \in [0,T]\,,\ \alpha > 0\,, \quad \text{and} \quad u \in U\,. \tag{3.35}$$

We now choose an arbitrary $r > 0$ which we keep fixed in Step 2 – Step 4 below.

Step 2. *Oscillations of $\eta = \mathfrak{p}_r[0,x]$.*

Let $\alpha > 0$, $x_0 \in [-c,c]$, and $u \in U$ be given, let $x \in X$ be a solution of (3.1), and put $\eta = \mathfrak{p}_r[0,x]$. Invoking Proposition 3.12 we construct a partition

$$0 = s_0 \leq t_0 < s_1 < t_1 < \ldots < s_m \leq t_m = T$$

such that η is

(i) monotone in $[t_{k-1}, s_k]$ for $k = 1,\ldots,m$,

(ii) constant in $[s_k, t_k]$ for $k = 0,\ldots,m$,

(iii) non-monotone in $[t_{k-1}, t_k + \delta]$ for any $\delta > 0$ and $k = 1,\ldots,m-1$.

We may assume that η is non-decreasing in $[s_0, s_1]$; the other case is analogous. Then $(-1)^{k-1}\eta$ is non-decreasing in $[s_{k-1}, s_k]$; hence, in particular,

$$(-1)^{k-1}(\eta(t) - \eta(s_k)) \;\leq\; 0 \qquad \text{for} \quad t \in [s_{k-1}, s_{k+1}]\,, \quad k = 1,\ldots,m-1\,. \tag{3.36}$$

By (3.31) we have

$$(-1)^{k-1}(x(s_k) - \eta(s_k)) \;=\; r, \qquad \text{for} \quad k = 1,\ldots,m-1\,, \tag{3.37}$$

and from (3.36) – (3.37) we obtain

$$\begin{aligned}(-1)^{k-1}(x(s_k) - x(s_{k-1})) &= (-1)^{k-1}(x(s_k) - \eta(s_k) + \eta(s_{k-1}) - x(s_{k-1})) \\ &\quad + (-1)^{k-1}(\eta(s_k) - \eta(s_{k-1})) \;\geq\; 2r \end{aligned} \tag{3.38}$$

for $k = 2, \dots, m-1$, similarly to (3.32). The elementary inequality $|x(t) - \eta(t)| \le r$ and (3.36) – (3.37) yield for $t \in [s_{k-1}, s_{k+1}]$, $k = 1, \dots, m-1$, that

$$(-1)^{k-1}(x(t) - x(s_k)) = (-1)^{k-1}(x(t) - \eta(t) + \eta(s_k) - x(s_k)) + (-1)^{k-1}(\eta(t) - \eta(s_k)) \;\le\; 0\,. \tag{3.39}$$

Integrating equation (3.1) we obtain from (3.39), for $h > 0$ sufficiently small, that

$$(-1)^{k-1}\frac{1}{h}\int_{s_k}^{s_k+h} (u(\tau) - g(x(\tau)))\,d\tau = (-1)^{k-1}\frac{\alpha}{h}(x(s_k+h) - x(s_k)) \;\le\; 0\,,$$

$$(-1)^{k-1}\frac{1}{h}\int_{s_k-h}^{s_k} (u(\tau) - g(x(\tau)))\,d\tau = (-1)^{k-1}\frac{\alpha}{h}(x(s_k) - x(s_k-h)) \;\ge\; 0\,,$$

and letting $h \to 0+$ we obtain, for $k = 1, \dots, m-1$, the inequalities

$$(-1)^{k-1}(u(s_k+) - g(x(s_k))) \le 0\,, \tag{3.40}$$

$$(-1)^{k-1}(u(s_k) - g(x(s_k))) \ge 0\,. \tag{3.41}$$

Step 3. *Oscillations of u.*
Set

$$\mu(r) = \inf\{|g_i(v) - g_i(w)|\,;\; v, w \in J_i\,,\; |v - w| \ge r\,,\; i = 1,2,3\}\,, \tag{3.42}$$

$$\kappa(r) = \frac{1}{2}\min\left\{G^+ - G^-, \mu(r)\right\}\,. \tag{3.43}$$

We claim that:

For every $k = 2, \dots m-3$ there exists an interval $[a_k, b_k] \subset [s_{k-1}, s_{k+2}]$ such that $|u(b_k) - u(a_k)| \ge \kappa(r)$. (3.44)

To prove the claim, we classify the points $s_1, \dots, s_{m-1}$ by introducing the sets

$$\Theta_i \;=\; \{s_k\,;\; k \in \{1, \dots, m-1\}\,,\; x(s_k) \in J_i\} \quad \text{for} \quad i = 1,2,3\,,$$

and fix $k \in \{2, \dots, m-3\}$. We say that the quadruple $\{s_{k-1}, s_k, s_{k+1}, s_{k+2}\}$ is of type

(ij) if there exists $\ell \in \{k, k+1, k+2\}$ such that $s_{\ell-1} \in \Theta_i$, $s_\ell \in \Theta_j$, $i, j = 1,2,3$,

(ijp) if there exists $\ell \in \{k, k+1\}$ such that $s_{\ell-1} \in \Theta_i$, $s_\ell \in \Theta_j$, $s_{\ell+1} \in \Theta_p$, $i, j, p = 1,2,3$,

(ijpq) if $s_{k-1} \in \Theta_i$, $s_k \in \Theta_j$, $s_{k+1} \in \Theta_p$, $s_{k+2} \in \Theta_q$, $i, j, p, q = 1,2,3$.

We distinguish the following categories **A** – **E**.

A. (ii) *for* $i = 1, 3$.
By (3.38), (3.40) – (3.42) we have

$$\begin{aligned}(-1)^{\ell-1}(u(s_\ell) - u(s_{\ell-1})) &= (-1)^{\ell-1}(u(s_\ell) - g(x(s_\ell)) + g(x(s_{\ell-1})) - u(s_{\ell-1})) \\ &\quad + (-1)^{\ell-1}(g(x(s_\ell)) - g(x(s_{\ell-1}))) \\ &\geq \ |g(x(s_\ell)) - g(x(s_{\ell-1}))| \geq \mu(2r)\,,\end{aligned}$$

and it suffices to put $a_k = s_{\ell-1}$, $b_k = s_\ell$.

B. (22i) *for* $i = 1, 2, 3$.
We again use (3.38), (3.40) – (3.42) to estimate

$$\begin{aligned}&(-1)^{\ell-1}(u(s_\ell+) - u(s_{\ell-1}+)) \\ &\quad = (-1)^{\ell-1}(u(s_\ell+) - g(x(s_\ell)) + g(x(s_{\ell-1})) - u(s_{\ell-1}+)) \\ &\qquad + (-1)^{\ell-1}(g(x(s_\ell)) - g(x(s_{\ell-1}))) \\ &\quad \leq -|g(x(s_\ell)) - g(x(s_{\ell-1}))| \ \leq \ -\mu(2r)\,.\end{aligned}$$

We obtain (3.44) by choosing $a_k \in]s_{\ell-1}, s_\ell]$ sufficiently close to $s_{\ell-1}$ and $b_k \in]s_\ell, s_{\ell+1}]$ sufficiently close to s_ℓ.

C. (1ij) *for* $i = 2, 3$, $j = 1, 2$.
We have $x(s_{\ell-1}) \leq x_- < x(s_\ell)$, hence ℓ is odd, and we find

$$\tau_\ell \ = \ \max\{\tau \in [s_{\ell-1}, s_\ell[\ ;\ x(\tau) = x_-\}\,.$$

From equation (3.1) it follows that

$$u(\tau_\ell+) \ \geq \ G_+ = g(x(\tau_\ell))\,, \tag{3.45}$$

and (3.40) – (3.41) yield

$$u(s_\ell+) \ \leq \ g(x(s_\ell))\,, \quad u(s_{\ell+1}) \ \leq \ g(x(s_{\ell+1}))\,. \tag{3.46}$$

We now make finer distinctions by considering the following subcategories.

C1. (121) .
We have $x(s_{\ell+1}) \leq x(\tau_\ell)$, and either

$$x(\tau_\ell) - x(s_{\ell+1}) \ \geq \ r\,, \tag{3.47}$$

or

$$x(\tau_\ell) - x(s_{\ell+1}) \ < \ r\,. \tag{3.48}$$

If (3.47) holds, then it follows from (3.45) – (3.46) that

$$u(\tau_\ell+) - u(s_{\ell+1}) \ \geq \ g(x(\tau_\ell)) - g(x(s_{\ell+1})) \ \geq \ \mu(r)\,,$$

and we choose $a_k \in]\tau_\ell, s_\ell]$ sufficiently close to τ_ℓ, and $b_k = s_{\ell+1}$.

If (3.48) holds, then (3.38) yields $x(s_\ell) - x(\tau_\ell) > r$, and it follows from (3.45) – (3.46) that

$$u(\tau_\ell+) - u(s_\ell+) \;\geq\; g(x(\tau_\ell)) - g(x(s_\ell)) \;\geq\; \mu(r)\,,$$

and we choose $a_k \in]\tau_\ell, s_\ell]$ sufficiently close to τ_ℓ, and $b_k \in]s_\ell, s_{\ell+1}]$ sufficiently close to s_ℓ.

C2. (122) .

We have $x(\tau_\ell) = x_- < x(s_{\ell+1}) < x(s_\ell)$, and $x(s_\ell) - x(\tau_\ell) > x(s_\ell) - x(s_{\ell+1}) \geq 2r$ by (3.38). From (3.45) – (3.46) it follows that

$$u(\tau_\ell+) - u(s_\ell+) \;\geq\; g(x(\tau_\ell)) - g(x(s_\ell)) \;\geq\; \mu(2r)\,,$$

and we choose suitable $a_k \in]\tau_\ell, s_\ell]$ and $b_k \in]s_\ell, s_{\ell+1}]$ as above.

C3. (13i) *for* $i = 1, 2$.

We have $x(s_{\ell-1}) \leq x_- < x_+ \leq x(s_\ell)$, $x(s_\ell) \geq x_+ > -x(s_{\ell+1})$. We find

$$\sigma_\ell \;=\; \max\{\tau \in [s_\ell, s_{\ell+1}[\;;\; x(\tau) = x_+\}\,.$$

Using equation (3.1) we argue as in (3.45), and obtain

$$u(\sigma_\ell+) \leq G_- = g(x(\sigma_\ell))\,.$$

We therefore have $u(\sigma_\ell+) - u(\tau_\ell+) \geq G_+ - G_-$, and it suffices to choose suitable $a_k \in]\tau_\ell, s_\ell]$ and $b_k \in]\sigma_\ell, s_{\ell+1}]$.

D. (3ij) *for* $i = 1, 2$, $j = 2, 3$.

This case is completely symmetric with respect to **C**, with the subcategories **D1: (323), D2: (322), D3: (31i)** for $i = 2, 3$. We leave the details to the reader.

E. (123), (321), (1332), (2113), (2331), (3112), (ijji) for $i, j = 1, 2, 3$, $i \neq j$.

These cases cannot occur, as this would mean $x(s_{\ell-1}) < x(s_\ell) < x(s_{\ell+1})$ or $x(s_{\ell-1}) > x(s_\ell) > x(s_{\ell+1})$ for some $\ell \in \{k, k+1\}$, in contradiction to (3.38).

Table 1 below shows that the above list exhausts all possible cases **(ijpq)**, $i, j, p, q = 1, 2, 3$.

ij \ pq	11	12	13	21	22	23	31	32	33
11	A	A	A	A	A	E	A	A	A
12	A	C1	C1	E	B	B	E	E	E
13	A	C3	C3	E	C3	C3	E	E	A
21	A	E	E	C1	C2	E	C3	C3	A
22	A	B	B	B	B	B	B	B	A
23	A	D3	D3	E	D2	D1	E	E	A
31	A	E	E	D3	D3	E	D3	D3	A
32	E	E	E	B	B	E	D1	D1	A
33	A	A	A	E	A	A	A	A	A

Table 1: *Distribution of cases* **(ijpq)** *into categories A – E*

Step 4. *Total variation of η.*

By hypothesis, the set U has uniformly bounded oscillation. Hence, the number m^* of intervals $]a_k, b_k[$ such that (3.44) holds is at most $N(\kappa(r))$. By (3.44), we may choose $]a_1, b_1[\subset [s_1, s_4]$, $]a_2, b_2[\subset [s_4, s_7], \ldots,]a_j, b_j[\subset [s_{3j-2}, s_{3j+1}]$ for $j = 1, \ldots m^*$, as long as $3j + 1 \le m - 1$. Putting

$$m^* = \max\{j \in \mathbb{N}\,;\, 3j + 1 \le m - 1\},$$

we have

$$m \le 3m^* + 4 \le 3N(\kappa(r)) + 4.$$

Using Step 1, Step 2 and Corollary 3.11, we obtain

$$\begin{aligned} \operatorname*{Var}_{[0,T]} \eta = \sum_{k=1}^{m} |\eta(s_k) - \eta(s_{k-1})| \;&\le\; \sum_{k=1}^{m} \|x(\cdot) - x(s_{k-1})\|_{[s_{k-1}, s_k]} \\ &\le m\,(B - A) \;\le\; (B - A)\,(3N(\kappa(r)) + 4). \end{aligned}$$

Since $\|x - \eta\|_{[0,T]} \le r$, by definition of the play, and since the above argument is valid for arbitrary $r > 0$, we conclude that X has uniformly bounded ε-variation, and the assertion of Proposition 3.4 completes the proof of Theorem 3.6. □

We now investigate the limit passage as $\alpha \to 0$. With the notation from Theorem 3.7, we first establish the following result.

Lemma 3.15. *Let $[a, b] \subset \mathbb{R}$ and $[s, t] \subset [0, T]$ be intervals such that*

(i) *g is increasing in $[a, b]$,*

(ii) *$x_\alpha(\tau) \in [a, b]$ for every $\tau \in [s, t]$ and for α sufficiently small.*

Then we have

$$u(\tau) \in [g(a), g(b)], \quad \lim_{\alpha \to 0+} x_\alpha(\tau) = \left(g|_{[a,b]}\right)^{-1}(u(\tau)) \quad \textit{for every } \tau \in]s, t].$$

Proof. Assume, for instance, that $u(\tau) < g(a)$ for some $\tau \in]s, t]$. We fix $h \in]0, \tau - s[$ and $\varepsilon > 0$ such that $u(\sigma) \le g(a) - \varepsilon$ for $\sigma \in [\tau - h, \tau]$. Integrating equation (3.1) from $\tau - h$ to τ we obtain that $\alpha(x_\alpha(\tau) - x_\alpha(\tau - h)) \le -h\varepsilon$, which is a contradiction for α sufficiently small. The same contradiction is obtained if $u(\tau) > g(b)$.

The system $\{x_\alpha\,;\, \alpha > 0\}$ is bounded and has uniformly bounded ε-variation, hence, according to Theorem 3.8 in [4], there exists a sequence $\alpha_n \to 0$ and a function $x^* \in G(0, T)$ such that

$$\lim_{n \to \infty} x_{\alpha_n}(\tau) = x^*(\tau), \qquad \forall \tau \in [0, T]. \tag{3.49}$$

We therefore have $x^*(\tau) \in [a, b]$ for $\tau \in [s, t]$. Lemma 3.15 will be proved if we check that

$$g(x^*(\tau)) = u(\tau) \qquad \text{for every } \tau \in]s, t]. \tag{3.50}$$

Indeed, then $x^*(\tau) = \left(g|_{[a,b]}\right)^{-1}(u(\tau))$ is the unique limit of $x_\alpha(\tau)$ as $\alpha \to 0+$ independently of the subsequence α_n.

For every test function $w \in \overset{o}{W}{}^{1,1}(0,T)$ we have

$$\alpha_n \int_0^T x_{\alpha_n}(\tau)\,\dot{w}(\tau)\,d\tau = \int_0^T (g(x_{\alpha_n}(\tau)) - u(\tau))\,w(\tau)\,d\tau\,,$$

and letting $\alpha_n \to 0$ we obtain

$$g(x^*(\tau)) = u(\tau), \qquad \text{for a. e. } \tau \in]0,T[\,.$$

To prove (3.50), we fix $\tau \in]s,t]$, and an arbitrary $\varepsilon > 0$. Let $\delta > 0$ and $h \in]0, \tau - s[$ be such that

$$\delta = \min\{|g(y) - g(z)|\,;\; y,z \in [a,b]\,,\; |y - z| \geq \varepsilon\}\,,$$

$$\sigma \in [\tau - h, \tau] \quad \Rightarrow \quad |u(\sigma) - u(\tau)| < \frac{\delta}{2}\,.$$

Let $\varrho_k \nearrow \tau$ be a sequence such that $0 < \tau - \varrho_k < h$, and $g(x^*(\varrho_k)) = u(\varrho_k)$ for every $k \in \mathbb{N}$. For all $\sigma \in [\varrho_k, \tau]$ and $k \in \mathbb{N}$, we then have

$$\begin{aligned} u(\sigma) &\in [g(a), g(b)] \cap [u(\varrho_k) - \delta, u(\varrho_k) + \delta] \\ &\subset [g(a), g(b)] \cap [g(x^*(\varrho_k) - \varepsilon), g(x^*(\varrho_k) + \varepsilon)]\,. \end{aligned}$$

For each fixed $k \in \mathbb{N}$, we find $n(k)$ such that, for $n \geq n(k)$, we have $|x_{\alpha_n}(\varrho_k) - x^*(\varrho_k)| < \varepsilon$. We now apply Lemma 3.14 for $\alpha = \alpha_n$ on intervals $[\varrho_k, \tau]$ instead of $[s,t]$ and $[a,b] \cap [x^*(\varrho_k) - \varepsilon, x^*(\varrho_k) + \varepsilon]$ instead of [a,b] and obtain

$$x_{\alpha_n}(\sigma) \in [a,b] \cap [x^*(\varrho_k) - \varepsilon, x^*(\varrho_k) + \varepsilon]\,, \qquad \forall \sigma \in [\varrho_k, \tau] \quad \forall n \geq n(k)\,.$$

Then (3.49) yields that $|x^*(\tau) - x^*(\varrho_k)| \leq \varepsilon$, and letting $k \to \infty$ we obtain that $|x^*(\tau) - x^*(\tau-)| \leq \varepsilon$. Since $\varepsilon > 0$ has been chosen arbitrarily, we conclude that $x^*(\tau) = x^*(\tau-)$ for every $\tau \in]s,t]$. Consequently, $g(x^*(\tau))$ and $u(\tau)$ are two left-continuous regulated functions which coincide almost everywhere, hence (3.50) holds and Lemma 3.15 is proved. □

We conclude this section by proving Theorems 3.7 and 3.8.

Proof of Theorem 3.7. The cases (i), (ii) are fully symmetric, and we therefore establish just (i). The crucial step consists in proving the following statement:

$$\forall t \in]t_1, t_2] \;\; \exists s \in]t_1, t] \;\; \exists \alpha_0 > 0 \quad \text{such that for } \alpha \in]0, \alpha_0[\quad \text{we have } x_\alpha(s) \in J_3\,. \quad (3.51)$$

We then obtain (3.21) and (3.22) immediately from Lemmas 3.14 and 3.15.

To prove (3.51), consider an arbitrary $t \in]t_1, t_2]$. We fix $\varepsilon > 0$, $\delta > 0$, and $\sigma \in]t_1, t[$ such that $u(\tau) > G_+ + \varepsilon$ for $\tau \in [\sigma - \delta, \sigma]$, and put

$$\alpha_0 = \frac{\varepsilon \delta}{B - A}\,,$$

where A, B are as in (3.35). For $\tau \in [\sigma - \delta, \sigma]$ we have

$$\alpha\dot{x}_\alpha(\tau) + g(x_\alpha(\tau)) > G_+ + \varepsilon\,. \tag{3.52}$$

If $g(x_\alpha(\tau)) \le G_+$ for all $\tau \in [\sigma - \delta, \sigma]$, then the inequality (3.52) would imply that $\alpha(B - A) \ge \alpha(x_\alpha(s) - x_\alpha(s - \delta) \ge \varepsilon\delta$, which is a contradiction for $\alpha < \alpha_0$. Hence, there exists $s \in [\sigma - \delta, \sigma]$ such that $x_\alpha(s) \in J_3$, and (3.51) follows. □

Proof of Theorem 3.8. The solution x_α does not depend on the value of $u(0)$ and we may assume throughout the proof that $u(0) = u(0+)$. In case (i) set

$$t_0 \;=\; \inf\{t \in]0, T]\,;\, u(t) \notin [G_-, G_+]\}\,, \tag{3.53}$$

and assume that $0 \le t_0 < T$ (otherwise the assertion follows from Lemmas 3.14, 3.15), and that, for instance, there exists a sequence $s_i \searrow t_0$ such that $u(s_i) > G_+$ for all i. For $j = 1, 2, \dots$ we put recursively

$$\begin{aligned} t_{2j-1} &= \max\{t \in]t_{2j-2}, T]\,;\, u(s) \ge G_- \text{ for } s \in]t_{2j-2}, t]\},\\ t_{2j} &= \max\{t \in]t_{2j-1}, T]\,;\, u(s) \le G_+ \text{ for } s \in]t_{2j-1}, t]\} \end{aligned}$$

until $t_m = T$ for some m. The sequence $\{t_k\}$ is finite as u is regulated, and the convergence in $[0, t_0]$, if $t_0 > 0$, follows from Lemmas 3.14, 3.15. For $k = 1, \dots, m$ and for any $t \in]t_{k-1}, t_k]$, we find $s \in]t_{k-1}, t]$ such that $u(s+) > G_+$ if k is odd and $u(s+) < G_-$ if k is even, and use Theorem 3.7 in the interval $]s, t_k]$. We know that the function $x(t) = \lim_{\alpha\to 0+} x_\alpha(t)$ is regulated by Theorem 3.6 and Propositions 3.4, 3.5. Moreover, x is left-continuous on each interval $[0, t_0]$, $]t_{k-1}, t_k]$, $k = 1, \dots, m$, hence $x \in G_L(0, T)$ and the assertion is proved.

In case (ii) assume, for instance, that $u(0+) > g(x_0)$; the other inequality is similar. We fix some $\tau \in]0, T]$ and $\delta > 0$ such that

$$u(t) - g(x_0) \;\ge\; \delta, \qquad \text{for } t \in]0, \tau]\,. \tag{3.54}$$

For $\alpha > 0$ set

$$\begin{aligned} s_\alpha &= \max\{t \in [0, \tau]\,;\, g(x_\alpha(t)) - g(x_0) \le \delta/2\}\,,\\ t_\alpha &= \sup\{t \in [0, \tau]\,;\, x_\alpha(t) \in J_2\}\,. \end{aligned}$$

We first check that $s_\alpha > t_\alpha$. Indeed, assuming $s_\alpha \le t_\alpha$, we obtain for $t \in]0, s_\alpha[$ that

$$\alpha\dot{x}_\alpha(t) + g(x_\alpha(t)) - g(x_0) \;=\; u(t) - g(x_0) \;\ge\; \delta\,,$$

and hence

$$x_\alpha(t) - x_\alpha(s) \;\ge\; \frac{\delta}{2\alpha}(t - s)\,, \qquad \forall 0 \le s < t \le s_\alpha\,. \tag{3.55}$$

The function g is decreasing in J_2, hence $g(x_\alpha(t)) - g(x(0)) < 0$ for all $t \in [0, s_\alpha]$, which is a contradiction. Inequality (3.55) therefore holds for all $0 \le s < t \le t_\alpha$, and in particular

$$t_\alpha \;\le\; \frac{2\alpha}{\delta}(x_+ - x_0) \;<\; \tau\,, \quad x_\alpha(t_\alpha) \;=\; x_+, \tag{3.56}$$

for α sufficiently small. As in (3.53), put

$$t_0 = \inf\{t \in]0,T] \,;\, u(t) < G_-\}\,.$$

By Lemma 3.14 we have $x_\alpha(t) \geq x_+$ for all $t \in [t_\alpha, t_0]$, and the convergence in each interval $[\bar{t}, t_0]$ for $\bar{t} \in]0, t_0[$ follows from Lemma 3.15. For $t > t_0$ we proceed as in case (i) and the proof is complete. □

Remark. The assertion of Theorem 3.8 does not hold in general if $x_0 \in J_2$ and $u(0+) = g(x_0)$. Let us consider the set

$$Y = \{u \in G_L(0,T)\,;\, u(0) = u(0+) = g(x_0)\,,\ u(t) \in [G_-, G_+]\ \ \forall t \in [0,T]\}\,,$$

and a sequence $\alpha_j \searrow 0$ as $j \to \infty$. Let x_j, for $j \in \mathbb{N}$, be the solution to the problem

$$\alpha_j \dot{x}_j + g(x_j(t)) = u(t)\,, \quad x_j(0) = x_0\,.$$

For a fixed $\bar{t} \in]0,T]$ and for $j \in \mathbb{N}$, we define the sets $A_+^j = \{u \in Y\,;\, x_j(\bar{t}) \leq x_+ - \alpha_j\}$, $A_-^j = \{u \in Y\,;\, x_j(\bar{t}) \geq x_- + \alpha_j\}$, $B_\pm^j = \bigcap_{i=j}^\infty A_\pm^i$. Then each A_+^j, A_-^j is closed in Y, and hence each B_+^j, B_-^j is also closed. The sets B_+^j (and analogously B_-^j) are nowhere dense in Y. Indeed, each $u \in B_+^j$ can be uniformly approximated by functions $u^{(r)} \in Y$ of the form

$$u^{(r)}(t) = \begin{cases} u(0)\,, & \text{for } t \in [0, t_r]\,, \\ \min\{u(t) + r, G_+\}\,, & \text{for } t \in]t_r, T]\,, \end{cases}$$

for $r \to 0+$, where $t_r \in]0, \bar{t}[$ is chosen in such a way that $|u(t) - u(0)| \leq r/2$ for $t \in [0, 2t_r]$. The solution $x_i^{(r)}$ to the problem

$$\alpha_i \dot{x}_i^{(r)} + g(x_i^{(r)}(t)) = u^{(r)}(t)\,, \quad x_i^{(r)}(0) = x_0$$

satisfies $g(x_i^{(r)}(t_r)) = g(x_0) = u(0) < u^{(r)}(t_r+)$. The argument (3.54) – (3.56) in the proof of Theorem 3.8, shifted from $t = 0$ to $t = t_r$ for each fixed $r > 0$, yields that, for $i \geq j$ sufficiently large, there exists $t_i \in]t_r, \bar{t}[$ such that $x_i^{(r)}(t_i) = x_+$. From Lemma 3.14 it follows that $x_i^{(r)}(\bar{t}) \geq x_+$, hence $u^{(r)} \in Y \setminus B_+^j$. By virtue of Baire's Theorem (see [6], Chapter II, §3), the set $Y_0 = Y \setminus \bigcup_{j=1}^\infty (B_+^j \cup B_-^j)$ is non-empty. By construction, for $u \in Y_0$ we have

$$\forall j \in \mathbb{N} \begin{cases} \exists i \geq j: & x_i(\bar{t}) > x_+ - \alpha_i\,, \\ \exists i' \geq j: & x_{i'}(\bar{t}) < x_- + \alpha_{i'}\,, \end{cases}$$

and hence $\limsup_{j\to\infty} x_j(\bar{t}) \geq x_+$, $\liminf_{j\to\infty} x_j(\bar{t}) \leq x_-$.

3.7 Stability of the Flow

The results of the previous sections suggest that a stable regime for $t \to \infty$ in Problem (3.14) – (3.18) can be expected only under rather restrictive conditions, cf. also Section 3 in [12]. Accordingly, we make the following hypothesis.

Hypothesis 3.16. *There exist $z \in \mathbb{R}$, $d > 0$ and $h > 0$ such that*

(i) $|\lambda| \; < \; \dfrac{d}{d+4}$,

(ii) $g(z) + \dfrac{2\lambda}{1-\lambda}\, z \; = \; 0$,

(iii) $\dfrac{g(x) - g(y)}{x - y} \; \geq \; d$ *for every* $x, y \in [z-h, z+h]$, $x \neq y$,

where λ is given by (3.12).

This will certainly be fulfilled if, for example, g is smooth, there exists $z_0 \in \mathbb{R}$ such that $g(z_0) = 0$, $g'(z_0) > 0$, and the valve parameter K in (3.5) is sufficiently close to the critical value $K = 1$. Physically, this means that the valve is adjusted with a prescribed accuracy so as to kill the reflected wave which otherwise may cause undesirable interactions with the pump. Under the above conditions, we have the following stability result.

Proposition 3.17. *Let Hypothesis 3.16 hold, let $\bar{p} : [0, \infty[\to \mathbb{R}$ be a function which is regulated on each bounded interval, and let there exist $\gamma > 0$ such that*

$$|\bar{p}(t)| \; \leq \; \gamma \; < \; hd\,\frac{(1-\varrho)(d+4)}{(1-\varrho)d + 4}\,, \qquad \forall t \geq 0\,,$$

where $\varrho := \dfrac{|\lambda|(d+4)}{d} \in [0,1[$. Assume that the initial conditions satisfy the inequalities

$$\begin{aligned} |q^0(0) - z| \quad &< \; h, \\ \|\psi_0 - g(z)\|_{[0,2]} \; &< \; dh - \gamma, \end{aligned} \tag{3.57}$$

with ψ_0 given by (3.13). Then for every $\varepsilon > \frac{(1-|\lambda|)\gamma}{(1-\varrho)d}$ and every $\alpha > 0$ there exists $\bar{t} > 0$ such that the functions y, ψ defined in (3.19) satisfy the estimate

$$\begin{aligned} |y(t) - z| \; &\leq \; \varepsilon, \\ |\psi(t) - g(z)| \; &\leq \; d\varepsilon - \gamma, \end{aligned} \qquad \text{for } t \geq \bar{t}\,.$$

In other words, Proposition 3.17 states that the values of $(y(t), \psi(t))$ remain in a small neighbourhood of the equilibrium point $(z, g(z))$ if the fluctuations $\bar{p}(t)$ are small, and $y(t) \to z$, $\psi(t) \to g(z)$ as $t \to \infty$ if $\bar{p} \equiv 0$.

Proof. We construct sequences $\{a_k\}$, $\{b_k\}$ recursively by the formula

$$a_0 \; = \; b_0 \; = \; h\,, \tag{3.58}$$

$$\begin{aligned} a_k \; &= \; \mu a_{k-1} + (1-\mu) b_{k-1}, \\ b_k \; &= \; \varrho\,(\nu a_{k-1} + (1-\nu) b_{k-1}) + \tfrac{1-|\lambda|}{d}\,\gamma, \end{aligned} \tag{3.59}$$

for $k = 1, 2, \ldots$, where $\mu = e^{-2d/\alpha}$, and $\nu = \frac{2}{d+4}$. The matrix $\begin{pmatrix} \mu & 1-\mu \\ \varrho\nu & \varrho(1-\nu) \end{pmatrix}$ has only real eigenvalues which belong to $]-1, 1[$, hence the sequence (a_k, b_k) converges to $(\bar{a}, \bar{b})$, where

$$\bar{a} = \bar{b} = \frac{(1-|\lambda|)\gamma}{(1-\varrho)d}, \tag{3.60}$$

and we have $0 \le a_k \le h$, $0 \le b_k \le h$ for every $k = 0, 1, 2, \ldots$.

We further prove by induction that the solutions y_k, ψ_k to (3.14) – (3.18) satisfy the inequalities

$$\|y_{k-1} - z\|_{[0,2]} \le \max\{a_{k-1}, b_{k-1}\}, \tag{3.61}$$

$$|y_k(0) - z| < a_k, \tag{3.62}$$

$$\|\psi_k - g(z)\|_{[0,2]} < d b_k - \gamma. \tag{3.63}$$

Assume that

$$\begin{aligned} |y_{k-1}(0) - z| &< a_{k-1}, \\ \|\psi_{k-1} - g(z)\|_{[0,2]} &< d b_{k-1} - \gamma, \end{aligned}$$

for some k. The induction hypothesis is fulfilled for $k = 1$, thanks to (3.57) and (3.15). From (3.14) it follows that

$$\alpha \dot{y}_{k-1}(t) + g(y_{k-1}(t)) - g(z) = \psi_{k-1}(t) - g(z) + \bar{p}_{k-1}(t), \qquad \text{for } t \in [0, 2]. \tag{3.64}$$

Let $\tau \in]0, 2]$ be any point such that $|y_{k-1}(t) - z| \le h$ for all $t \in [0, \tau]$. Testing equation (3.64) by $\operatorname{sign}(y_{k-1}(t) - z)$ we obtain

$$\alpha \frac{d}{dt} |y_{k-1}(t) - z| + d\, |y_{k-1}(t) - z| < d b_{k-1}, \qquad \text{for } t \in]0, \tau[\,,$$

and hence the inequality

$$|y_{k-1}(t) - z| < e^{-td/\alpha} a_{k-1} + \left(1 - e^{-td/\alpha}\right) b_{k-1} \tag{3.65}$$

holds for all $t \in [0, \tau]$. Consequently, we may take $\tau = 2$ and obtain immediately (3.61) – (3.62) from (3.59) and (3.65). To check that (3.63) holds, we use (3.18) which yields that

$$\psi_k(t) - g(z) = \lambda\,(\psi_{k-1}(t) - g(z)) - 2\lambda\,(y_{k-1}(t) - z), \qquad \text{for } t \in [0, 2],$$

and hence

$$\begin{aligned} \|\psi_k - g(z)\|_{[0,2]} &\le |\lambda| \left(\|\psi_{k-1} - g(z)\|_{[0,2]} + 2\, \|y_{k-1} - z\|_{[0,2]} \right) \\ &< |\lambda|\,(d b_{k-1} - \gamma + 2\,(a_{k-1} + b_{k-1})) \\ &= d b_k - \gamma. \end{aligned}$$

We have thus proved that (3.61) – (3.63) hold for all $k = 1, 2, \ldots$, and the assertion follows from (3.60). □

3.8 Acknowledgements

This work was supported by the Grant Agency of the Czech Republic under Grant no. 201/02/1058 and by the Boole Centre for Research in Informatics, University College Cork.

Bibliography

[1] G. Aumann, *Reelle Funktionen*, Springer-Verlag, Berlin – Göttingen – Heidelberg, 1954 (in German).

[2] M. Brokate and P. Krejčí, *Duality in the space of regulated functions and the play operator*, Math. Z., 245(4) (2003), pp. 667–688.

[3] M. Brokate and J. Sprekels, *Hysteresis and phase transitions*, Appl. Math. Sci., 121, Springer-Verlag, New York, 1996.

[4] D. Fraňková, *Regulated functions*, Math. Bohem., 119 (1991), pp. 20–59.

[5] W. Kolarčík, P. Krejčí, V. Lovicar, and I. Straškraba, *On the dynamics of pump surge*, in: Proceedings of the 5th International Meeting of IAHR, Section "The behaviour of hydraulic machinery under steady oscillatory conditions", M. Fanelli, ed., Milano, Sept. 16–18, 1991, pp. 1–17.

[6] A. N. Kolmogorov and S. V. Fomin, *Elements of the Theory of Functions and Functional Analysis*, Graylock Press, Albany, New York, 1961.

[7] M. A. Krasnosel'skii and A. V. Pokrovskii, *Systems with Hysteresis*, Nauka, Moscow, 1983 (English edition: Springer, 1989).

[8] P. Krejčí, *Hysteresis, convexity and dissipation in hyperbolic equations*, Gakuto Int. Ser. Math. Sci. Appl., 8, Gakkōtosho, Tokyo, 1996.

[9] P. Krejčí and Ph. Laurençot, *Hysteresis filtering in the space of bounded measurable functions* Boll. Unione Mat. Ital., 5-B (2002), pp. 755–772.

[10] P. Krejčí and Ph. Laurençot, *Generalized variational inequalities*, J. Convex Anal., 9 (2002), pp. 159–183.

[11] V. Lovicar, I. Straškraba, and P. Krejčí, *Hysteresis in singular perturbation problems with nonuniqueness in limit equation*, in Models of Hysteresis, A. Visintin, ed., Longman, Harlow 1993, pp. 91–101.

[12] I. Straškraba and V. Lovicar, *Qualitative behavior of solutions to* 2×2 *linear hyperbolic system with nonlinear boundary conditions*, SAACM, 3 (1993), pp. 277–290.

[13] G. Tronel and A. Vladimirov, *On BV-type hysteresis operators*, Nonlinear Anal., 39 (2000), pp. 79–98.

[14] A. Visintin, *Differential Models of Hysteresis*, Springer, Berlin - Heidelberg, 1994.

Chapter 4

Combined Asymptotic Expansions

E. Benoît, A. Fruchard, and A. El Hamidi

A structured and synthetic presentation of Vasil'eva's [7] combined expansions is proposed. These expansions simultaneously take into account the limit layer and the slow motion of solutions of a singularly perturbed differential equation. An asymptotic formula is established which gives the distance between two exponentially close solutions. An "input-output" relation around a *canard* solution is carried out in the case of a turning point. Finally, the distance between two *canard* values of differential equations with a parameter is given.

We illustrate this study on the Liouville equation and the splitting of energy levels in the one dimensional steady Schrödinger equation in the symmetric double well case. The structured nature of our approach allows us to give effective symbolic algorithms.

This paper is a short version of [1].

4.1 Introduction

Combined asymptotic expansions are studied in the book [7]. They are used to study singularly perturbed ordinary differential equations. Due to their complexity, turning points are avoided in the basic literature. In this paper, we use combined asymptotic expansions to study the global behaviour of solutions around a turning point. The domain of a combined asymptotic expansion is divided into two parts: the boundary, or inner, layer with its inner expansion, and the outer expansion valid outside the layer. In some of the literature, the two domains are connected using matching techniques, combined asymptotic expansions give approximation formulae valid in large domains containing the boundary layers and the turning point.

Our presentation of combined asymptotic expansions is more algebraic and algorithmic than the classical one. It seems to us that it allows the more efficient use of this theory. In the presentation here some technical proofs are not given as the details are given in [1].

For this presentation, we will use Nonstandard Analysis (with the axioms of Internal Set Theory: see [6]). In fact, all the proofs could be translated into classical language, but the translation would give more complex formulae: with nonstandard techniques, we can take a fixed infinitesimal positive real number ε, then the symbol ε is a parameter and not a variable, and we can write $f(t)$ for a function depending on t and ε. An object not depending on ε is, in our context, a standard object. We will give some examples of translations from nonstandard language to the classical one. In addition, this will fix some notation (see [5]):
The formula $A = \emptyset$ (we read *"A is infinitesimal"*), is translated by

$$\forall \alpha > 0 \quad \exists \varepsilon_0 \quad \forall \varepsilon \quad 0 < \varepsilon < \varepsilon_0 \implies |A(\varepsilon)| < \alpha \ .$$

The formula $A = \pounds$ (we read *"A is limited"*), is translated by

$$\exists M \quad \exists \varepsilon_0 \quad \forall \varepsilon \quad 0 < \varepsilon < \varepsilon_0 \implies |A(\varepsilon)| < M \ .$$

The formula $A = \not\infty$ (we read *"A is unlimited"*), is translated by

$$\forall M \quad \exists \varepsilon_0 \quad \forall \varepsilon \quad 0 < \varepsilon < \varepsilon_0 \implies |A(\varepsilon)| > M \ .$$

The formula $A = @$ (we read *"A is appreciable"*), is translated by

$$\exists \alpha > 0 \quad \exists \varepsilon_0 \quad \forall \varepsilon \quad 0 < \varepsilon < \varepsilon_0 \implies \alpha < A(\varepsilon) < 1/\alpha \ .$$

The formula *"for all positive non limited x, $A(x) = \pounds e^{-@x}$"* (we say *"A has S-exponential decay at infinity"*), is translated by

$$\exists \alpha > 0 \;\; \exists M \;\; \exists \varepsilon_0 \;\; \exists x_0 \;\; \forall x > x_0 \;\; \forall \varepsilon \;\; 0 < \varepsilon < \varepsilon_0 \implies |A(x,\varepsilon)| < M e^{-\alpha x} \ .$$

In the first part of this paper, we introduce combined asymptotic expansions and we write algorithms for computations with such expansions. These algorithms are written in Maple and can be downloaded from

http://www.univ-lr.fr/labo/lmca/publications/02-01/02-01.mws

In the second part we give the main lemma, which is very similar to that given in [7], for the existence of a combined asymptotic expansion for a solution of a singularly perturbed ordinary differential equation. The lemma is given in the one-dimensional case; we did not study the n-dimensional case.

In the third part we extend the theory with the introduction of turning points (canards) and we give an application to compute the exponentially small difference between the first two energy levels of the stationary Schrödinger equation with a symmetric double well potential (a problem already studied, for example, in [4] or [3]).

4.2 Computations with Combined Asymptotic Expansions

We give first the nonstandard definition of a combined asymptotic expansion, and after that its translation to classical language.

Definition 4.1. *A function* $\varphi : [t_1, t_2] \to \mathbb{R}^d$ *has a* combined asymptotic expansion *if there are two standard sequences of* C^∞ *functions* $(\varphi_n)_{n\in\mathbb{N}}$, $(\psi_n)_{n\in\mathbb{N}}$, $\varphi_n : [t_1, t_2] \to \mathbb{R}^d$, $\psi_n : \mathbb{R}^+ \to \mathbb{R}^d$ *such that*

$$\bullet \forall^{st} N \in \mathbb{N}\ \forall t \in [t_1, t_2]\ ,\quad \varphi(t) = \sum_{n=0}^{N-1} \left(\varphi_n(t) + \psi_n\left(\tfrac{t-t_1}{\varepsilon}\right)\right)\varepsilon^n + \pounds\varepsilon^N,$$

$$\bullet \forall^{st} n \in \mathbb{N}\ \forall x \in \mathbb{R}^+,\quad \psi_n(x) = \pounds e^{-@x}.$$

The translation into classical language is:

Definition 4.2. *A function* $\varphi : [t_1, t_2]\times]0, \varepsilon_0] \to \mathbb{R}^d$ *has a* combined asymptotic expansion *if there are two sequences of* C^∞ *functions* $(\varphi_n)_{n\in\mathbb{N}}$, $(\psi_n)_{n\in\mathbb{N}}$, $\varphi_n : [t_1, t_2] \to \mathbb{R}^d$, $\psi_n : \mathbb{R}^+ \to \mathbb{R}^d$ *such that*

$$\bullet \forall N \in \mathbb{N}\ \exists C_N\ \ \forall \varepsilon \in]0, \varepsilon_0]\ \ \forall t \in [t_1, t_2]\ ,$$
$$\left|\varphi(t) - \sum_{n=0}^{N-1} \left(\varphi_n(t) + \psi_n\left(\tfrac{t-t_1}{\varepsilon}\right)\right)\varepsilon^n\right| < C_N\varepsilon^N,$$

$$\bullet \forall n \in \mathbb{N}\quad \limsup_{x\to+\infty} \frac{\ln|\psi_n(x)|}{x} < 0.$$

The formal sum $\sum \varphi_n(t)\varepsilon^n$ is called the *slow expansion*, and the formal sum $\sum \psi_n\left(\frac{t-t_1}{\varepsilon}\right)\varepsilon^n$ the *fast expansion*.

We note that the approximation is valid for $t = t_1 + \varepsilon\pounds$ (usually called the boundary layer), for $t = t_1 + @$ (usually called the exterior domain), but also between these two domains. In the exterior domain, the functions ψ_n are exponentially small, hence smaller than ε^N, and the fast expansion is negligible. However, in contrast, the slow expansion is not negligible in the boundary layer.

Proposition 4.3. *Let* Φ *be a* C^∞ *standard function from an open subset* U *of* $\mathbb{R}^d$ *to* $\mathbb{R}^p$ *and* f *a function from* $[t_1, t_2]$ *to* $\mathbb{R}^d$ *having a combined expansion* (φ_n, ψ_n). *Suppose that for all* $t \in [t_1, t_2]$, $\varphi_0(t) \in U$ *and for all* $x \in \mathbb{R}^+$, $\varphi_0(t_1) + \psi_0(x) \in U$. *Then* $\Phi \circ f$ *is well defined and has a computable combined expansion.*

Corollary 4.4. *If* f *and* g *have combined asymptotic expansions on* $[t_1, t_2]$, *the same is true for the sum* $f + g$, *the product* fg, *the scalar product, etc. If* $g(t) \not\simeq 0$, *the same is true for the quotient* f/g, *etc.*

Corollary 4.5. *If $\Phi(u)$ has an asymptotic expansion for u in a standard open domain U, if f has a combined asymptotic expansion (φ_n, ψ_n) on $[t_1, t_2]$, if $\forall t \in [t_1, t_2]$ $\varphi_0(t) \in U$ and if $\forall x \in \mathbb{R}^+$ $\varphi_0(t_1) + \psi_0(x) \in U$, then $\Phi \circ f$ has a combined asymptotic expansion.*

One proves the corollary 4.4 by the application of the proposition to $\Phi(f, g) = f + g$ (resp; fg, $\langle f, g\rangle$, f/g,...). To prove corollary 4.5, we apply the proposition to each function Φ_n of the expansion of Φ. All the computations of these corollaries are implemented in Maple (in the one dimensional case).

The proof of the proposition 4.3 is technical, and details can be found in [1]. An outline is as follows: First, we claim that the slow part of the result is the formal sum $\Phi(\sum \varphi_n \varepsilon^n)$. Then, we study the difference $\Phi(f(t)) - \Phi(\sum_{n=0}^{N-1} \varphi_n \varepsilon^n)$. For that purpose, we have to decompose Φ with $\Phi(a+h) = \Phi(a) + \Delta\Phi(a,h).h$ where $\Delta\Phi$ is a C^∞ standard function defined on a subdomain of $U \times \mathbb{R}^d$. Then, we replace t by $t_1 + \varepsilon x$, we expand each term using Taylor's formula, and we can prove that the remainders are bounded using the exponential decay of the functions ψ_n. The technical difficulties come from the complicated notation in Taylor's formulae.

Proposition 4.6. *If f has a combined asymptotic expansion (φ_n, ψ_n), then $F : [t_1, t_2] \to \mathbb{R}^d, t \mapsto \int_{t_1}^t f(\tau)d\tau$ has a combined asymptotic expansion (Φ_n, Ψ_n) given by*

$$\Phi_0(t) = \int_{t_1}^t \varphi_0(\tau)d\tau \ , \quad \Psi_0(x) = 0,$$

and for $n \geq 1$:

$$\Phi_n(t) = \int_{t_1}^t \varphi_n(\tau)d\tau + \int_0^{+\infty} \psi_{n-1}(x)dx \ ,$$

$$\Psi_n(x) = -\int_x^{+\infty} \psi_{n-1}(\xi)d\xi \ .$$

In particular, we have

$$\int_{t_1}^{t_2} f(t)dt \sim \int_{t_1}^{t_2} \varphi_0(t)dt + \sum_{n\geq 1} \left(\int_{t_1}^{t_2} \varphi_n(t)dt + \int_0^{+\infty} \psi_{n-1}(x)dx \right) \varepsilon^n \ .$$

The proof is straightforward using the exponential decay of the ψ_n. The implementation of this proposition is very easy in Maple, but the computations of integrals give many difficulties for effective computations in applications.

4.3 Main Lemma

In this section, we study a one-dimensional singularly perturbed equation

$$\varepsilon\dot{u} = f(t, u).$$

We don't assume that f is analytic or C^∞ in ε (in our nonstandard presentation ε is a given infinitesimal number, so this kind of regularity with respect to ε is not relevant).

HYPOTHESIS 1 — *The function f is C^∞ and all its derivatives of standard order (f included) have ε-expansions in a standard open subset U of $\mathbb{R}^2$.* If we write $f(t,u) \sim \sum f_n(t,u)\varepsilon^n$, the asymptotic expansions of the derivatives are given by the derivatives of the f_n.

HYPOTHESIS 2 — *There is a standard slow curve $u = u_0(t)$ in U defined and C^∞ on a standard compact interval $[t_1, t_2]$, i.e.*

$$\forall t \in [t_1, t_2] \ , \ (t, u_0(t)) \in U \text{ and } f_0(t, u_0(t)) = 0 \ .$$

Let c_0 standard be such that the segment joining $(t_1, u_0(t_1))$ and (t_1, c_0) is in U.

HYPOTHESIS 3 — (attractiveness) *The function $a_0(t,u) := \frac{\partial f_0}{\partial u}(t,u)$ is bounded above by some standard negative constant on U.*

Proposition 4.7. *We consider the differential equation*

$$\varepsilon \dot{u} = f(t,u), \tag{4.1}$$

where the function f satisfies hypotheses 1,2 and 3. Then, for every number $c \sim \sum_{n\geq 0} c_n\varepsilon^n$, the solution u of (4.1) with the initial condition $u(t_1) = c$ is defined and has a combined asymptotic expansion in $[t_1, t_2]$:

$$u(t) \sim \sum_{n\geq 0} u_n(t)\varepsilon^n + \sum_{n\geq 0} y_n(x)\varepsilon^n \quad , \qquad t = t_1 + \varepsilon x \ .$$

Sketch of the proof (details are given in [1]):

- The slow expansion is well known in this situation (even with the weak hypothesis 1.): It is the asymptotic expansion of a solution $u^\natural$ of (4.1), with $u^\natural(t) = u_0(t) + \emptyset$ for all t in some standard open interval containing $[t_1, t_2]$,

$$u^\natural(t) \ = \ \sum_{n=0}^{N-1} u_n(t)\varepsilon^n \ + \ \pounds\varepsilon^N \ .$$

  It is also the unique formal solution of (4.1), and the u_n can be computed using formal identification.

- Let $\tilde{y}$ and y be defined by

$$\tilde{y}(t) \ = \ u(t) - u^\natural(t) \ \ , \ \ y(x) \ = \ \tilde{y}(t_1 + \varepsilon x) \ .$$

  The function $y(x)$ is the solution of

$$\begin{cases} y'(x) &= \ g(\varepsilon x, y(x))\, y(x), \\ y(0) &= \ c - u^\natural(0), \end{cases} \tag{4.2}$$

  where $g(t,y) := \left[f(t, u^\natural(t) + y) \ - \ f(t, u^\natural(t)) \right] / y$ is a function with an asymptotic expansion.

- Then, we formally solve the problem (4.2) to obtain a formal series $\sum y_n(x)\varepsilon^n$, where $y_n(x)$ is given in terms of $y_0, y_1, \ldots, y_{n-1}$.
- The formula shows recursively the exponential decay of the y_n.
- Finally, we have to study the remainder r_N defined by

$$y(x) = \sum_{n=0}^{N-1} y_n(x)\varepsilon^n + r_N(x)\varepsilon^N .$$

We can prove that r_N is the solution of a linear differential equation

$$r'_N(x) = a_N(x)\, r_N(x) + b_N(x)$$

$$\text{with} \begin{cases} a_N(x) = g(\varepsilon x, Y_N(x)) + \dfrac{g(\varepsilon x, y(x)) - g(\varepsilon x, Y_N(x))}{y(x) - Y_N(x)} , \\ b_N(x) = \varepsilon^{-N} \{g(\varepsilon x, Y_N(x))Y_N(x) - Y'_N(x)\} , \\ Y_N(x) = \displaystyle\sum_{n=0}^{N-1} y_n(x)\varepsilon^n . \end{cases}$$

We can prove that $a_N(x) = -@$ for x unlimited, and (after many technical difficulties) that $b_N(x) = \pounds e^{-@x}$. That is enough to prove that $r_N(x) = \pounds e^{-@x}$.

4.4 Improvement to Turning Points with Canards

In this section, we replace hypothesis 3 by a weaker hypothesis allowing canards at a point $(t_0, u_0(t_0))$.

Let us define the function

$$A_0(t) = \int_{t_1}^{t} \frac{\partial f_0}{\partial u}(\tau, u_0(\tau))d\tau .$$

Its derivative is the function a_0 controlling the attractivity of the slow curve. Assume that there is a turning point at time t_0, i.e. $A(t)$ is decreasing for t in $[t_1, t_0]$, and increasing for t in $[t_0, t_2]$. We know (by differentiation with respect to the initial condition) that the order of magnitude of the distance between two solutions of the equation (4.1) is $\exp(A_0(t)/\varepsilon)$.

HYPOTHESIS 4 — *There exists a canard $u^\natural(t)$, i.e. a solution of (4.1) satisfying $u^\natural(t) \simeq u_0(t)$ on an open interval containing $[t_1, t_2]$. Moreover, $A_0(t) < 0$ for all t in $]t_1, t_2]$, and $a_0(t_1, u) < 0$ for u in $[u_0(t_1), c]$.*

Proposition 4.8. *With hypotheses 1, 2 and 4, proposition 4.7 is valid.*

For a proof, if $t \in [t_1, (t_0 + t_1)/2]$, we can apply proposition 4.7, and for bigger t, we know that the difference $u(t) - u^\natural(t)$ and the fast expansion are both exponentially small.

4.5 Application

In this section, we apply the above theory to the following classical problem: The energy levels of the stationary Schrödinger equation:

$$\varepsilon^2 \ddot{\psi} = \left(\varphi(t)^2 - E\right) \psi \qquad \text{with} \qquad \varphi(t) = 1 - t^2 .$$

The energy levels are the values of the parameter E such that there exists a nontrivial L^2-solution. It is known that the first two energy levels $E^\flat$ and $E^\sharp$ have the same asymptotic expansion. Formal computation yields the first terms of this expansion:

$$E^\flat \sim E^\sharp \sim 2\varepsilon - \frac{1}{2}\varepsilon^2 - \frac{9}{32}\varepsilon^2 - \frac{89}{256}\varepsilon^3 + \pounds\varepsilon^4 .$$

The problem is to evaluate the difference $E^\sharp - E^\flat$.

Proposition 4.9. *There is a Maple program to compute an asymptotic expansion for $E^\sharp - E^\flat$ (see*

http://www.univ-lr.fr/labo/lmca/publications/02-01/02-01.mws).

The first terms of this expansion are

$$E^\sharp - E^\flat = \frac{32}{\sqrt{2\pi}} \sqrt{\varepsilon} \exp\left(\frac{-4}{3}\frac{1}{\varepsilon}\right)\left(1 - \frac{71}{96}\varepsilon - \frac{6299}{18432}\varepsilon^2 - \frac{2691107}{5308416}\varepsilon^3 + \pounds\varepsilon^4\right) .$$

Sketch of the proof:

With the change of variable $u = \varepsilon\dot{\psi}/\psi$, we transform the linear Schrödinger equation to the Riccati equation

$$\varepsilon\dot{u} = \varphi(t)^2 - E - u^2 .$$

This equation has two slow curves $u = \pm\varphi(t)$ intersecting at points $t = -1$ and $t = 1$. These two points are turning points. It is known (see [2, 4]) that an L^2-solution of the Schrödinger equation corresponds to a canard at both points $t = \pm 1$. Therefore, we look for a solution u such that

$$\forall t = -@ , \; u(t) \simeq -\varphi(t), \qquad \text{and} \qquad \forall t = +@ , \; u(t) \simeq \varphi(t) .$$

Due to the symmetry, there are two possibilities: a solution $u^\sharp$ with $u^\sharp(0) = \infty$, and another $u^\flat$ with $u^\flat(0) = 0$. The first one has a pole for $t = 0$.

A difficulty appears due to the pole. To cancel it, we have to do another change of variables, for example $u = \frac{v+\delta}{1-\delta v}$ which corresponds to a rotation of the cylinder (the natural phase space of a Riccati equation) of angle $-\arctan\delta$. In the computations below, we do not write formulae with this new variable v to avoid long and complicated formulae, instead we write all with the variable u, knowing that effective computations need the variable v.

Let us denote by $y(t)$ the difference $u^\sharp(t) - u^\flat(t)$. It is a solution of the equation

$$\varepsilon\dot{y} = c(t)y - (E^\sharp - E^\flat)b(t) \quad , \quad c(t) = -(u^\sharp(t) + u^\flat(t)) \quad , \quad b(t) = 1 ,$$

$$\text{with} \qquad y(0) = \infty \ , \quad y(\infty) = 0 \ .$$

We can solve this equation:

$$y(t_1) = y(t_2)\exp\left(\frac{-1}{\varepsilon}\int_{t_1}^{t_2} c(\tau)d\tau\right) + \frac{E^\sharp - E^\flat}{\varepsilon}\int_{t_1}^{t_2} b(s)\exp\left(\frac{-1}{\varepsilon}\int_{t_1}^{s} c(\tau)d\tau\right)ds \ .$$

We can then write

$$E^\sharp - E^\flat \;=\; \varepsilon\frac{y(t_1) - y(t_2)\exp\left(\frac{-1}{\varepsilon}\int_{t_1}^{t_2} c(\tau)d\tau\right)}{\int_{t_1}^{t_2} b(s)\exp\left(\frac{-1}{\varepsilon}\int_{t_1}^{s} c(\tau)d\tau\right)ds} \ .$$

We now apply this formula for $t_1 = 0$, and $t_2 = 2$.

Using the theory above, we can compute the combined asymptotic expansions of $u^\sharp(t)$ and $u^\flat(t)$ for $t \in [0, 2]$ (here we need the other chart v), then of $c(t)$ and $b(t)$, then of the integral $A(t) := \int_{t_1}^{t} c(\tau)d\tau$. We isolate the standard part $A_0(t)$ of this integral (which has the same meaning as in Section 4.4) and then compute the asymptotic expansion of $\exp\{(A(t) - A_0(t))/\varepsilon\}$.

Doing an asymptotic expansion of a Laplace integral $\int_0^2 f(t)\exp(-A_0(t)/\varepsilon)dt$, the term $\sqrt{\varepsilon}\,\exp\left(\frac{-4}{3}\,\frac{1}{\varepsilon}\right)$ appears. All the computations can be done with algorithms on combined asymptotic expansions. But we have to remember that we are studying a Riccati equation on the cylinder $(t, u) \in \mathbb{R} \times S^1$, or better $(t, v) \in \mathbb{R} \times S^1$.

Bibliography

[1] E. Benoît, A. El Hamidi and A. Fruchard, *On combined asymptotic expansions in singular perturbations*, Electronic J. of Differential Equations, 51 (2002), pp. 1–27.

[2] J. L. Callot, *Solutions visibles de l'équation de Schrödinger*, Mathématiques finitaires et Analyse Non Standard. Publications mathématiques de l'Université Paris 7(31–1) (1985), pp. 105–119.

[3] E. Delabaere, H. Dillinger, and F. Pham, *Exact semiclassical expansions for one-dimensional quantum oscillators*, J. Math. Phys., 38(12) (1997), pp. 6126–6184.

[4] A. Gaignebet, *Équation de Schrödinger unidimensionnelle stationnaire. Quantification dans le cas d'un double puits de potentiel symétrique*, C. R. Acad. Sci. Paris, 315, Série I (1992), pp. 113–118.

[5] F. Koudjeti and I. P. van der Berg, *Neutrices, external numbers, and external calculus*, in Nonstandard Analysis in Practice, F. Diener and M. Diener, eds., pp. 145–170, Springer, 1995.

[6] E. Nelson, *Internal set theory*, Bull. Amer. Math. Soc. 83 (1977), pp. 1165–1198.

[7] A. B. Vasil'eva and V. F. Butuzov, *Asymptotic Expansions of the Solutions of Singularly Perturbed Equations*, Nauka, Moscow, 1973 (in Russian).

Chapter 5

Contrast Structures of Alternating Type

A. Vasilieva

The paper is devoted to spike-type solutions (contrast structures) of singularly perturbed parabolic equations.

5.1 Equation without Explicit Dependence on u_x

We consider the singularly perturbed parabolic equation (ε is a small parameter)

$$\varepsilon^2(u_{xx} - u_t) = F(u, x, t), \quad (x, t) \in \Omega := \{x \in [0, 1], t \in \mathbb{R}\}, \tag{5.1}$$

under the boundary conditions

$$u(0, t, \varepsilon) = 0, \qquad u(1, t, \varepsilon) = 0, \tag{5.2}$$

and the 2π-periodicity condition

$$u(x, t, \varepsilon) = u(x, t + 2\pi, \varepsilon). \tag{5.3}$$

Remark. We assume that the boundary conditions (5.2) are homogeneous for simplicity of presentation of our results. However, one can consider $-\infty < a < b < +\infty$, $u \in [a, b]$ and the boundary conditions can have the form $u(0, t, \varepsilon) = u^0$, $u(1, t, \varepsilon) = u^1$ (see [1]).

Let the following assumptions hold:

1) The function $F(u, x, t) \in C^2(G)$, *where* $G := \{\mathbb{R} \times \Omega\}$.

2) The degenerate ($\varepsilon = 0$) *equation*

$$0 = F(u, x, t)$$

has three roots $\varphi_1(x,t) < \varphi_2(x,t) < \varphi_3(x,t)$ *in the domain* Ω. *Moreover*

$$F_u(\varphi_i(x,t),x,t) > 0, (i=1,3), \quad F_u(\varphi_2(x,t),x,t) < 0. \tag{5.4}$$

We introduce the auxiliary equation

$$\frac{d^2\tilde{u}}{d\tau_\xi^2} = F(\tilde{u},\xi,t), \quad \tau_\xi = (x-\xi)/\varepsilon, \tag{5.5}$$

where $0 \le \xi \le 1$ and t are parameters. The phase plane $(\tilde{u}, \frac{d\tilde{u}}{d\tau_\xi})$ of this equation is represented in Fig. 5.1. This picture may be easily obtained, because equation (5.5) has a first integral in the explicit form

$$\frac{du}{d\tau} = \pm\left(2\int_{\varphi_1}^{u} F(u,\xi,t)\,du + C\right)^{1/2},$$

where the indices u and τ are omitted to keep the last expression from becoming too involved. We can distinguish the following three cases

a)

$$\int_{\varphi_1}^{\varphi_2} F(u,\xi,t)\,du > 0;$$

b)

$$\int_{\varphi_1}^{\varphi_2} F(u,\xi,t)\,du < 0;$$

c)

$$\int_{\varphi_1}^{\varphi_3} F(u,\xi,t)\,du = 0.$$

Notice that Fig. 5.1(a) and 5.1(b) contain trajectories corresponding to $C \neq 0$, where arrows indicate increasing time.

By (5.4) the points $(\varphi_i(\xi,t),0), (i=1,3)$ are saddles, and the point $(\varphi_2(\xi,t),0)$ is a center.

We determine functions

$$u_1^{(+)}(\tau_0), u_3^{(+)}(\tau_0), (\xi=0, \tau_0=x/\varepsilon);\; u_1^{(-)}(\tau_1), u_3^{(-)}(\tau_1), (\xi=1, \tau_\xi=(x-1)/\varepsilon),$$

as the solutions of the equation (5.5) with the appropriate boundary conditions:

$$u_1^{(+)}(0)=0, u_1^{(+)}(\infty)=\varphi_1(0); \qquad u_3^{(+)}(0)=0, u_3^{(+)}(\infty)=\varphi_3(0);$$
$$u_1^{(-)}(0)=0, u_1^{(-)}(-\infty)=\varphi_1(1); \quad u_3^{(-)}(0)=0, u_3^{(-)}(-\infty)=\varphi_3(1).$$

From Fig. 5.1, where $\tau_\xi = \tau_0$, we can see that for the existence of the function $u_1^{(+)}(\tau_0)$ (for instance) the vertical line $u=0$ must intersect the separatrix entering to the saddle $(\varphi_1(0,t),0)$ as $\tau_0 \to \infty$. Thus the existence of $u_1^{(+)}(\tau_0)$ depends on the position of the value $u=0$ on the u-axis:

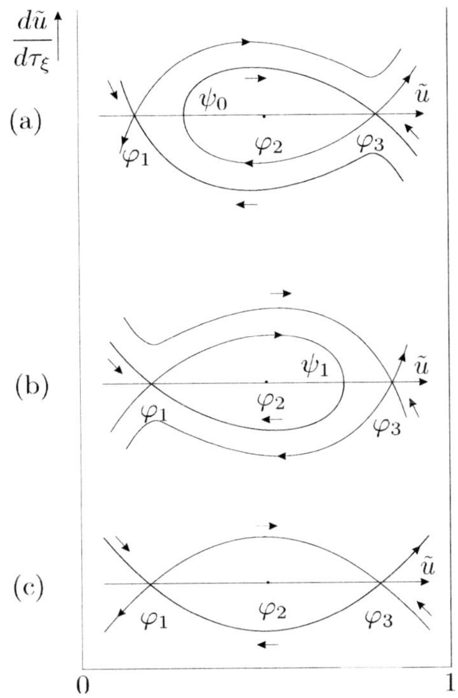

Figure 5.1.

in case a), $u=0$ can be any point of the u-axis,

in case b), the inequality $0<\psi_1$ must be fulfilled,

in case c), the inequality $0<\varphi_3$ must be fulfilled.

In general: The function $u_1^{(+)}(\tau_\xi)$ (or $u_3^{(+)}(\tau_\xi)$) exists if the vertical line $u=0$ in the phase plane intersects the separatrix entering the saddle $(\varphi_1(\xi,t),0)$ as $\tau_\xi \to \infty$ (or $(\varphi_3(\xi,t),0)$). The function $u_1^{(-)}(\tau_\xi)$ (or $u_3^{(-)}(\tau_\xi)$) exists if the vertical line $u=0$ in the phase plane intersects the separatrix entering the saddle $(\varphi_1(\xi,t),0)$ (or $(\varphi_3(\xi,t),0)$) as $\tau_\xi \to 0$.

Remark. Near $x=0$ we need consider only $u_1^{(+)}(\tau_0)$ and $u_3^{(+)}(\tau_0)$; near $x=1$ we need consider only $u_1^{(+)}(\tau_1)$ and $u_3^{(+)}(\tau_1)$.

We now introduce so called boundary layer functions:

$$\Pi_1^{(+)}(\tau_0,t) = u_1^{(+)}(\tau_0) - \varphi_1(0,t), \qquad \Pi_1^{(-)}(\tau_1,t) = u_1^{(-)}(\tau_1) - \varphi_1(1,t),$$

$$\Pi_3^{(+)}(\tau_0,t) = u_3^{(+)}(\tau_0) - \varphi_3(0,t), \quad \Pi_3^{(-)}(\tau_1,t) = u_3^{(-)}(\tau_1) - \varphi_3(1,t),$$

and add the further assumptions:

3,1) *Let* $u_1^{(+)}(\tau_0), u_1^{(-)}(\tau_1)$ *exist*, (i. e. let the line $u=0$ intersect the separatrix in the phase plane $(\tilde{u}, d\tilde{u}/d\tau_0)$ entering the saddle $(\varphi_1(0,t), 0)$ as $\tau_0 = +\infty$ and let the line $u = 1$ intersect the separatrix in the phase plane entering the saddle $(\tilde{u}, d\tilde{u}/d\tau_1)$ as $\tau_1 = -\infty$).

3,3) *Let* $u_3^{(+)}(\tau_0), u_3^{(-)}(\tau_1)$ *exist.*

Theorem 5.1. *Let assumptions 1), 2) and 3,1) hold. Then for sufficiently small* ε *there exists a solution* $u_1(x,t,\varepsilon)$ *of problem (5.1)–(5.3) such that, for* $0 \le x \le 1$ *and* $-\infty < t < +\infty$, *we have the asymptotic expansion*

$$u_1(x,t,\varepsilon) = \varphi_1(x,t) + \Pi_1^{(+)}(\tau_0,t) + \Pi_1^{(-)}(\tau_1,t) + O(\varepsilon).$$

We call the solution $u_1(x,t,\varepsilon)$ *the lower boundary layer solution*. The following limit relation holds

$$\lim_{t\to 0} u_1(x,t,\varepsilon) = \varphi_1(x,t), \quad 0 \le x \le 1, \ -\infty < t < +\infty.$$

Remark. A similar theorem is valid for the *upper boundary layer solution* $u_3(x,t,\varepsilon)$ if we replace the condition 3,1) by 3,3):

$$u_3(x,t,\varepsilon) = \varphi_3(x,t) + \Pi_3^{(+)}(\tau_0,t) + \Pi_3^{(-)}(\tau_1,t) + O(\varepsilon).$$

To study the solution which has not only boundary layers but also some interior layers we add the assumptions 4) and 5)

4) *The equation*

$$I(x,t) := \int_{\varphi_1(x,t)}^{\varphi_3(x,t)} F(u,x,t)\,du = 0 \tag{5.6}$$

defines a periodic function $x = x_0(t)$ *such that*

$$0 < x_0(t) < 1. \tag{5.7}$$

Moreover, suppose $I_x(x_0(t),t) \neq 0$.

Since $\xi = x_0(t)$, we have a cell (i.e. heteroclinic orbits) as in Fig. 5.1(c) in the phase plane. The assumption 4) guarantees the existence of the functions $u_1^{(-)}(\tau), u_3^{(-)}(\tau), u_1^{(+)}(\tau), u_3^{(+)}(\tau)$ (the index ξ on τ is omitted for notational simplicity). Moreover a) the function $u_1^{(-)}(\tau)$ is smoothly connected with $u_3^{(+)}(\tau)$ (i. e. there exists a separatrix going from the saddle $(\varphi_3(x_0,t),0)$ as $\tau \to -\infty$ to the saddle $(\varphi_1(x_0,t),0)$ as $\tau \to +\infty$), b) the function $u_3^{(-)}(\tau)$ is smoothly connected with $u_1^{(+)}(\tau)$.

By means of $u_k^{(\pm)}$ we obtain $\Pi_k^{(\pm)}$, for instance $\Pi_k^{(-)}(\tau,t) = u_1^{(-)}(\tau,t) - \varphi_1(x_0(t),t)$, (similarly for $\Pi_3^{(+)}, \Pi_3^{(-)}, \Pi_1^{(-)}$) which we naturally call *the interior layer functions*.

We must introduce one more assumption, because the theorem guaranteeing a solution with an interior layer is proved by the method of differential inequalities.

The foundation of this method is given in [2, 3]. For applications of this method in studying singularly perturbed problems, an important role was played by works of Nefedov (see [4, 5, 6]). Theorem 5.1 is proved by the method of differential inequalities as well, but without any additional assumptions. So, we add a further assumption

5) *The inequality*

$$\int_{\varphi_3(x_0(t),t)}^{\varphi_1(x_0(t),t)} F_x(u, x_0(t), t)\, du < 0 \tag{5.8}$$

holds. This inequality is called the stability condition.

Theorem 5.2. *If assumptions 1), 2), 3,3) for $u_3^{(+)}(\tau_0)$, 3,1) for $u_1^{(-)}(\tau_1)$, 4) and 5) all hold, then for sufficiently small ε there exists a periodic solution $u_3(x,t,\varepsilon)$ such that*

$$u_{31}(x,t,\varepsilon) = \begin{cases} \varphi_3(x,t) + \Pi_3^{(+)}(\tau_0,t) + \Pi_3^{(-)}(\tau^*,t) + O(\varepsilon), & 0 \le x \le x*, \\ \varphi_1(x,t) + \Pi_1^{(+)}(\tau^*,t) + \Pi_1^{(-)}(\tau_0,t) + O(\varepsilon), & x^* \le x \le 1, \end{cases} \tag{5.9}$$

where $x^* = x_0 + O(\varepsilon)$, $\tau^* = (x - x^*)/\varepsilon$.

A more exact asymptotic approximation is constructed in [4, 6].

The solution $u_{31}(x,t,\varepsilon)$ is called *a contrast structure of step type (CSST)* with the passage (transition) from $\varphi_3(x,t)$ to $\varphi_1(x,t)$ and with transition point $x_0(t)$. This solution has a boundary layer near $x = 0$, a boundary layer near $x = 1$ and an interior layer near $x = x_0(t)$. The graph of this solution in the (x,u) plane moves periodically, with $x_0(t)$, forward and backward along the x-axis. The transition point does not reach the boundaries $x = 0$ and $x = 1$. It is a moving CSST.

The following limit relation is the result of (5.9)

$$\lim_{\varepsilon\to 0} u_{31}(x,t,\varepsilon) = \begin{cases} \varphi_3(x,t), & 0 < x < x_0(t), \\ \varphi_1(x,t), & x_0(t) < x < 1. \end{cases} \tag{5.10}$$

A similar theorem is valid for the solution $u_{13}(x,t,\varepsilon)$ which has a transition direction from $\varphi_1(x,t)$ to $\varphi_3(x,t)$. We must, however, interchange the indices 1 and 3 in assumption 5) and in the limit relation (5.10), and use the assumption 3,1) for $u_1^{(+)}(\tau_0)$ and 3,3) for $u_3^{(-)}(\tau_1)$.

5.2 Equation with Weak Dependence on u_x

We now consider the case when the first derivative, u_x, is present with a small parameter in the right hand side of (5.1), see [7, 8]:

$$\varepsilon^2(u_{xx} - u_t) = F(\varepsilon u_x, u, x, t). \tag{5.11}$$

The first integral of the equation

$$\frac{d^2\tilde{u}}{d\tau_\xi^2} = F(\frac{d\tilde{u}}{d\tau_\xi}, \tilde{u}, \xi, t) \tag{5.12}$$

cannot be obtained in explicit form. But for some special cases, for instance when

$$F = \varepsilon a u_x + (u^2 - 1)(u - \varphi(x,t)\,), \tag{5.13}$$

we can obtain the separatrix connecting the saddle $(-1,0)$ with the saddle $(1,0)$ (as is necessary for the existence of an interior layer) in the form of a parabola (we omit the sign " $\sim$ ")

$$\frac{du}{d\tau} = A(u^2 - 1). \tag{5.14}$$

The equation (5.12) becomes

$$\frac{d^2u}{d\tau^2} = a\frac{du}{d\tau} + (u^2 - 1)(u - \varphi(\xi,t)\,). \tag{5.15}$$

Substituting (5.14) into (5.15) and equating terms with the same degree of u we obtain $2A^2 = 1$ (i.e. $A = \pm 1/\sqrt{2}$), $aA = \varphi(\xi,t)$. The transition point ξ_{13} for the transition from $(-1,0)$ to $(1,0)$ is obtained from the equation $\varphi(\xi,t) = -a/\sqrt{2}$; the equation for the separatrix is $du/d\tau = -1/(\sqrt{2})(u^2-1)$ and we can solve it explicitly. For the transition point ξ_{31} from $(1,0)$ to $(-1,0)$ we have $\varphi(\xi_3,t) = a/\sqrt{2}$, $du/d\tau = 1/(\sqrt{2})(u^2-1)$. Thus the separatrix corresponding to ξ_{13} is unique and it is different from the separatrix corresponding to ξ_{31}. Note that in the case (5.1), we have two separatrices for each transition point ξ.

5.3 The Case when Degenerate Equation Has Several Roots

When the derivative u_x appears in the right hand side of (5.11) without any small parameter, the behavior of the solution changes radically. In this case we can consider only the quasilinear equation because the derivative u_x tends to infinity as $\varepsilon \to 0$.

Thus we consider the quasilinear problem

$$\varepsilon^2(u_{xx} - u_t) = A(u,x,t)u_x + B(u,x,t), \tag{5.16}$$

$$u(a,t,\varepsilon) = 0, \qquad u(b,t,\varepsilon) = 0, \tag{5.17}$$

$$u(x,t,\varepsilon) = u(x,t+T,\varepsilon). \tag{5.18}$$

We impose the following five conditions:

1) *A and B are T periodic functions with respect to t and*

$$A \in C^3(\mathbb{R}\times\bar{\Omega}),\ B \in C^2(\mathbb{R}\times\bar{\Omega}),\ \Omega := \{\,a < x < b,\ t \in \mathbb{R}\,\}.$$

2) *The degenerate equation $Au_x + B = 0$ has two solutions $\bar{u}^{(\pm)}(x,t)$ such that*

$$\bar{u}^{(-)}(a,t) = 0, \qquad \bar{u}^{(+)}(b,t) = 0.$$

Moreover

$$A(\bar{u}^{(-)}(x,t),x,t) > 0, \quad A(\bar{u}^{(+)}(x,t),x,t) < 0, \qquad (x,t) \in \Omega.$$

3) *The equation*

$$I(x,t) = \int\limits_{\bar{u}^{(+)}(x,t)}^{\bar{u}^{(-)}(x,t)} A(u,x,t)\,du = 0 \tag{5.19}$$

has a solution $x = x_0(t)$ *such that* $a < x_0(t) < b$ *and* $I'_x(x_0(t),t) \neq 0$ *for* $\forall t \in \mathbb{R}$.

4) *The function*

$$F(u,t) := \int\limits_{\bar{u}^{(-)}(x_0(t),t)}^{u} A(\eta, x_0(t), t)\,d\eta$$

has no roots between $\bar{u}^{(-)}(x_0(t),t)$ *and* $\bar{u}^{(+)}(x_0(t),t)$ *for* $\forall t \in \Omega$.

5) *The inequality*

$$\frac{I'_x(x_0(t),t)}{(\bar{u}^{(+)}(x_0(t),t) - \bar{u}^{(-)}(x_0(t),t))} < 0$$

is satisfied $\forall t \in \mathbb{R}$.

Remark. Condition 3) corresponds to the previous assumption 4) and condition 5) corresponds to the stability condition (5.8).

Theorem 5.3. *If assumptions 1) – 5) hold, then for sufficiently small* ε *there exists a* T*-periodic solution* $u(x,t,\varepsilon)$ *of the problem (5.16) – (5.18),* $u \in C^{2,1}_{x,t}$ *satisfying the limit relation*

$$\lim_{\varepsilon \to 0} u(x,t,\varepsilon) = \begin{cases} \bar{u}^{(-)}(x,t), & a \le x < x_0(t), \\ \bar{u}^{(+)}(x,t), & x_0(t) < x \le b. \end{cases}$$

This solution has no boundary layers, only an interior layer near $x = x_0(t)$. It is the CSST.

A uniform asymptotic representation similar to (5.9) can also be obtained, see [9].

5.4 Periodic Contrast Structures of Step Type

We have considered a periodic CSST. The important condition for the existence of such a contrast structure is (5.7): $0 < x_0(t) < 1$. The following question arises: what happens to the solution $u(x,t,\varepsilon)$ when the transition point $x_0(t)$ approaches the boundary $x = 0$ or $x = 1$. Up until now this question has not been analytically resolved. We propose a number of examples for which one can give a rough asymptotic description, which coincides, in a certain sense, with the results of numerical computations.

A particular case, for which one can give an exact analysis, will be described below in Section 5.5.

Let us consider the equation

$$\varepsilon^2(u_{xx} - u_t) = F(u, x, t) := (u^2 - 1)(u - \varphi(x, t)) \tag{5.20}$$

with the additional conditions (5.2), (5.3). In this case equation (5.6) for the transition point $x_0(t)$ reduces to

$$\varphi(x_0(t), t) = 0.$$

To see this, it is sufficient to note that for equation (5.20) we have

$$I(x, t) = \int_{-1}^{+1} (u^2 - 1)(u - \varphi)\, du = -\varphi \int_{-1}^{+1} (u^2 - 1)\, du,$$

and $I = 0$ implies $\varphi = 0$.

Consider the solution $u_{31}(x, t, \varepsilon)$. Let us construct, for this solution, the function $u_{31}(\tau)$ which is the smooth connector of $u_3^{(-)}(\tau)$ and $u_1^{(+)}(\tau)$ (see assumption 4) in Section 1 and the subsequent explanatory comments). We have

$$u_{31}(\tau) = \frac{1 - \exp\{\frac{\sqrt{2}}{\varepsilon}(x - x_0(t))\}}{1 + \exp\{\frac{\sqrt{2}}{\varepsilon}(x - x_0(t))\}}. \tag{5.21}$$

From (5.21) we get the relation:

$$\lim_{\varepsilon \to 0} u_{31}(\tau) = \begin{cases} +1, & 0 < x < x_0(t), \\ -1, & x_0(t) < x < 1. \end{cases}$$

The stability condition (5.8) gives

$$-\int_{+1}^{-1} (u^2 - 1)\varphi_x(x_0(t), t)\, du < 0$$

or, equivalently

$$\varphi_x(x_0(t), t)\, du > 0. \tag{5.22}$$

The transition point $x_0(t)$ varies when t varies. The function (5.21) has a jump at the point $x_0(t)$ as $\varepsilon \to 0$ and we call the solution $u_{31}(x, t, \varepsilon)$ the moving or travelling step. The solution $u_{31}(x, t, \varepsilon)$ has not only an interior layer described by $u_{31}(\tau)$, but also boundary layers at $x = 0$ and $x = 1$.

Let be $x_0(0) = x^0 \in (0, 1)$ (for instance $x_0(0) = 0.5$). Let $x_0(t)$ increase for increasing t. What happens if for a certain $t = t_0$ the relation $x(t_0) = 1$ holds? Strictly speaking, it is not possible to use the expression (5.21) at $t = t_0$, but we postulate that the value found from the relation $x(t_0) = 1$ is the moment when the solution $u_{31}(x, t, \varepsilon)$ becomes the purely boundary layer solution $u_3(x, t, \varepsilon)$ (see the remark after the Theorem 5.1).

Numerical experiments confirm that this is possible. The step u_{31} at $t = t_0$ is transformed into the the purely boundary layer solution u_3. But the solution u_3 does not exist in general for all $t \le t_0$ since the requirement 3,1) for $u_3^{(-)}(\tau_1)$ can be violated for a certain $t = t_1 > t_0$. The solution u_3 blows up. After that, a travelling step can again arise which moves back to $x = 0$.

Similar phenomena can arise when the transition point $x_0(t)$ hits the boundary $x = 0$.

Remark. We will investigate the transformation of the solution u_{31} using the assumption that the $u_3^{(+)}(\tau_0)$ exists during the transformation of u_{31} (from u_{31} to u_3 and to a moving step again) after its exit from the boundary $x = 1$, i.e. the left boundary layer exists all the time.

Contrast structures which at a certain value of t change their form from a moving step to a purely boundary layer solution and back to a moving step again are called *the contrast structures of alternating type (CSAT).*

A number of analytical results helps us to study the CSAT. For the existence of the step u_{31} with boundary layers it is necessary that the vertical $u = 0$ meets the separatrix entering the saddle $(\varphi_3(0,t), 0) = (1, 0)$ as $\tau_0 \to \infty$ and the vertical $u = 0$ arrives at the separatrix entering the saddle $(\varphi_1(1,t), 0) = (-1, 0)$ as $\tau_1 \to -\infty$. For brevity we will speak about the joining of $u = 0$ with " + 1" and the joining of $u = 0$ with " − 1".

The possibility of joining $u = 0$ with "+ 1" at the boundary $x = 0$ depends on the value $\varphi_0 := \varphi(0,t)$. If $\varphi_0 < 0$, we have the phase picture shown in Fig. 5.1 (b); if $\varphi_0 = 0$, we have Fig. 5.1 (c); if $\varphi_0 > 0$, we have Fig. 5.1 (a). In the case $\varphi_0 \le 0$ the value $u = 0$ can be any point of interval $(-1, +1)$, and in the case $\varphi_0 > 0$ the value $u = 0$ should not be less than ψ_0. Therefore if $\varphi_0 > 0$ varies with t then, to join $u = 0$ with " + 1", the maximum possible value φ_0 is a value for which $\psi_0 = 0$. Denote it by $\varphi_0^{(+)}$. From the explicit expression for the separatrix we can directly calculate that

$$\varphi_0^{(+)} = \frac{3}{8} = 0.375. \tag{5.23}$$

Similarly, to reveal the possibility of joining $u = 0$ with " − 1", we introduce the notation $\varphi_0^{(-)}$ which is the minimum possible value of φ_0 for joining u with " − 1". We have

$$\varphi_0^{(-)} = -\frac{3}{8} = -0.375. \tag{5.24}$$

The same formulas can be written for the boundary $x = 1$. We denote by $\varphi_1^{(+)}$ the maximum possible value of φ_1 for the joining of $\varphi_1 = \varphi(1,t)$ with " + 1" and we denote by $\varphi_1^{(-)}$ the minimum possible value of φ_1 for the joining of $u = 0$ with " − 1". We have

$$\varphi_1^{(+)} = \frac{3}{8} = 0.375, \qquad \varphi_1^{(-)} = -\frac{3}{8} = -0.375. \tag{5.25}$$

It is convenient to use the formulas (5.23) – (5.25) to study the existence of the purely boundary layer solutions u_1 or u_3 and the steps u_{31} (or u_{13}).

In Fig. 5.1 where $\varphi_1 = \varphi_3 = 1$ and $\xi = 1$, we see that if $\varphi(1,t)$ increases, then the picture c) is transformed to the picture a) and $\varphi(1,t) = 0.375$ is the last value of $\varphi(1,t)$ when the vertical $u=0$ intersects the separatrix entering the saddle $(1,0)$. If $\varphi(1,t)$ is greater than the number 0.375, the vertical $u=0$ does not intersect the separatrix entering the saddle $(0,1)$ and the boundary function $u_3^{(-)}(\tau_1)$ does not exist. Thus the boundary layer phase ends at $t=t_1$ where t_1 is obtained from solving the equation $\varphi(1,t_1)=0.375$.

A. P. Petrov [10] proposed a method to describe a process that arises after $t=t_1$. Let us introduce the differential equation

$$\varepsilon\frac{dr}{dt}=-\sqrt{2}\varphi(r,t), \tag{5.26}$$

where r is the value of x for which $u(x,t,\varepsilon)=0$ and $\phi(r,t)$ is the known root of $F=0$. The initial condition is taken as

$$r(t_1,\varepsilon)=1, \tag{5.27}$$

since during the boundary layer phase the argument x is near the value 1.

Equation (5.26) is a singularly perturbed first order equation. According to Tikhonov's theorem, (see [11]), together with the stability condition (5.22), the solution $r(t,\varepsilon)$ of the Cauchy problem (5.26), (5.27) exponentially approaches the transition point $x_0(t)$ which is the solution of the degenerate equation $\varphi(r,t)=0$.

Substituting $r(t,\varepsilon)$ in the formula (5.21) we obtain

$$\hat{u}_{31}=\frac{1-\exp\{\frac{\sqrt{2}}{\varepsilon}(x-r(t,\varepsilon))\}}{1+\exp\{\frac{\sqrt{2}}{\varepsilon}(x-r(t,\varepsilon))\}}.$$

It turns out that $\hat{u}_{31}$ satisfies the equation (5.20) with an error of order $O(\varepsilon)$.

The following four stages are distinguished in the character of the behavior of CSAT.

a) The solution u_{31} has the form of a travelling step with an interior layer function (5.21) from $t=0$ to $t=t_0$, where t_0 is obtained from the equation $x_0(t_0)=0$. The graph of u_{31} moves with $x_0(t)$, and we call this stage the *"moving"* or the *"travelling"* stage.

b) From t_0 to t_1 (at t_1 the function $u_3^{(-)}(\tau_1)$ no longer exists) we have the stage which is called the *"halt"*. The solution has a purely boundary layer form and varies slowly. This stage may be rather long, because t_1-t_0 may be large.

c) During the interval (t_0,t_1) the value $x_0(t)$ is greater than 1 and is next equal to 1 at $t=t_2>t_0$. If $t_2=t_1$ then a new moving stage is starting. The solution has the form of step u_{31} and moves to the left with the motion of $x_0(t)$.

d) If the inequality $t_2<t_1$ holds, then at $t=t_1$, the solution has a form close to $(\hat{u}_{31})$, as if it overtakes the solution u_{31} described by $u_{31}(\tau)$ in the region of the interior layer. This stage is called the *"run"*.

If $x_0(t)$ attains the boundary $x=0$, there may be a similar transformation as for $x=1$. In general the collapse of the $u_3^{(+)}(\tau_0)$ may arise before $x_0(t)$ arrives at $x=0$ (we assumed above that this did not occur). Then a more complicated regime may arise.

We can illustrate all the above phenomena by some numerical results. These numerical results were obtained (see [10]) by A. A. Plotnikov (MSU, Moscow) and confirmed by M. Radziunas (WIAS, Berlin). The computations were performed for the problem

$$\varepsilon^2(u_{xx} - u_t) = (u^2 - 1)(u - 0.5x + 0.73 \sin t + 0.25)\,),$$

$$u(0, t, \varepsilon) = u(1, t, \varepsilon), \qquad \varepsilon^2 = 10^{-3}.$$

This problem was solved with initial conditions $u(x, 0, \varepsilon) = \sin 2\pi x$. The solution rapidly arrives at the CSST u_{31}.

For this case our rough analytical calculations give: $\varphi(x, t) = 0.5x - 0.73 \sin t - 0.25$, the value t_0 may be found from the equation $x_0(t_0) = 2(0.73 \sin t_0 + 0.25) = 1$ and is equal to 0.34. The value t_1 is equal 3.31 and may be obtained from the equation $\varphi(1, t_1) = -0.73 \sin t_1 + 0.25 = 0.375$.

The inequality $t_2 < t_1$ is satisfied (the value t_2 is the other root of the equation $x_0(t) = 1$). During the time interval $t_1 - t_2$, $x_0(t)$ goes to the left and arrives at the value 0.25 at $t = t_1$. Starting from the instant of time t_1, we have a run to the left according to the initial value problem

$$\varepsilon \frac{dr}{dt} = -\sqrt{2}(0.5r - 0.73 \sin t - 0.25)\,),$$
$$r(3.31, \varepsilon) = 1, \qquad \varepsilon^2 = 10^{-3}. \tag{5.28}$$

The function $r(t, \varepsilon)$ approaches rapidly the value $x_0(t) = 2(0.73 \sin t + 0.25)\,)$ which moves and arrives at the boundary $x = 0$ at the value $t = t_3 = 3.48$ ($x_0(3.48) = 0$).

After $t = t_3$ we have the stage of the halt in the form of the lower purely boundary layer solution. By $t = t_4$ (the value is found from the equation $\varphi(0, t_4) = -0.375$) there arises a run to the right in direction of $x = 1$. The run is described by the equation (5.28) with initial condition $r(t_4, \varepsilon) = 0$. The boundary $x = 1$ is attained at $t_5 = 6.62 = t_0 + 2\pi$.

The results of the numerical computations are in good agreement with our analytical results.

Other examples of a CSAT can be found in [1].

Remark. A CSAT can arise for the equation (5.11) as well. But even in the case (5.13), for which the transition point is obtainable, we cannot get the value t_1 where the purely boundary layer solution collapses. Unlike the case (5.20), when the relation $\varphi(t_1, 1) = \varphi_1^{(+)} = 0.375$ for t_1 is obtained by considering the phase picture in Fig.5.1, for the case (5.13) we do not construct the phase picture because we do not have the first integral of (5.15) in explicit form.

We must obtain the phase picture by computation. Then we look for the value t_1 as that value of t when the line $u = 0$ is tangent to the separatrix leaving the focus $(\varphi, 0)$ and entering the saddle $(0, 1)$.

5.5 Periodic Contrast Structures of Alternating Type

In this section we give a new result. We present the case when the existence of CSAT can be rigorously proved [12, 13].

We return to the problem (5.16) – (5.18). Suppose the following conditions are satisfied:

1) and 2) are the same as in the Section 3.

3) there are two values t_1 and t_2, $t_1 < t_2$, such that

i) the equation (5.19) has the solution $x = x_0(t)$, which satisfies the inequality $a < x_0(t) < b$, if $t \in [0, t_1] \cup [t_2, T]$, and $x_0(t) = b$, if $t = t_1$ and $t = t_2$.

ii) $I(x,t) \neq 0$ for all $(x,t) \in [a,b] \times (t_1, t_2)$.

4) The function

$$F(u,t) = \int\limits_{\bar{u}^{(-)}(x_0(t),t)}^{u} A(v, x_0(t), t)\, dv$$

has no roots between $\bar{u}^{(-)}(x_0(t),t)$ and $\bar{u}^{(+)}(x_0(t),t)$ for $t \in [0, t_1] \cup [t_2, T]$. The function

$$\tilde{F}(u,t) = \int\limits_{\bar{u}^{(-)}(b,t)}^{u} A(v, b, t)\, dv$$

has no roots between $\bar{u}^{(-)}(b,t)$ and zero for $t \in (t_1, t_2)$.

5) The inequality 5) from Section 3 is satisfied on $[0, t_1] \cup [t_2, T]$.

Theorem 5.4. *If the assumptions 1) – 5) hold, then for sufficiently small ε there exists a T-periodic solution $u(x,t,\varepsilon)$ of the problem (5.16) – (5.18), $u \in C^{2,1}_{x,t}$, and following limit relation is valid*

$$\lim_{\varepsilon\to 0} u(x,t,\varepsilon) = \begin{cases} \bar{u}^{(-)}(x,t), & a \le x < x_0(t),\ t \in [0,t_1] \cup [t_2, T], \\ \bar{u}^{(+)}(x,t), & x_0(t) < x \le b,\ t \in [0,t_1] \cup [t_2, T], \\ u^{(-)}(x,t), & a \le x < b,\ t \in (t_1, t_2). \end{cases} \tag{5.29}$$

The theorem is proved by means of the method of differential inequalities [5]. The upper and lower solutions (barriers), β and α, are constructed by modifying the barriers for the case $a < x_0(t) < b$ (see [9]). These barriers satisfy the limit relation (5.29), the difference between α and β tends to 0 as $\varepsilon \to 0$, but this difference is not $O(\varepsilon)$ for all x and t.

In [13] we obtain the remainder to be of order $O(\varepsilon)$ for all x and t, except in a small vicinity of the points (b, t_1) and (b, t_2).

Remark. The moving step for our problem does not have boundary layers near $x = a$ and $x = b$, and has only an interior layer. This step solution transforms into the boundary layer solution with one boundary layer, near the boundary $x = b$. The run-stage does not appear after the halt-stage, but the moving-stage follows immediately. The presence of only two stages raises the possibility of constructing a proof for Theorem 5.4.

It is probable, that such a concept is applicable for a CSAT for some cases in Sections 5.1, 5.2, 5.4.

5.6 Acknowledgements

This work was supported in part by the Boole Centre for Research in Informatics, University College Cork.

Bibliography

[1] A. B. Vasil'eva, *Contrast structures of alternating type*, J. of Math. Sciences, 121(1) (2004), pp. 2080–2116.

[2] M. Nagumo, *Ueber die Differentialgieichung* $y'' = f(x, y, y')$, Proc. Phys. Math. Soc. Jpn., 19 (1937), pp. 861–866.

[3] H. Amann, *Periodic solutions of semilinear parabolic equatuons*, in Nonlinear Analysis (A Collection of Papers in Honour of Erich Rothe), L. Cesari, R. Kannan and H. F. Weinberger, eds., Academic Press. New York, 1978, pp. 1–29.

[4] V. F. Butuzov, A. B. Vasil'eva, and N. N. Nefedov, *Asymptotic theory of contrast structures*, Automat. Remote Control, 58(7-1) (1997), pp. 1068–1091.

[5] N. N. Nefedov, *Asymptotic method of differential inequalities for studying periodic contrast structures, asymptotics and stability*, Differ. Equat., 36(10) (2000), pp. 1544–1550.

[6] A. B. Vasil'eva, V. F. Butuzov, and N. N. Nefedov, *Contrast structures in singularly perturbed problems*, Fundam. Prikl. Mat., 4(3) (1998), pp. 799–851.

[7] A. B. Vasil'eva and M. A. Davydova, *On a step-like contrast structure for a class of second order nonlinear singularly perturbed equations*, Comput. Math. Math. Phys., 38(6) (1998), pp. 900–908.

[8] A. B. Vasil'eva, *Periodic solutions of a parabolic problem with a small parameter multiplying the derivatives*, Comp. Math. and Math. Physics, 67(3) (2003), pp. 346–348.

[9] A. B. Vasil'eva and O. E. Omel'chenko, *Periodic step-like contrast structures for a singularly perturbed parabolic equation*, Differ. Equat., 36(2) (2000), pp. 236–246.

[10] A. B. Vasil'eva, A. P. Petrov, and A. A. Plotnikov, *Theory of contrast structures of alternating type*, Comput. Math. Math. Phys., 38(9) (1998), pp. 1471–1480.

[11] A. B. Vasil'eva, V. F. Butuzov, and L. V. Kalachev, *The Boundary Function Method for Singular Perturbation Problems*, SIAM, Philadelphia, 1995.

[12] A. B. Vasil'eva and O. E. Omel'chenko, *Contrast structures of alternating type in singularly perturbed quasilinear equations*, Russian Acad. Sci. Dokl. Math., 390(3) (2003), pp. 298–300.

[13] ——, *Contrast structures of alternating type in singularly perturbed parabolic equations*, Differ. Uravn, to appear.

Chapter 6

Multi-Dimensional Internal Layers for Spatially Inhomogeneous Reaction-Diffusion Equations

N. Nefedov

In this paper singularly perturbed partial differential equations are considered and typical problems of the asymptotic theory of contrast structures are discussed.

6.1 Introduction

Mathematical problems concerning reaction–diffusion problems are of increasing interest because of many applications of practical importance. Some important cases when the reaction term is dominated can be described by nonlinear singularly perturbed parabolic equations. Physically, this problem may be interpreted as a model for reaction-diffusion systems in chemical kinetics, synergetics, astrophysics, biology, *et.al.* (see, for example, [5] and references therein). The solutions of these problems often feature a narrow boundary layer region of rapid change as well as internal layers of different structures (contrast structures), and it requires the development of new asymptotic methods to treat it formally as well as rigorously. In the present work we demonstrate some recent results of rigorous investigations which are based on the asymptotic method of differential inequalities and on the SLEP-method (Singular Limit Eigenvalue Problem Method). Using these methods we consider some classes of singularly perturbed equations. For all problems considered here we prove the existence of the solutions, estimate the accuracy of the asymptotics and investigate the stability of the stationary and periodic solutions as solutions of the corresponding initial boundary value problems for parabolic equations. To define contrast structures we consider the following problem

$$\begin{gathered}\varepsilon^2 \Delta u = f(u, x, \varepsilon)\,, \quad x \in \mathcal{D} \subset R^2, \\ u|_{\partial\mathcal{D}} = g(x)\,,\end{gathered} \tag{6.1}$$

where $\varepsilon > 0$ is small parameter and Δ is the Laplacian. The spike–type contrast structure is the solution $u(x, \varepsilon)$ of the problem (6.1), which is close to some solution $u = \varphi(x)$ of the degenerate equation $f(u, x, 0) = 0$ everywhere inside the domain D with the exception of a small neighborhood of some curve C where the solution $u(x, \varepsilon)$ differs significantly from $\varphi(x)$ (i.e. the solution $u(x, \varepsilon)$ has a spike).

The step-type contrast structure is the solution $u(x, \varepsilon)$ of the problem (6.1), which is close to two different solutions of the degenerate equation on different sides of some curve C.

In what follows we concentrate our attention on multi-dimensional problems. A relevant consideration of some one-dimensional problems can be found in [16], [18], [21], [22].

6.2 Step-type Contrast Structures in Partial Differential Equations

6.2.1 The noncritical case

Consider the equation

$$\varepsilon^2 \Delta u = f(u, x, \varepsilon), \quad x \in \mathcal{D} \subset R^2, \tag{6.2}$$

with the Dirichlet boundary condition

$$u|_{\partial\mathcal{D}} = g(x). \tag{6.3}$$

Here Δ is the Laplace operator, $\mathcal{D}$ is a smooth bounded domain. The functions f, g and the boundary $\partial\mathcal{D}$ are assumed to be sufficiently smooth.

For the reduced equation

$$f(u, x, 0) = 0 \tag{6.4}$$

we assume that

I. *Equation* (6.4) *has three solutions* $u = \varphi_i(x)$ $(i = 1, 2, 3)$ *such that*

a) $\varphi_1(x) < \varphi_2(x) < \varphi_3(x)$ *for* $x \in \overline{\mathcal{D}}$ *and there are no other solutions of* (6.4) *in the intervals* (φ_1, φ_2) *and* (φ_2, φ_3).

b) $f_u(\varphi_i(x), x, 0) > 0$, $i = 1, 3$; $f_u(\varphi_2(x), x, 0) < 0$ *for* $x \in \overline{\mathcal{D}}$.

We will construct the asymptotics of a contrast step-type structure for the problem (6.2), (6.3). A contrast step-type structure is defined (analogously to Section 6.1) as a solution of (6.2), (6.3) which is close to $\varphi_1(x)$ and $\varphi_3(x)$ of the reduced equation (6.4) on different sides of some closed curve Γ located inside $\mathcal{D}$, and thus, the interior transition layer lies in a small neighborhood of the curve Γ. The location of the curve Γ is not known a priori; it is defined during the construction of the asymptotics.

We introduce the function $I(x)$:

$$I(x) = \int_{\varphi_1(x)}^{\varphi_3(x)} f(u, x, 0)\, du\,.$$

II. *Let the equation* $I(x) = 0$ *define some closed smooth simple curve* Γ_0 *located inside* $\mathcal{D}$ *and let*

$$\frac{\partial}{\partial n} I(x) < 0 \quad for \quad x \in \Gamma_0, \tag{6.5}$$

where $\frac{\partial}{\partial n}$ *is the derivative along the inward unit normal vector to* Γ_0.

Such a case, when condition (6.5) holds, is called *noncritical.* The critical case $I(x) \equiv 0$ for $x \in \overline{\mathcal{D}}$ is called *the case with balanced nonlinearities.* This case will be discussed in the next subsection.

The curve Γ_0 will be the main "term" of the asymptotic representation of the curve Γ mentioned above. We define the location of Γ by the condition

$$u(x, \varepsilon) = \varphi_2(x) \quad \text{for} \quad x \in \Gamma. \tag{6.6}$$

The closed curve Γ (yet unknown) divides the domain $\mathcal{D}$ into two subdomains: $\mathcal{D}_i$ (interior to Γ) and $\mathcal{D}_e$ (exterior to Γ).

The following condition is a standard one which assumes that the boundary conditions belong to the domain of attraction of the solution of the reduced equation (it is used to construct the boundary functions).

III. *Let*

$$\int\limits_{\varphi_1(x)}^{s} f(u, x, 0)\, du > 0 \quad for \quad x \in \partial\mathcal{D}, \quad s \in (\varphi_1(x), g(x)] .$$

Under the conditions I – III, we seek an asymptotic expansion of the solution in the form

$$u(x,\varepsilon) = \begin{cases} \bar{u}^i + Q^i = \sum\limits_{k=0}^{\infty} \varepsilon^k [\,\bar{u}_k^i(x) + Q_k^i(\tau,\theta)], & \text{for } x \in \overline{\mathcal{D}}_i; \\ \bar{u}^e + Q^e + \Pi = \sum\limits_{k=0}^{\infty} \varepsilon^k [\,\bar{u}_k^e(x) + Q_k^e(\tau,\theta) + \Pi_k(\rho, l)], & \text{for } x \in \overline{\mathcal{D}}_e. \end{cases} \tag{6.7}$$

Here $\bar{u}_k^i$, $\bar{u}_k^e$ are the regular terms of the asymptotics, Q_k^i and Q_k^e are transition layer functions, Π_k are boundary functions. The variables ρ, l are local coordinates near the boundary. They are defined in the standard way. The variables τ, θ are introduced as follows. At first we introduce new (local) coordinates (r, θ) in a small neighborhood of the curve Γ_0: r is the distance from a given point M to the curve Γ_0 along the normal to Γ_0 with the plus sign if $M \in \mathcal{D}_i$ and with the minus sign if $M \in \mathcal{D}_e$, θ is the coordinate along this normal on Γ_0 (we assume that $0 \le \theta \le 2\pi$).

We will seek the equation for Γ in the form

$$r = \lambda(\theta, \varepsilon) = \varepsilon\lambda_1(\theta) + \varepsilon^2\lambda_2(\theta) + \cdots,$$

where the functions $\lambda_i(\theta)$ must be found during the construction of the asymptotics. We now introduce the transition layer variable

$$\tau = \frac{r - \lambda(\theta, \varepsilon)}{\varepsilon}.$$

The operator $\varepsilon^2\Delta$ in terms of the new variables τ, θ has the form

$$\varepsilon^2\Delta = (1+\varepsilon^2\alpha)\frac{\partial^2}{\partial\tau^2} + \sum_{j=1}^{\infty}\varepsilon^j L_j\,,$$

where $\alpha = \alpha(\varepsilon\tau, \theta, \varepsilon)$ is a known function, L_j are linear differential operators containing the differentiations $\frac{\partial}{\partial\tau}, \frac{\partial}{\partial\theta}, \frac{\partial^2}{\partial\theta^2}$; for example, $L_1 = k(\theta)\frac{\partial}{\partial\tau}$ and $k(\theta)$ is a known function (the curvature of Γ_0).

Let us substitute the series (6.7) into the equation (6.2) and change the right-hand side f into an expression of the type (6.7): $\bar{f}^i + Q^i f$ for $\overline{\mathcal{D}}_i$ and $\bar{f}^e + Q^e f + \Pi f$ for $\overline{\mathcal{D}}_e$. Now, expanding $\bar{f}^i, \ldots, \Pi f$ into power series in ε and equating coefficients of powers of ε separately for regular terms, for transition layer functions and for boundary functions, we obtain the equations for the terms of the asymptotic expansion of the solution.

For $\bar{u}_0(x)$, we have the degenerate equation:

$$f(\bar{u}_0, x, 0) = 0.$$

We choose

$$\begin{aligned} \bar{u}_0 &= \bar{u}_0^i = \varphi_3(x) \text{ for } x \in \overline{\mathcal{D}}_i, \\ \bar{u}_0 &= \bar{u}_0^e = \varphi_1(x) \text{ for } x \in \overline{\mathcal{D}}_e. \end{aligned}$$

In a similar way we can find the next terms of the regular part of the asymptotics, for example

$$\bar{u}_1^i = f_u^{-1}(\varphi_3(x), x, 0) f_\varepsilon(\varphi_3(x), x, 0),\ \bar{u}_1^e = f_u^{-1}(\varphi_1(x), x, 0) f_\varepsilon(\varphi_1(x), x, 0).$$

To determine $Q_0^i(\tau,\theta)$ and $Q_0^e(\tau,\theta)$ we obtain the equations (the functions $\varphi_i(x)$ and $f(u,x,\varepsilon)$ after change the variables x to the variables (r,θ) are denoted by $\varphi_i(r,\theta)$ and $f(u,r,\theta,\varepsilon)$):

$$\frac{\partial^2 Q_0^i}{\partial\tau^2} = f(\varphi_3(0,\theta) + Q_0^i, 0, \theta, 0), \quad \tau > 0, \tag{6.8}$$

$$Q_0^i(0,\theta) = \varphi_2(0,\theta) - \varphi_3(0,\theta), \quad Q_0^i(+\infty,\theta) = 0, \tag{6.9}$$

$$\frac{\partial^2 Q_0^e}{\partial\tau^2} = f(\varphi_1(0,\theta) + Q_0^e, 0, \theta, 0), \quad \tau < 0, \tag{6.10}$$

$$Q_0^e(0,\theta) = \varphi_2(0,\theta) - \varphi_1(0,\theta), \quad Q_0^e(-\infty,0) = 0. \tag{6.11}$$

Here θ is considered as a parameter. The first condition in (6.9) and (6.11) follows from (6.6).

We now introduce the function

$$\tilde{Q}_0(\tau,\theta) = \begin{cases} \varphi_1(0,\theta) + Q_0^e(\tau,\theta), & \tau \le 0; \\ \varphi_3(0,\theta) + Q_0^i(\tau,\theta), & \tau \ge 0. \end{cases}$$

From (6.8) – (6.11) $\tilde{Q}_0$ satisfies the following

$$\begin{aligned} \frac{\partial^2 \tilde{Q}_0}{\partial\tau^2} &= f(\tilde{Q}_0, 0, \theta), \quad -\infty < \tau + \infty, \\ \tilde{Q}_0(0,\theta) &= \varphi_2(0,\theta),\ \tilde{Q}_0(-\infty,\theta) = \varphi_1(0,\theta),\ \tilde{Q}_0(\infty,\theta) = \varphi_3(0,\theta). \end{aligned} \tag{6.12}$$

From conditions I – II it follows that for each fixed θ in the phase plane ($\tilde{Q}_0$, $\frac{\partial \tilde{Q}_0}{\partial \tau}$) of equation (6.12) there is a cell: the saddle points ($0, \varphi_1$) and ($0, \varphi_3$) are connected by separatrices. Thus, the problem (6.12) and, hence, the problems (6.8) – (6.11) are solvable. The functions Q_0^i and Q_0^e satisfy the exponential estimate

$$|\, Q_0^i(\tau,\theta)\,| \le C\, \exp\{\, æ\,\tau\,\};\ |\, Q_0^e(\tau,\theta)\,| \le C\, \exp\{\, æ\,\tau\,\}. \tag{6.13}$$

Here C and æ denote some suitable positive constants independent of ε. Note that

$$\frac{\partial Q_0^i}{\partial \tau}(0,\theta) = \frac{\partial Q_0^e}{\partial \tau}(0,\theta) = \frac{\partial \tilde{Q}_0}{\partial \tau}(0,\theta)\,,$$

i.e. the derivatives of zeroth approximation are matched on the curve Γ.

To determinate $Q_1^i(\tau,\theta)$, $Q_1^e(\tau,\theta)$ we have the equations

$$\begin{aligned} \frac{\partial^2 Q_1^i}{\partial \tau^2} &= f_u(\tau,\theta)\, Q_1^i + f_1^i(\tau,\theta)\,, \quad \tau > 0, \\ Q_1^i(0,\theta) &= -\lambda_1(\theta)\tfrac{\partial \varphi_3}{\partial r}(0,\theta),\ Q_1^i(\infty,\theta) = 0; \\ \frac{\partial^2 Q_1^e}{\partial \tau^2} &= f_u(\tau,\theta)\, Q_1^e + f_1^e(\tau,\theta), \quad \tau < 0, \\ Q_1^e(0,\theta) &= -\lambda_1(\theta)\tfrac{\partial \varphi_1}{\partial r}(0,\theta),\ Q_1^e(-\infty,\theta) = 0, \end{aligned}$$

where

$$f_1^i = [\, f_u(\tau,\theta)\frac{\partial \varphi_3}{\partial r}(0,\theta) + f_r(\tau,\theta)\,]\,(\tau + \lambda_1(\theta)\,) + f_\varepsilon(\tau,\theta) - k(\theta)\frac{\partial \tilde{Q}_0}{\partial \tau},$$

f_1^e has a similar expression with φ_1 replacing φ_3, while $f_u(\tau,\theta)$, $f_r(\tau,\theta)$ and $f_\varepsilon(\tau,\theta)$ are evaluated at the point ($\tilde{Q}_0, 0, \theta$). The solutions to these equations can be represented in an explicit form using the function $\Phi(\tau,\theta) = \frac{\partial \tilde{Q}_0}{\partial \tau}(\tau,\theta)$. For example, the expression for Q_1^i has the form:

$$\begin{aligned} Q_1^i(\tau,\theta) &= -\lambda_1(\theta)\,\tfrac{\partial \varphi_3}{\partial r}(0,\theta)\,\Phi^{-1}(0,\theta)\,\Phi(\tau,\theta) \\ &\quad -\Phi(\tau,\theta)\int_0^\tau \Phi^{-2}(\xi,0)\int_\xi^\infty \Phi(\sigma,\theta)\, f_1^i(\sigma,\theta)\, d\sigma d\xi. \end{aligned} \tag{6.14}$$

The matching condition for the derivatives at the first approximation gives the equality

$$\frac{\partial \varphi_1}{\partial r}(0,\theta) + \frac{\partial Q_1^e}{\partial \tau}(0,\theta) = \frac{\partial \varphi_3}{\partial r}(0,\theta) + \frac{\partial Q_1^i}{\partial \tau}(0,\theta).$$

Using (6.14) and an analogous expression for $Q_1^e(\tau,\theta)$, we can rewrite this equality in the form

$$\frac{\partial \varphi_3}{\partial r}(0,\theta) - \frac{\partial \varphi_1}{\partial r}(0,\theta) - \Phi^{-1}(0,\theta)\int_{-\infty}^{\infty} \Phi(\tau,\theta)\, f_1(\tau,\theta)\, d\tau = 0.$$

Next, using the expression for $f_1(\tau,\theta)$, which is equal to $f_1^i(\tau,\theta)$ for $\tau \ge 0$, and to $f_1^e(\tau,\theta)$ for $\tau \le 0$, after some transformations we get the equation

$$\left[\frac{\partial}{\partial r}\int_{\varphi_1(r,\theta)}^{\varphi_2(r,\theta)} f(u,r,\theta)\, du\right]\Bigg|_{r=0} \lambda_1(\theta) = \int_{-\infty}^{\infty} [\, k(\theta)\,\Phi(\tau,\theta) - f_r(\tau,\theta)\,\tau \ - f_\varepsilon(\tau,\theta)]\,\Phi(\tau,\theta)\, d\tau. \tag{6.15}$$

The coefficient multiplying $\lambda_1(\theta)$ is the derivative $\frac{\partial}{\partial n}I(x)$ for $x \in \Gamma_0$ from condition (6.5). According to assumption II this coefficient is not equal to zero and therefore $\lambda_1(\theta)$ is uniquely defined by equation (6.15). In a similar way, the transition layer functions Q_k^i, Q_k^e can be found successively for $k = 2, 3, \ldots$. Using the C^1-matching condition we get for the functions $\lambda_k(\theta)(k = 2, 3, \ldots)$ equations which are similar to equation (6.15). All functions Q_k^i, Q_k^e have estimates of the type (6.13).

For the boundary function $\Pi_0(\rho, l)$ we have the following equations where l enters as a parameter (notation for the functions f, g in terms of the variables ρ, l is analogous to the notation in terms of the variables τ, θ):

$$\begin{aligned}\frac{\partial^2 \Pi_0}{\partial \rho^2} &= f(\varphi_1(0,l) + \Pi_0, 0, l), \quad \rho > 0,\\ \Pi_0(0,l) &= g(0,l) - \varphi_1(0,l), \quad \Pi_0(\infty, l) = 0.\end{aligned}$$

By virtue of the condition III, this problem has a unique monotone solution (see [2]) and $\Pi_0(\rho, l)$ satisfies the exponential estimate

$$|\,\Pi_0(\rho, l)\,| \le C \exp\{æ\rho\}. \tag{6.16}$$

The functions $\Pi_k(\rho, l)$ $(k = 1, 2, \ldots)$ are defined similarly to $Q_k(\tau, \theta)$. Functions $\Pi_k(\rho, l)$ are given by formulae analogous to (6.14). All $\Pi_k(\rho, l)$ possess estimates of the type (6.16).

The functions Q_k^i, Q_k^e, Pi_k must be multiplied by cut-off functions to continue them on the full domain $\mathcal{D}$.

We set $\tau = [\,r - (\,\varepsilon\,\lambda_1(\theta) + \cdots + \varepsilon^{n+1}\lambda_{n+1}(\theta)\,)\,]/\varepsilon$ and denote the n-th partial sum of the series (6.7) by $U_n(x, \varepsilon)$. The main result for problem (6.2), (6.3) is the following theorem.

Theorem 2.1. *Under conditions I – III and for sufficiently small ε, problem (6.2), (6.3) has a solution $u(x, \varepsilon)$ with an internal layer (step-type contrast structure) which satisfies the limit relations*

$$\lim_{\varepsilon \to 0} u(x, \varepsilon) = \begin{cases} \varphi_3(x), & \text{for } x \in \mathcal{D}_i, \\ \varphi_1(x), & \text{for } x \in \mathcal{D}_e. \end{cases}$$

Moreover the series (6.7) are the asymptotic expansion of the solution $u(x, \varepsilon)$ in the domain $\overline{\mathcal{D}}$, i.e. the estimate

$$\max_{\overline{\mathcal{D}}} |\,u - U_n\,| = O(\varepsilon^{n+1}) \tag{6.17}$$

holds.

The limit relation from Theorem 2.1 was first obtained in [3] by using the generalized implicit function theorem. The estimate (6.17) and the stability result have been obtained in [7], [8] by the asymptotic method of differential inequalities.

Remark. *If instead of the Dirichlet boundary condition (6.3) we have the Neumann boundary condition*

$$\left.\frac{\partial u}{\partial n}\right|_{\partial \mathcal{D}} = 0,$$

the asymptotic expansion for a step-type solution is constructed in the same way. For this case $\Pi_0(\rho, l) = 0$, so we don't need the condition III.

6.2.2 The case of balanced nonlinearity

The results of Theorem 2.1 can be extended to the case when problem (6.2), (6.3) has any finite dimension. In what follows we demonstrate it by considering the special multi-dimensional case of problem (6.2), (6.3) which we call *the case of balanced nonlinearity*. Our results can be found in [10], [20] and [11]. The one-dimensional case was considered in [21].

We consider a spatially inhomogeneous reaction-diffusion equation with the homogeneous Neumann boundary condition

$$\begin{cases} \dfrac{\partial u}{\partial t} = \epsilon^2 \Delta u - f(u, x, \epsilon) & (x \in \mathcal{D} \subset R^N,\ t > 0), \\ \dfrac{\partial u}{\partial n} = 0 & (x \in \partial\mathcal{D},\ t > 0), \end{cases} \tag{6.18}$$

and investigate the existence of equilibrium internal layer solutions. Our first assumption is as in Section 6.2

(A1) *The equation* $f(u, x, 0) = 0$ *has exactly three solutions* $u = \phi^{(\pm)}(x)$, $\phi^{(0)}(x)$ *such that*

$$\phi^{(-)}(x) < \phi^{(0)}(x) < \phi^{(+)}(x), \qquad x \in \overline{\mathcal{D}}$$

and

$$\overline{f}_u^{\pm}(x) \equiv f_u(\phi^{(\pm)}(x), x, 0) > 0, \qquad x \in \overline{\mathcal{D}}.$$

In contrast to assumption II of Section 6.2, we assume that the function $I(x)$

$$I(x) := \int_{\phi^{(-)}(x)}^{\phi^{(+)}(x)} f(u, x, 0)\, du,$$

satisfies

(A2) $I(x) \equiv 0$ *for* $x \in \overline{\mathcal{D}}$.

In order to state our problem succinctly, we define a set $\mathcal{S}$ of interfaces:

$$\mathcal{S} = \Big\{ \Gamma \subset \mathcal{D} \mid \Gamma \text{ is an } N-1 \text{ dimensional, smooth, connected, closed manifold} \Big\}.$$

Now the equation $I(x) = 0$ does not define a surface Γ_0 (in contrast to the noncritical case), and the main terms of the asymptotics in $\overline{\mathcal{D}}^{(-)}$ and $\overline{\mathcal{D}}^{(+)}$ are matched for any choice of Γ_0 (for a given interface Γ, we denote by $\mathcal{D}^{(-)}$ and $\mathcal{D}^{(+)}$ subdomains of $\mathcal{D}$ outside and inside Γ, respectively, and let $\nu(x; \Gamma)$ be the unit normal vector on Γ pointing into $\mathcal{D}_\Gamma^{(+)}$).

Let us define a function $V_1(x, \Gamma)$ for surfaces $\Gamma \subset \mathcal{S}$ by

$$V_1(x, \Gamma) \equiv -\kappa(x, \Gamma) m(x) + J(x; \Gamma),$$

where $\kappa(x,\Gamma)$ is the mean curvature of Γ,

$$m(x) = \int_{-\infty}^{\infty} \left(\frac{\partial \tilde{Q}_0(\tau;x)}{\partial \tau} \right)^2 d\tau,$$

$$J(x;\Gamma) = \int_{-\infty}^{\infty} \left[\tau \left(\nabla_x f(u,x,0) \Big|_{u=\tilde{Q}_0(\tau;x)} \cdot \nu(x;\Gamma) \right) + f_\epsilon(\tilde{Q}_0(\tau;x),x,0) \right] \frac{\partial \tilde{Q}_0(\tau;x)}{\partial \tau} d\tau, \qquad x \in \Gamma,$$

and $\tilde{Q}_0(\tau;x)$ is the unique solution of the boundary value problem for the zero order internal layer function (c.f. problem (6.12) in Section 6.2):

$$\begin{cases} \dfrac{d^2\tilde{Q}_0}{d\tau^2} - f(\tilde{Q}_0,x,0) = 0, & \tau \in R, \\ \lim_{\tau\to\pm\infty} \tilde{Q}_0(\tau;x) = \phi^{(\pm)}(x), & \tilde{Q}_0(0;x) = \phi^{(0)}(x). \end{cases}$$

The matching condition for the derivatives of the first approximation leads to the following assumption (compare with (6.15))

(A3) *There exists a* Γ *such that*

$$V_1(x,\Gamma) \equiv 0, \qquad x \in \Gamma.$$

It turns out that the interface Γ in (A3) does not necessarily give rise to equilibrium solutions of (6.18) with transition layers on Γ. We need an extra *non-degeneracy condition.* In order to state the non-degeneracy condition on Γ, we define an elliptic operator $\mathcal{A}^\Gamma$ by

$$\begin{aligned} \mathcal{A}^\Gamma R(x) :=& m(x) \left(\Delta^\Gamma + \sum_{j=1}^{N-1} \kappa_j(x)^2 \right) R(x) + \nabla_\Gamma m(x) \cdot \nabla_\Gamma R(x) \\ & - \kappa(x;\Gamma) \frac{\partial m(x)}{\partial \nu(x;\Gamma)} R(x) + J_r(x;\Gamma) R(x), \\ =& \operatorname{div}_\Gamma \Big(m(x) \nabla_\Gamma R(x) \Big) + \left(\sum_{j=1}^{N-1} \kappa_j(x)^2 \right) R(x) \\ & - \kappa(x;\Gamma) \frac{\partial m(x)}{\partial \nu(x;\Gamma)} R(x) + J_r(x;\Gamma) R(x), \qquad x \in \Gamma, \end{aligned}$$

where Δ^Γ, $\operatorname{div}_\Gamma$, and ∇_Γ stand, respectively, for the *Laplace-Beltrami, divergence* and *gradient* operators on the manifold Γ , and $\kappa_j(x)$ ($j = 1,\ldots,N-1$) are the principal curvatures of Γ at x. The function $J_r(x;\Gamma)$ is defined by

$$J_r(x;\Gamma) \equiv \frac{d}{dr} J(x + r\nu(x;\Gamma);\Gamma_r) \Big|_{r=0}.$$

We note that $\mathcal{A}^\Gamma R$ is nothing but the *linearization* of $V_1(x,\Gamma)$ in the direction of

$$\{x + R(x)\nu(x,\Gamma) \mid x \in \Gamma\}.$$

Since $\mathcal{A}^\Gamma$ is self-adjoint, its eigenvalues are real. We denote them by

$$\sigma(\mathcal{A}^\Gamma) = \{\lambda_j\}_{j=0}^{\infty}, \qquad \lambda_0 > \lambda_1 > \ldots > \lambda_j \to -\infty,$$

where only *distinct eigenvalues* are listed. The non-degeneracy condition on Γ is:

(A4) *The spectrum $\sigma(\mathcal{A}^\Gamma)$ does not contain 0.*

Theorem 2.2. *Assume that the conditions (A1), (A2), (A3), (A4) are satisfied. Then for sufficiently small $\epsilon_0 > 0$ there exists an equilibrium solution $u(x,\epsilon)$ of (2.18) such that for each $d_0 > 0$ fixed*

$$\lim_{\epsilon\to 0} u(x,\epsilon) = \begin{cases} \phi^{(-)}(x), & x \in \mathcal{D}_\Gamma^{(-)}\backslash\Gamma^{(d_0)}, \\ \phi^{(+)}(x), & x \in \mathcal{D}_\Gamma^{(+)}\backslash\Gamma^{(d_0)}, \end{cases}$$

uniformly, where $\Gamma^{(d_0)}$ stands for the d_0-neighborhood of Γ. Moreover, if the principal eigenvalue of $\mathcal{A}^\Gamma$ in (A4), λ_0, is negative, $u(x,\epsilon)$ is asymptotically stable. If there are some positive eigenvalues of $\mathcal{A}^\Gamma$, then $u(x,\epsilon)$ is unstable with an instability index equal to the number of positive eigenvalues.

6.3 Spike-type Contrast Structures in Partial Differential Equations

In this section internal layers of spike type are considered. We consider the periodic case where we illustrate the algorithm for constructing the asymptotics and the multi-dimensional stationary case. A rigorous asymptotic treatment of this type of solution for second order ODE's was developed in [17] for the autonomous case, and in [1] for the nonautonomous case. The two-dimensional case for elliptic problems has been treated in [6].
Periodic moving spikes were considered in [9].

6.3.1 Moving spikes

In what follows we consider the scalar singularly perturbed problem

$$\begin{aligned} &\varepsilon^2\left(\frac{\partial^2 u}{\partial x^2} - \frac{\partial u}{\partial t}\right) = f(u,x,t,\varepsilon) \text{ in } Q_T = (0<x<1)\times(-\infty<t<\infty), \\ &\qquad u(0,t,\varepsilon) = g_0(t), \ u(1,t,\varepsilon) = g_1(t), \end{aligned} \tag{6.19}$$

assuming

(A_0) : *functions f, g_0, and g_1 are sufficiently smooth T-periodic functions of t in the domain of interest.*

We are looking for T-periodic solutions of (6.19) satisfying

$$u(x,t,\varepsilon) = u(x,t+T,\varepsilon). \tag{6.20}$$

Algorithm for the construction of the asymptotics

We construct the asymptotics of the solution of (6.19), (6.20) which are close to some solution of the degenerate (reduced) equation

$$f(u,x,t,0) = 0 \tag{6.21}$$

in Q_T, except in a small neighborhood of some T-periodic smooth curve C in which a spike solution arises. The location of the curve C is unknown beforehand and is found in the process of constructing the asymptotics. To this end, we assume that the curve C can be defined by the equation

$$x = h(t,\varepsilon) = h_0(t) + \varepsilon h_1(t) + \dots, \tag{6.22}$$

where the $h_i(t)$ $(i = 0,1,\dots)$ are smooth T-periodic functions. To define the curve C, we also impose the additional condition that, for every fixed t, the solution $u(x,t,\varepsilon)$ has an extremum at $x = h(t,\varepsilon)$, i.e.,

$$\frac{\partial u}{\partial x}(h(t,\varepsilon),t,\varepsilon) = 0. \tag{6.23}$$

We look for the asymptotic expansion of a spike type solution in the following form

$$\begin{aligned} u(x,t,\varepsilon) &= \bar{u}(x,t,\varepsilon) + \pi(\tau,t,\varepsilon) + \pi^1(\tau_1,t,\varepsilon) + Q(\xi,t,\varepsilon) \\ &= \textstyle\sum_{i=0}^{\infty} \varepsilon^i (\bar{u}_i(x,t) + \pi_i(\tau,t) + \pi_i^1(\tau_1,t) + Q_i(\xi,t)), \end{aligned} \tag{6.24}$$

where $\bar{u}(x,\varepsilon)$ is the regular part of the asymptotics, $\pi(\tau,t,\varepsilon)$ and $\pi^1(\tau_1,t,\varepsilon)$ are boundary layer functions in the neighborhood of $t = 0$ and $t = 1$, respectively, and $Q(\xi,t,\varepsilon)$ is spike type function; the stretched variables are $\tau = x/\varepsilon$, $\tau_1 = (x-1)/\varepsilon$, and $\xi = (x - h(t,\varepsilon))/\varepsilon$.

Suppose that the following conditions are fulfilled.

(A_1). *The degenerate equation (6.21) has two solutions* $u = \varphi(x,t)$ *and* $u = \varphi_0(x,t)$, *and, for definiteness, suppose* $\varphi(x,t) < \varphi_0(x,t)$ *in* $\bar{Q}_T$.

(A_2). *The inequalities*

$$f_u(\varphi(x,t),x,t,0) := \bar{f}_u(x,t) > 0,$$
$$f_u(\varphi_0(x,t),x,t,0) < 0 \quad \text{in } \bar{Q}_T$$

are satisfied.

(A_3). *There exists a function* $\psi(x,t)$ *such that*

$$\int_{\varphi(x,t)}^{\psi(x,t)} f(u,x,t,0)\,du = 0 \qquad in\ \bar{Q}_T.$$

In order to find the coefficients of the asymptotic expansion (6.22), (6.24) we substitute (6.22), (6.24) into (6.19), (6.20), (6.23) and represent the right hand side of (6.19) in a form similar to (6.24). By equating expressions with the same power of ε (separately for unstretched and for the same stretched variables), we obtain equations to determine the unknown coefficients of the asymptotic expansion. In particular under our conditions, $\bar{u}_0(x,t)$ is determined by the degenerate equation (6.21) and $\bar{u}_0(x,t) = \varphi(x,t)$, whereas $\pi_0(\tau,t)$ is determined by the boundary value problem

$$\begin{aligned} \frac{\partial^2 \pi_0}{\partial \tau^2} &= f(\varphi(0,t) + \pi_0(\tau,t), 0, t, 0)\ ,\ \tau > 0, \\ \pi_0(0,t) &= g_0(t) - \varphi(0,t), \quad \pi_0(\infty,t) = 0. \end{aligned} \tag{6.25}$$

An analogous problem is obtained for $\pi_0^1(\tau_1,t)$. To ensure the solvability of (6.25) and the problem for $\pi_0^1(\tau_1,t)$, we assume (for definiteness with $\varphi(i,0) \le g_i(t)$, i = 0,1) that

(A_4). $\int_{\varphi(i,t)}^{s} f(u,i,t,0)\,du > 0$ *for all* $s \in (\varphi(i,t), g_i(t)]$, $i = 0,1$.

Assumption (A_4) guarantees that the initial conditions for π_0 and π_0^1 are in the domain of attraction of the steady-state solution $u = \varphi(0,t)$. The problem (6.25) is an autonomous second order equation (t is a parameter) and can be investigated by using phase plane arguments (see also [2]). It follows that under assumptions $(A_1), (A_2), (A_4)$ problem (6.25) has a unique monotonic (in τ) solution which satisfies the exponential estimate

$$|\pi_0(\tau,t)| \le c\exp(-\kappa\tau) \text{ for all } t, \tag{6.26}$$

where c and κ are some positive constants. The boundary layer function $\pi_0^1(\tau_1,t)$ has an estimate

$$|\pi_0^1(\tau_1,t)| \le c\exp(-\kappa\tau_1) \text{ for all } t$$

similar to (6.26). The zeroth order spike term $Q_0(\xi,t)$ satisfies the boundary value problem

$$\frac{\partial^2 Q_0}{\partial \xi^2} = f(\varphi_0(t) + Q_0, h_0(t), t, 0), -\infty < \xi < \infty, \tag{6.27}$$

$$\frac{\partial Q_0}{\partial \xi}(0,t) = 0\ ,\ Q_0(\pm\infty,t) = 0, \tag{6.28}$$

where $\varphi_0(t) = \varphi(h_0(t),t)$.
The first condition in (6.28) follows from (6.23), and second one follows from the

requirement that the Q_0-function be a spike layer function of the stretched variable ξ. It follows from (A_2) that for every t the rest point $Q_0 = 0$ of (6.27) is a saddle point, and it follows from (A_3) that the separatrix forms a loop. The upper and lower parts of the separatrix are respectively described by the equations

$$\frac{\partial Q_0}{\partial \xi} = \pm \left[\int_{\varphi_0(t)}^{\varphi_0(t)+Q_0} f(u, h_0(t), t, 0)\, du \right]^{\frac{1}{2}} . \tag{6.29}$$

The solution of equation (6.29) with the "+"-sign on the right hand side and the initial condition $Q_0(0,t) = \psi(h_0(t), t) - \varphi_0(t)$ and the solution of equation (6.29) with the "−"-sign on the right hand side and the same initial condition form the nontrivial solution of the problem (6.27), (6.28) which satisfies the estimate

$$|Q_0(\xi, t)| \leq c \, \exp(-\kappa|\xi|) \text{ for all } t,$$

where c and κ are some positive constants. We note that the function $h_0(t)$ is still unknown and its equation will be obtained at the next step of the asymptotic construction.

For higher-order regular terms, we get

$$\bar{u}_1(x,t) = -\bar{f}_u^{-1}(x,t)\bar{f}_\varepsilon(x,t)$$

and

$$\bar{u}_k(x,t) = -\bar{f}_u^{-1}(x,t)\bar{f}_k(x,t) \ , \ k \geq 2,$$

where $\bar{f}_k(x,t)$ is a known function depending on $\bar{u}_1(x,t), \ldots, \bar{u}_{k-1}(x,t)$ (in these relations, the bar means that functions are evaluated at the point $(\varphi(x,t), x, t, 0)$). For higher-order boundary layer functions $\pi_i(\tau, t)$, we obtain linear problems of the form

$$\begin{aligned} \frac{\partial^2 \pi_i}{\partial \tau^2} &= f_u(\varphi(0,t) + \pi_0, 0, t, 0) + f_i(\tau, t), \tau > 0, \\ \pi_i(0,t) &= -\bar{u}_i(0,t) \ , \ \pi_i(\infty, t) = 0, \end{aligned} \tag{6.30}$$

where the $f_i(\tau, t)$ are completely determined by preceding terms in the expansion for the boundary layer correction. The solution of (6.30) has the form

$$\begin{aligned} \pi_i(\tau,t) = &-\bar{u}_i(0,t)\phi(\tau,t)/\phi(0,t) - \\ &\phi(\tau,t)\int_0^\tau \phi^{-2}(p,\tau)\int_p^\infty \phi(\sigma,t)f_i(\sigma,t)\,d\sigma dp, \end{aligned} \tag{6.31}$$

where $\phi(\tau, t) = \frac{\partial \pi_0}{\partial \tau}$ (note that $\phi(\tau, t) \neq 0$ because we have chosen a monotonic solution of the problem (6.25)). Here $f_1(\tau, t)$ has the form

$$f_1(\tau,t) = \left[f_u(\tau,t)\frac{\partial \varphi}{\partial x}(0,t) + f_x(\tau,t) \right] \tau + f_u(\tau,t)\bar{u}_1(0,t) + f_\varepsilon(\tau,t),$$

where $f_u(\tau,t)$, $f_x(\tau,t)$ and $f_\varepsilon(\tau,t)$ are evaluated at the point $(\varphi(0,t)+\pi_0(\tau,t),0,t,0)$. From the estimate (6.26) it follows that $f_1(\tau,t)$ decays exponentially to zero as $\tau\to\infty$. By using (6.31) we can show that $\pi_1(\tau,t)$ is exponentially decaying. By induction, we can then show that all $\pi_i(\tau,t)$ are exponentially decaying, that is

$$|\pi_i(\tau,t)|\le c\ \exp(-\kappa\tau)\quad \text{for all } t,$$

where c and κ are positive constants. The functions $\pi_i^1(\tau_1,t)$ satisfy a similar estimate

$$|\pi_i(\tau,t)|\le c\ \exp(-\kappa\tau)\quad \text{for all } t.$$

For higher-order spike layer terms we obtain linear problems. For $Q_1(\xi,t)$ we get

$$\begin{aligned}\frac{\partial^2 Q_1}{\partial\xi^2}&=f_u(\xi,t)Q_1+f_1(\xi,t),\ -\infty<\xi<\infty,\\ \frac{\partial Q_1}{\partial\xi}(0,t)&=-\frac{\partial\varphi}{\partial x}(h_0(t),t),\ Q_1(\pm\infty,t)=0,\end{aligned}\tag{6.32}$$

where

$$\begin{aligned}f_1(\xi,t)=&\left[f_u(\xi,t)\frac{\partial\varphi}{\partial x}(h_0(t),t)+f_x(\xi,t)\right](h_1(t)+\xi)\\ &-f_u(\xi,t)\bar u_1(h_0(t),t)+f_\varepsilon(\xi,t)-\frac{\partial Q_0}{\partial\xi}h_0'(t),\end{aligned}\tag{6.33}$$

and partial derivatives $f_u(\xi,t)$, $f_x(\xi,t)$ and $f_\varepsilon(\xi,t)$ are evaluated at the point $(\varphi(h_0(t),t)+Q_0(\xi,t),h_0(t),t,0)$. The homogeneous problem corresponding to (6.32) has the nontrivial solution $\frac{\partial Q_0}{\partial\xi}(\xi,t)$. It can be easily checked that $\frac{\partial Q_0}{\partial\xi}$ is an eigenfunction corresponding to the zeroth eigenvalue which is simple. Therefore equation (6.32) is solvable if and only if the function $f_1(\xi,t)$ is orthogonal to $\frac{\partial Q_0}{\partial\xi}(\xi,t)$. The solvability condition has the form

$$\int_{-\infty}^{\infty} f_1(\xi,t)\frac{\partial Q_0}{\partial\xi}(\xi,t)\,d\xi=0.\tag{6.34}$$

By using the expression (6.33) for $f_1(\xi,t)$, the fact that $Q_0(\xi,t)$ is an even function of ξ, $\frac{\partial Q_0}{\partial\xi}(\xi,t)$ is an odd function of ξ, and that integration by parts yields

$$\begin{aligned}&\int_{-\infty}^{\infty} f_u(\xi,t)\xi\frac{\partial Q_0}{\partial\xi}\,d\xi=\int_{-\infty}^{\infty}\frac{\partial^3 Q_0}{\partial\xi^3}\xi\,d\xi=\int_{-\infty}^{\infty}d\Big(\frac{\partial^2 Q_0}{\partial\xi^2}\Big)\xi=\\ &-\int_{-\infty}^{\infty}\frac{\partial^2 Q_0}{\partial\xi^2}\,d\xi=-\int_{-\infty}^{\infty}d\Big(\frac{\partial Q_0}{\partial\xi}\Big)=0,\end{aligned}$$

(6.34) implies the following solvability condition

$$h_0'(t)\int_{-\infty}^{\infty}\left(\frac{\partial Q_0}{\partial \xi}\right)^2 d\xi - \int_{-\infty}^{\infty} f_x(\xi,t)\xi\frac{\partial Q_0}{\partial \xi}\,d\xi = 0. \tag{6.35}$$

Now we assume

(A_5). *Equation (6.35) has a unique T-periodic smooth solution $h_0(t)$ satisfying $0 < h_0(t) < 1$ for all t.*

Then the problem (6.32) has a solution which can be written in the form

$$Q_1 = -q_1(\xi,t)\int_0^{\xi} q_2(s,t)f_1(s,t)\,ds + q_2(\xi,t)\int_{-\infty}^{\xi} q_1(s,t)f_1(s,t)\,ds \\ +\gamma(t)\frac{\partial Q_0}{\partial \xi}(\xi,t), \tag{6.36}$$

where q_1 and q_2 are a fundamental system of solutions of the homogeneous equation corresponding to (6.32)

$$q_1(\xi,t) = \frac{\partial Q_0}{\partial \xi}(\xi,t) \quad \text{and} \quad q_2(\xi,t) = q_1(\xi,t)\int_{-1}^{T}\frac{ds}{q_1^2(s,t)},$$

while

$$\gamma(t) = -\left(\frac{\partial \varphi}{\partial x}(h_0(t),t) + \alpha(t)\right)/f(\psi(h_0(t),t),h_0(t),t,0),$$

and

$$\alpha(t) = \frac{\partial q_2}{\partial \tau}(0,t)\int_{-\infty}^{0} q_1(s,t)f_1(s,t)\,ds.$$

From the representation (6.36) it follows that $Q_1(\xi,t)$ satisfies the exponential estimate

$$|Q_1(\xi,t)| \le c\,\exp(-\kappa|\xi|) \text{ for all } t,$$

where c and κ are positive constants. We note that $Q_1(\xi,t)$ depends on an unknown function $\lambda_1(t)$ which can be obtained at the next step. The key step to constructing a higher-order spike layer function is the following. We introduce the function $\tilde{Q}_0(\xi,t,x)$ as a spike type solution of the problem (6.27), (6.28) where the function f is defined for all x in a small neighborhood of the curve $x = h_0(t)$, i.e.

$$\frac{\partial^2 \tilde{Q}_0}{\partial \xi^2} = f(\varphi(h_0(t)+x)+\tilde{Q}_0, h_0(t)+x,t,0), -\infty < \xi < \infty,$$

$$\frac{\partial \tilde{Q}_0}{\partial \xi}(0,t) = 0\ , \tilde{Q}_0(\pm\infty,t) = 0.$$

By using this function $\tilde{Q}_0$ and also considering the left hand side of (6.35) for all x in this neighborhood, we can introduce the function $G(x,t)$ by

$$G(x,t) = h_0'(t)\int_{-\infty}^{\infty}\left(\frac{\partial \tilde{Q}_0}{\partial \xi}\right)^2 d\xi - \int_{-\infty}^{\infty} f_x(\varphi(h_0(t)+x)+\tilde{Q}_0, h_0(t)+x,t,0)\xi\frac{\partial \tilde{Q}_0}{\partial \xi}\,d\xi.$$

It follows from (6.35) that $G(0,t) = 0$.
We also need the assumption:

(A_6). $G_x(0,t) > 0$ *for all* t.

Continuing our procedure, we can show that all $Q_i(\xi,t)$ are solutions of the problems

$$\frac{\partial^2 Q_i}{\partial \xi^2} = f_u(\xi,t)Q_i + f_i(\xi,t), \ -\infty < \xi < \infty,$$

$$\frac{\partial Q_i}{\partial \xi}(0,t) = g_i(0,t)\ ,\ Q_i(\pm\infty,t) = 0,$$

where, for each step, $f_i(\xi,t)$ and $g_i(0,t)$ are known functions. By using the solvability condition and some lengthy calculations, we can show that all $h_i(t)(i \geq 1)$ are solutions of the following problems

$$h_i'(t)\int_{-\infty}^{\infty}\left(\frac{\partial Q_0}{\partial \xi}\right)^2 - h_i(t)G_x(0,t) = \phi_i(t), \tag{6.37}$$

where the $\phi_i(t)$ are known T-periodic functions at each step.

It follows from assumption (A_6) that problem (6.37) has a unique T-periodic solution; moreover it can be written in an explicit form. If we choose $h_i(t)$ satisfying (6.37), then, by induction, we can show that all $Q_i(\xi,t)$ decay exponentially, that is, they satisfy an estimate of exponential type. We can then construct $U_n(x,t,\varepsilon)$ which is the truncated sum of the asymptotic expansion (6.24) where ξ is substituted by $\left(x - \sum_{i=0}^{n+1}\varepsilon^i h_i(t)\right)/\varepsilon$. By this asymptotic construction, we get the following result:

Theorem 3.1. *Assume the hypotheses $(A_1) - (A_6)$ to be valid. Then there exists a sufficiently small ε_0 such that for $0 < \varepsilon \leq \varepsilon_0$ the problem (6.19), (6.20) has a spike type solution $u(x,t,\varepsilon)$ satisfying*

$$\max_{\bar{Q}_T}|u(x,t,\varepsilon) - U_n(x,t,\varepsilon)| = O(\varepsilon^{n+1}).$$

6.3.2 Stationary spikes

We will consider again a spatially inhomogeneous reaction-diffusion equation with homogeneous Neumann boundary condition

$$\begin{cases} \dfrac{\partial u}{\partial t} = \epsilon^2\Delta u - f(u,x,\epsilon), & \left(x \in \mathcal{D} \subset R^N,\ t > 0\right), \\ \dfrac{\partial u}{\partial n} = 0 & \left(x \in \partial\mathcal{D},\ t > 0\right), \end{cases} \tag{6.38}$$

and investigate the existence of spike-type equilibrium internal layer solutions.

Our first assumption is as in Section 6.3.

(B1) *The equation* $f(u,x,0)=0$ *has a solution* $u=\varphi(x)$ *such that*

$$f_u(\varphi(x),x,)>0, \qquad x\in\overline{\mathcal{D}}.$$

We also assume

(B2) *There exists a function* $\psi(x)$ *such that*

$$\int_{\varphi(x)}^{\psi(x)} f(u,x,0)\,du=0\,,$$
$$\int_{\varphi(x)}^{s} f(u,x,0)\,du>0 \text{ for } x\in\overline{\mathcal{D}}, s\in(\varphi(x),\psi(x))\,.$$

Similarly to Section 6.2, we define a function $V_1(x,\Gamma)$ for closed surfaces $\Gamma\subset\mathcal{S}$ by

$$V_1(x,\Gamma)\equiv-\kappa(x,\Gamma)m(x)+J(x;\Gamma),$$

where $\kappa(x,\Gamma)$ is the mean curvature of Γ,

$$m(x)=\int_{-\infty}^{\infty}\left(\frac{\partial\tilde{Q}_0(\tau;x)}{\partial\tau}\right)^2 d\tau,$$

$$J(x;\Gamma)=\int_{-\infty}^{\infty}\left[\tau\left(\nabla_x f(u,x,0)\Big|_{u=\tilde{Q}_0(\tau;x)}\cdot\nu(x;\Gamma)\right)\right]\frac{\partial\tilde{Q}_0(\tau;x)}{\partial\tau}\,d\tau, \qquad x\in\Gamma.$$

Here $\tilde{Q}_0(\tau;x)$ is the solution of the boundary value problem for the zero order spike-type function (see also problem (6.27), (6.28) in Section 6.3):

$$\begin{cases}\dfrac{d^2\tilde{Q}_0}{d\tau^2}-f(\tilde{Q}_0,x,0)=0, & \tau\in R,\\[2ex] \lim_{\tau\to\pm\infty}\tilde{Q}_0(\tau;x)=\phi(x), & \frac{d\tilde{Q}_0}{d\tau}=0.\end{cases}$$

It is known that this problem has a unique nontrivial solution.

The matching condition for the asymptotics of the first approximation leads to the assumption

(B3) *There exists a* Γ *such that*

$$V_1(x,\Gamma)\equiv0, \qquad x\in\Gamma.$$

As in Section 6.2, the interface Γ in (B3) does not necessarily give rise to equilibrium solutions of (6.38) with a spike layer on Γ. We need an extra *non-degeneracy condition.* Introducing the operator $\mathcal{A}^\Gamma$ by the same formula as in Section 6.2, we assume

(B4) *The spectrum* $\sigma(\mathcal{A}^\Gamma)$ *does not contain* 0.

Theorem 3.2. *Assume that the conditions* $(B1),(B2),(B3),(B4)$ *are satisfied. Then for sufficiently small* ϵ *there exists an equilibrium spike type solution* $u(x,\epsilon)$ *of (6.38) such that*

$$\lim_{\epsilon\to 0} u(x,\epsilon) = \begin{cases} \varphi(x), & x\in\mathcal{D}\backslash\Gamma, \\ \psi^*(x), & x\in\Gamma \quad (\psi^*(x)\neq\phi(x)). \end{cases}$$

The spike-type solution $u(x,\epsilon)$ *is unstable.*

Remark. The instability of the spike-type solution does not follow from the spectral properties of the operator $\mathcal{A}^\Gamma$. It can be proved by using differential inequality techniques.

6.4 Applications

6.4.1 Phase transitions models

Such models are often in the form

$$\frac{\partial T}{\partial t} + \lambda\frac{\partial u}{\partial t} = \Delta T + g(x)\,, \tag{6.39}$$

$$\varepsilon^2\Delta u - \varepsilon^2\alpha\,\frac{\partial u}{\partial t} = f(u,T,\varepsilon)\,. \tag{6.40}$$

Here T is the temperature of a substance, and u is the so-called structural parameter which shows that the substance is in the solid phase if u is close to -1 and in the liquid phase if u is close to 1. The parameters ε, λ and α are a dimensionless interaction length, a latent heat and a relaxation time respectively. A discussion of the physics in such models is given in [4].

The system $(6.39), (6.40)$ is usually considered in some bounded domain $\mathcal{D}$ with impermeability conditions

$$\left.\frac{\partial T}{\partial n}\right|_{\partial\mathcal{D}} = 0\,, \quad \left.\frac{\partial u}{\partial n}\right|_{\partial\mathcal{D}} = 0$$

on the boundary $\partial\mathcal{D}$ of $\mathcal{D}$. Here $\frac{\partial}{\partial n}$ is the derivative along the outward normal to $\partial\mathcal{D}$.

Under some standard conditions ε is a small parameter and $f(u,T,\varepsilon)$ is represented in the form

$$f(u,T,\varepsilon) = u^3 - u + \varepsilon T\,.$$

If we consider a stationary process, then $\frac{\partial u}{\partial t} = 0$ and the problem for T is uncoupled from that for u. Having found T, we obtain the following problem for u from (6.40):

$$\begin{aligned} \varepsilon^2\Delta u &= u^3 - u + \varepsilon T(x)\,, \\ &\left.\tfrac{\partial u}{\partial n}\right|_{\partial\mathcal{D}} = 0\,. \end{aligned} \tag{6.41}$$

For physical applications it is of interest to find the location of the phase transition zone.

The degenerate equation for (6.41) has three solutions: $\varphi_1 = -1$, $\varphi_2 = 0$, $\varphi_3 = 1$, and the conditions $(A1)$ and $(A2)$ of Section 6.3 hold.

In what follows we assume that $\mathcal{D}$ is a two-dimensional domain, and the interface Γ can be described by the equation $R = R(\theta) = \sigma^{-1}(\theta)$. The leading term describing the phase transition (function $\tilde{Q}_0(\tau,\theta)$) satisfies the problem (see also (6.12) in Section 6.2):

$$\frac{\partial^2 \tilde{Q}_0}{\partial \tau^2} = \tilde{Q}_0^3 - \tilde{Q}_0\,, \quad -\infty < \tau < \infty\,,$$
$$\tilde{Q}_0(0,\theta) = 0\,, \ \tilde{Q}_0(-\infty,\theta) = -1\,, \ \tilde{Q}_0(+\infty,\theta) = 1\,.$$

The solution can be found in the explicit form:

$$\tilde{Q}_0(\tau,\theta) = \frac{\exp\{\sqrt{2}\,\tau\} - 1}{\exp\{\sqrt{2}\,\tau\} + 1}\,.$$

The equation for the curve Γ has the form (see assumption $(A3)$ and the definition of function $V_1(x)$):

$$\frac{\sigma'' + \sigma}{\left[1 + \left(\frac{\sigma'}{\sigma}\right)^2\right]^{3/2}} = -\left[\int_{-\infty}^{\infty}\left(\frac{\partial \tilde{Q}_0}{\partial \tau}\right)^2 d\tau\right]^{-1}\left[\int_{-\infty}^{\infty}\frac{\partial \tilde{Q}_0}{\partial \tau}\,d\tau\right]\cdot T(0,\theta),$$

where

$$\frac{\sigma'' + \sigma}{\left[1 + \left(\frac{\sigma'}{\sigma}\right)^2\right]^{3/2}} = k(0,\theta)$$

is the curvature of the closed curve Γ and the prime refers to differentiation with respect to θ. Denoting $\left[\int_{-\infty}^{\infty}\left(\frac{\partial \tilde{Q}_0}{\partial \tau}\right)^2 d\tau\right]^{-1}$ by γ, and using $\int_{-\infty}^{\infty}\frac{\partial \tilde{Q}_0}{\partial \tau}\,d\tau = 2$, we get the equation

$$\frac{\sigma'' + \sigma}{\left[1 + \left(\frac{\sigma'}{\sigma}\right)^2\right]^{3/2}} = -2\,\gamma\, T(0,\theta)\,, \tag{6.42}$$

where $T(0,\theta)$ is the temperature on the curve Γ. Solving the equation (6.42) we find the curve Γ. In the particular case when T depends on R and does not depend on θ (radially symmetric case: $T = T(R)$), the equation (6.42) may have solutions independent of θ which are defined by the equation

$$\frac{1}{R} = -2\,\gamma\, T(R)\,.$$

In such a case, Γ is a circle.

6.4.2 Stationary spikes in Fisher's equation

Consider the equation with the quadratic nonlinearity

$$\varepsilon^2 \Delta u = u\,[\,a(x) - u\,]\,, \quad x \in \mathcal{D}, \tag{6.43}$$

under the boundary condition

$$u|_{\partial\mathcal{D}} = 0\,,$$

where $a(x) > 0$. Equation (6.43) is often used in biology and is called *Fisher's equation.* This problem, under some conditions on $a(x)$, has a spike-type solution for which $\bar{u}_0(x,y) = 0$. We consider again the two-dimensional case. For this case $Q_0(\tau,\theta)$ is defined as the nontrivial solution of the problem (see also Section 6.4):

$$\begin{aligned} \frac{\partial^2 Q_0}{\partial\tau^2} &= Q_0\,[\,a(0,\theta) - Q_0\,]\,, \quad -\infty < \tau < \infty\,, \\ \frac{\partial Q_0}{\partial\tau}(0,\theta) &= 0\,, \quad Q_0(-\infty,\theta) = Q_0(+\infty,\theta) = 0\,. \end{aligned}$$

The solution can be found in the explicit form:

$$Q_0(\tau,\theta) = 6\,a_0 \frac{\exp\{\,\sqrt{a_o}\,\tau\,\}}{\left(1 + \exp\{\,\sqrt{a_0}\,\tau\,\}\right)^2}\,, \qquad a_0 = a(0,\theta)\,.$$

Using this expression we obtain

$$\begin{aligned} \int_{-\infty}^{+\infty} \left(\frac{\partial Q_0}{\partial\tau}\right)^2 d\tau &= \tfrac{1}{30}\,a^{5/2}(0,\theta)\,; \\ \int_{-\infty}^{+\infty} f_r(\tau,\theta)\,\tau\,\frac{\partial Q_0}{\partial\tau}\,d\tau &= \int_{-\infty}^{+\infty} a_r(0,\theta)\,Q_0\,\tau\,\frac{\partial Q_0}{\partial\tau}\,d\tau = -\tfrac{1}{12}\,a_r(0,\theta)\,a^{3/2}(0,\theta)\,, \end{aligned}$$

where $f_r = \frac{\partial f}{\partial r}$, $a_r = \frac{\partial a}{\partial r}$. Therefore, the equation for the curve Γ is:

$$\frac{\sigma'' + \sigma}{\left[1 + \left(\frac{\sigma'}{\sigma}\right)^2\right]^{3/2}} = \frac{5}{2}\,\frac{a_r(0,\theta)}{a(0,\theta)}\,.$$

6.4.3 Moving spikes in Fisher's equation

We consider the following special case of system (6.19).

$$\begin{gathered} \varepsilon^2\left(\frac{\partial^2 u}{\partial x^2} - \frac{\partial u}{\partial t}\right) = u(a(x,t) - u), \\ \text{in } Q_T = (-2 < x < 2) \times (-\infty < t < \infty), \\ u(0,t,\varepsilon) = u(1,t,\varepsilon) = 0, u(x,t,\varepsilon) = u(x,t+T,\varepsilon), \end{gathered}$$

which is known as a generalized Fisher's equation, assuming that $a(x,t) > 0$ is a smooth T-periodic function of t. Here $\varphi(x,t) = 0$, $\psi(x,t) = (3/2)a(x,t)$ and $\pi(\tau,t) = \pi^1(\tau_1,t) = 0$, so problem (6.27), (6.28) has the form

$$\begin{aligned} \frac{\partial^2 Q_0}{\partial\xi^2} &= Q_0(a_0(t) - Q_0), \ -\infty < \xi < \infty, \\ \frac{\partial Q_0}{\partial\xi} &= 0\,,\ Q_0(\pm\infty,t) = 0, \end{aligned}$$

where $a_0(t) = a(h_0(t), t)$. It has the nontrivial solution (the zeroth order spike term $Q_0(\xi, t)$)

$$Q_0(\xi, t) = 6a_0(t)\exp(a_0^{1/2}(t)\xi)/(1 + \exp a_0^{1/2}(t)\xi)^2. \tag{6.44}$$

The equation (6.35) for $h_0(t)$ can be written as

$$h_0'(t) = a_x(h_0(t), t)\int_{-\infty}^{\infty}(Q_0(\xi, t))^2\,d\xi/2\int_{-\infty}^{\infty}\left(\frac{\partial Q_0}{\partial \xi}\right)^2 d\xi. \tag{6.45}$$

Using (6.44), after some lengthy calculations, we can rewrite (6.45) as

$$h_0'(t) = 5/2(\ln a)_x(h_0(t), t). \tag{6.46}$$

For the case when $a(x, t) = \exp(-(x - \sin t)^2)$, for example, equation (6.46) becomes linear

$$h_0'(t) = 5(h_0(t) - \sin t),$$

and has the periodic solution

$$h_0(t) = 25/24\sin t + 5/24\cos t.$$

6.5 Asymptotic Method of Differential Inequalities

6.5.1 Basic ideas

Let $D \subset R^2$ be an open bounded simply connected region with a smooth boundary $\partial\mathcal{D}$, let I_1 be the interval $I_1 := \{\varepsilon \in R : 0 < \varepsilon \le \varepsilon_1\}$ with $\varepsilon_1 << 1$. We consider the singularly perturbed nonlinear boundary value problem

$$\begin{aligned} \varepsilon^2\Delta u &= f(u, x, \varepsilon), \quad \text{for} \quad x \in D, \\ \frac{\partial u}{\partial n} - q(x)u &= 0, \quad \text{for} \quad x \in \partial\mathcal{D}, \end{aligned} \tag{6.47}$$

where $\partial/\partial n$ denotes the derivative along the inner normal of $\partial\mathcal{D}$.
Our approach is based on the classical concept of lower and upper solutions of problem (6.47), and we recall the following definition:

Definition *Let $\mathcal{C} \subset \mathcal{D}$ be a smooth closed curve, $\mathcal{D}_1$ and $\mathcal{D}_2$ are inner and outer subdomains of $\mathcal{D}$ with respect to $\mathcal{C}$. The functions $\alpha(x, \varepsilon)$ and $\beta(x, \varepsilon)$, which are defined in $\overline{\mathcal{D}} \times I$ where I is some subset of I_1, are called lower and upper solutions, respectively, to the boundary value problem (6.47) if for all $\varepsilon \in I$ they satisfy the following conditions*

(i) α and β are continuously differentiable with respect to $x \in \overline{\mathcal{D}}_1$ and twice continuously differentiable with respect to x in $\mathcal{D}_1 \cup \mathcal{C}$ and in $\overline{\mathcal{D}}_2$.

(ii) $\frac{\partial \alpha}{\partial r}(x)\Big|_{+0} - \frac{\partial \alpha}{\partial r}(x)\Big|_{-0} \ge 0,\ \frac{\partial \beta}{\partial r}(x)\Big|_{+0} - \frac{\partial \beta}{\partial r}(x)\Big|_{-0} \le 0$ *for* $x \in \mathcal{C}$
where $\partial/\partial r$ denotes the differentiation with respect to the

inner normal of $\mathcal{C}$.

$(iii) L_\varepsilon \alpha(x,\varepsilon) := \Delta\alpha(x,\varepsilon) - f(\alpha(x,\varepsilon), x, \varepsilon) \geq 0, \;\; L_\varepsilon \beta(x,\varepsilon) \leq 0,$

for $x \in \mathcal{D}_1 \cup \mathcal{C}$ *and for* $x \in \overline{\mathcal{D}}_2$.

$(iv) \dfrac{\partial \alpha}{\partial n} - \lambda(x)\alpha \geq 0, \;\; \dfrac{\partial \beta}{\partial n} - q(x)\beta \leq 0, \qquad$ *for* $x \in \partial\mathcal{D}$.

It is known (see, for example, [19]) that if there exist ordered lower and upper solutions to (6.47) i.e., they satisfy the inequality

$$\alpha(x,\varepsilon) \leq \beta(x,\varepsilon) \quad \text{for } (x,\varepsilon) \in \overline{\mathcal{D}} \times I,$$

then the problem (6.47) has a solution $u(x,\varepsilon)$ satisfying

$$\alpha(x,\varepsilon) \leq u(x,\varepsilon) \leq \beta(x,\varepsilon) \quad \text{for } (x,\varepsilon) \in \overline{\mathcal{D}} \times I.$$

The main idea of our approach is to construct lower and upper solutions to the boundary value problem (6.47) by using formal asymptotics. Formal asymptotics of order n in ε for the problem (6.47) were constructed in [6] for solutions with boundary layers, and in [7] for solutions with internal layers. The modification of these asymptotics to construct the lower and upper solutions $\alpha_n(x,\varepsilon)$ and $\beta_n(x,\varepsilon)$ satisfying

$$L_\varepsilon \alpha_n(x,\varepsilon) = \varepsilon^n g(x,\varepsilon), L_\varepsilon \beta_n(x,\varepsilon) = -\varepsilon^n g(x,\varepsilon),$$

where $g(x,0) > 0 \not\subset \overline{\mathcal{D}}$, and

$$\beta_n - \alpha_n = O(\varepsilon^n) > 0, \beta_n - U_n = O(\varepsilon^n),$$

was given in [7], [8]. From these last relations, by the classical differential inequality theorem, it immediately follows that problem (6.47) has a solution and it satisfies

$$|u(x,\varepsilon) - U_{n-1}(x.\varepsilon)| = O(\varepsilon^n).$$

The lower and upper solutions, $\alpha_n(x,\varepsilon)$ and $\beta_n(x,\varepsilon)$, are used to prove local uniqueness of the solution, which is important for computations, and to investigate the stability of the solution (in the Lyapunov sense) as a stationary solution of the corresponding parabolic problem

$$\begin{aligned} \varepsilon^2 \Delta u - \frac{\partial u}{\partial t} &= f(u,x,\varepsilon), \;\; \text{for} \quad x \subset D, \\ \frac{\partial u}{\partial n} - q(x)u &= 0, \;\; \text{for} \quad x \in \partial\mathcal{D}, \end{aligned} \tag{6.48}$$

with prescribed initial conditions

$$u(x,0,\varepsilon) = u^0(x,\varepsilon).$$

From the fact that $\alpha_n(x,\varepsilon)$ and $\beta_n(x,\varepsilon)$ are the lower and upper solutions for the problem (6.48), it follows that a solution $u_s(x,\varepsilon)$ is a stable solution of problem

(6.48) with an appropriate choice of u^0. In order to get a more precise result we can construct lower and upper solutions in the form

$$\alpha(x,\varepsilon) = u_s - (u_s - \alpha_n)e^{-\lambda(\varepsilon)t},$$

$$\beta(x,\varepsilon) = u_s + (\beta_n - u_s)e^{-\lambda(\varepsilon)t},$$

where $\lambda(\varepsilon) > 0$ can be chosen in such way that the corresponding differential inequalities for problem (6.48) are satisfied. We get the following result:

Theorem 5.1. *The solution $u_s(x,\varepsilon)$ of problem (6.47) is locally unique and asymptotically stable as a solution of the corresponding initial boundary value problem (6.48) with domain of stability $[\alpha_2(x,\varepsilon), \beta_2(x,\varepsilon)]$.*

We can prove this theorem by verifying the corresponding differential inequalities for the parabolic problem (6.48).

6.5.2 SLEP-method

In a series of works [12, 13, 14, 15] on one dimensional reaction-diffusion systems, Nishiura and his co-workers have established a powerful method called the *singular limit Eigenvalue Problem* method (SLEP-method, for short) to determine the stability property of equilibrium transition layer solutions. The basic structure of the method is concisely expressed in the following diagram.

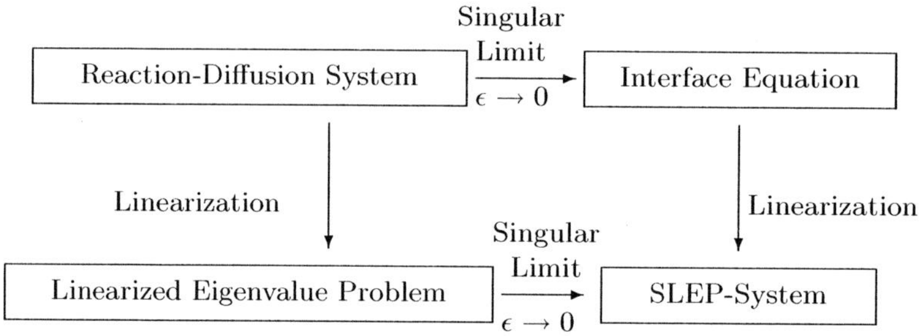

Figure 6.1. *Relationship between reaction-diffusion system and SLEP-system*

First construct an equilibrium transition layer solution to the reaction-diffusion system and linearize the system around it to obtain an eigenvalue problem. The singular limit of the eigenvalue problem is called the SLEP-system and contains full information on the stability of the equilibrium. Moreover, Nishiura et al. show that the SLEP-system is also obtained by first passing to the singular limit of the reaction diffusion system to obtain an associated system of interface equations and then linearizing the latter around its equilibrium.

Our results fit precisely into the same framework. We first find an equilibrium solution to the system of interface equations and linearize around it to obtain a

SLEP-system. Our assertion is that if the SLEP-system thus obtained is non-degenerate, then the equilibrium of the system of interface equations gives rise to an equilibrium transition layer solution of the reaction-diffusion system. Moreover, the SLEP-system also carries the full information on the stability of the transition layer solution. The case of the spike type internal layer is completed by the investigation of the principal eigenvalue which is positive and not critical. We also point out the following facts which are guiding principles in our proof in both cases.

The interface equations are nothing but the lowest order C^1-matching conditions.

The SLEP-system is the principal part of the higher order C^1-matching conditions.

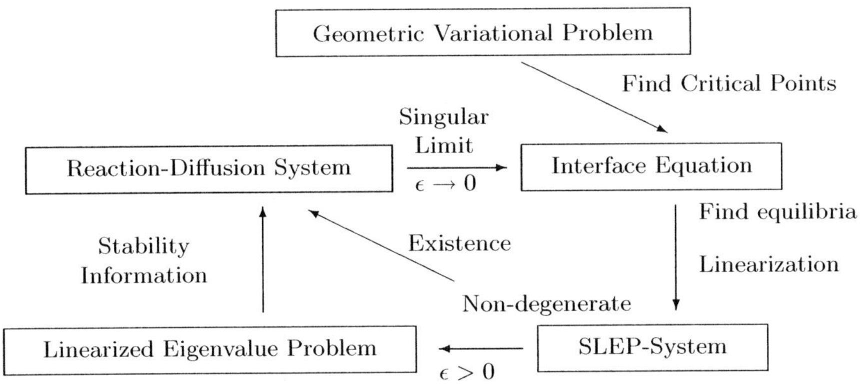

Figure 6.2. *Non-degenerate equilibria of the interface equations give rise to transition layer solutions of reaction-diffusion system and their stability properties are completely determined by the SLEP-system*

6.5.3 Outline of proof

The approach which is based on the SLEP-method was applied to the problems in Sections 6.2 and 6.3. The proof of these results consists of three steps:

(1) **Construction of highly accurate approximate solutions** U^ϵ_{app} via the method of matched asymptotic expansion. The conditions **A3** and **A4** (resp. **B3** and **B4**) allow us to find C^1-matched approximate solutions to arbitrarily high order of accuracy. As pointed out at the end of the previous section, **A3** (**B3**) is the lowest order C^1-matching condition, and **A4** (**B4**) allows us to find higher order C^1-matched approximations. Once an approximate solution is constructed, the original problem is then written as

$$\mathcal{L}^\epsilon \varphi + \mathcal{N}^\epsilon(\varphi) + \mathcal{R}^\epsilon = 0, \tag{6.49}$$

where $\mathcal{L}^\epsilon\varphi$ is obtained from the original problem by linearization around the approximate solution, $\mathcal{N}^\epsilon(\varphi)$ stands for nonlinear terms containing quadratic and higher order terms in φ, and the remainder $\mathcal{R}^\epsilon$ measures how well the approximation satisfies the original problem.

(2) The spectral analysis of $\mathcal{L}^\epsilon$. The linear operator $\mathcal{L}^\epsilon$ in general has small eigenvalues that go to zero as $\epsilon \to 0$, called critical eigenvalues. In the case of the problems considered in sections 6.2 and 6.3, these critical eigenvalues are of order $O(\epsilon^2)$, and when divided by ϵ^2 they are essentially the eigenvalues of the SLEP-system. Therefore, **A4** (**B4**) guarantees that the linear operator $\mathcal{L}^\epsilon$ is invertible, although it has small eigenvalues that converge to zero as $\epsilon \to 0$.

(3) To establish the solvability of (6.49). Since the linear part $\mathcal{L}^\epsilon\varphi$ is small, $O(\epsilon^2)$, one needs to make the contribution of the nonlinear term $\mathcal{N}^\epsilon(\varphi)$ smaller than the linear part. This, in turn, is possible if the remainder term $\mathcal{R}^\epsilon$ is small enough, say, $\|\mathcal{R}^\epsilon\| = O(\epsilon^8)$ in the present situation. By using a fixed point theorem one obtains the true solution U^ϵ very close to the approximate one. Now the linearization of the original problem around the genuine solution U^ϵ is a very small perturbation of $\mathcal{L}$, and hence the stability properties of U^ϵ are completely determined by the SLEP-system, which has already been analyzed in Step (2).

The strategy described above seems to have a wide range of applicability in dealing with transition layers and interfaces.

6.6 Acknowledgements

This work has been supported by the Russian Foundation for Basic Research under grant no. 01-01-00753.

Bibliography

[1] V. F. Butuzov and A. B. Vasiljeva, *On asymptotics of contrast structure solutions*, USSR Math. Notes, 42(6) (1987), pp. 831–841.

[2] P. C. Fife, *Semilinear elliptic boundary value problems with small parameters*, Arch.Rational Mech. Anal., 52 (1973), pp. 205–232.

[3] P. C. Fife and W. M. Greenlee, *Internal transition layers for elliptic boundary value problems with small parameters*, Uspehi Mat. Nauk., 29(4) (1974), pp. 103–131.

[4] P. C. Fife, *Dynamics of Internal Layers and Diffusive Interfaces*, SIAM, Philadelphia, Pennsylvania, 1988.

[5] D. Henry, *Geometric Theory of Semilinear Parabolic Equations*, Lect. Notes in Math., 840, Spring-Verlag, Berlin, 1983.

[6] N. N. Nefedov, *Contrast structures of spike type in nonlinear singularly perturbed elliptic equations*, Russian Acad. Sci. Dokl. Math., 46(3) (1993), pp. 410–413.

[7] , *Method of differential inequalities for some classes of nonlinear singularly perturbed problems with internal layers*, Differ. Equat., 31(7) (1995), pp. 1077–1085

[8] ——, *Two-dimensional contrast structures of step type: asymptotics, existence, stability*, Russian Acad. Sci. Dokl. Math. 349(5) (1996), pp. 603–605 (in Russian, MR 98d:35016).

[9] ——, *On moving spike type internal layer in nonlinear singularly perturbed problems*, J. Mathematical Analysis and Applications, 221(1) (1998), pp. 1–12.

[10] N. N. Nefedov and K. Sakamoto, *Geometric and spectral problems arising in reaction-diffusion systems*, in Progress in Nonlinear Science, 1 (Nizhny Novgorod, 2001), pp. 320–322, RAS, Inst. Appl. Phys., Nizhniĭ Novgorod, 2002.

[11] ——, *Multi-dimensional stationary internal layers for spatially Inhomogeneous Reaction-Diffusion Equations with Balanced nonlinearity*, Hiroshima Math. J., 33 (2003), pp. 391–432.

[12] Y. NISHIURA, *Coexistence of infintely many stable solutions to reaction diffusion systems in the singular limit*, Dynamics Reported, 3 (1994), pp. 25–103.

[13] ——, *Nonlinear Problems 1 – Mathematics for Pattern Formation*, 7, Iwanami Series on Developments of Modern Mathematics, Tokyo, 1998.

[14] Y. NISHIURA AND H. FUJII, *Stability of singularly perturbed solutions to reaction diffusion equations*, SIAM J. Math. Anal., 18 (1987), pp. 1726–1770.

[15] Y. NISHIURA AND M. MIMURA, *Layer oscillations in reaction-diffusion systems*, SIAM J. Appl. Math., 49 (1989), pp. 481–514.

[16] R. E. O'MALLEY, *Introduction to Singular Perturbations*, Academic Press, New York, 1974.

[17] ——, *Phase - plane solutions to some singular perturbation problems*, J.Math. Anal. Appl., 54 (1976), pp. 449–456.

[18] ——, *Singular Perturbation Methods for Ordinary Differential Equations*, Springer - Verlag, New York, 1991.

[19] C. V. PAO, *Nonlinear Parabolic and Elliptic Equations*, Plenum Press, New York and London, 1992.

[20] K. SAKAMOTO AND N. N. NEFEDOV, *Geometric variational problems arising in reaction-diffusion systems*, in Free Boundary Problems, RIMS 1210, pp. 29–43, Kyoto, 2001.

[21] A. B. VASIL'EVA AND V. F. BUTUZOV, *Asymptotic Expansions of Solutions of Singularly Perturbed Equations*, Nauka, Moscow, 1973 (in Russian, MR 57 #16876).

[22] ——, *Asymptotic Methods in the Theory of Singular Perturbation*, Vyshaja Shkola, Moscow, 1990 (in Russian, MR 92i:34072).

Chapter 7

Geometry of Singular Perturbations: Critical Cases

V. Sobolev

The paper is a contribution to advancing the geometrical approach to the investigation of singularly perturbed systems in cases when the hypotheses of the famous Tikhonov theorem are violated.

7.1 Introduction. Elements of the Geometric Theory of Singularly Perturbed Systems

7.1.1 Slow integral manifolds

It is common knowledge that a wide range of processes in various aspects of nature are characterized by extreme differences in the rates of change of variables, so singularly perturbed ordinary differential systems are used as models of such processes [21, 25, 43, 44, 46, 47]).

Consider the ordinary differential system

$$\begin{aligned} \frac{dx}{dt} &= f(x, y, t, \varepsilon), \\ \varepsilon \frac{dy}{dt} &= g(x, y, t, \varepsilon), \end{aligned} \tag{7.1}$$

with vector variables x and y, and a small positive parameter ε. The usual approach in the qualitative study of (7.1) is to consider first the degenerate system

$$\frac{dx}{dt} = f(x, y, t, 0),$$
$$0 = g(x, y, t, 0),$$

and then to draw conclusions about the qualitative behavior of the full system (7.1) for sufficiently small ε. A special case of this approach is the quasi–steady state assumption. A mathematical justification of that method can be given by means of the theory of integral manifolds for singularly perturbed systems (7.1) (see e.g. [2, 3, 6, 13, 14, 15, 17, 20, 24, 34, 43, 50, 49]). The integral manifolds method has been applied to a wide range of problems (see e. g. [5, 7, 8, 9, 15, 23, 33, 37, 38, 39, 41, 40, 45]), and in particular, to some applications in chemical kinetics problems which are discussed in Chapters *8* and *9*. In order to recall a basic result of the geometric theory of singularly perturbed systems we introduce the following terminology and assumptions.

The system of equations

$$\frac{dx}{dt} = f(x, y, t, \varepsilon) \tag{7.2}$$

represents the slow subsystem, and the system of equations

$$\varepsilon\frac{dy}{dt} = g(x, y, t, \varepsilon) \tag{7.3}$$

the fast subsystem, so it is natural to call (7.2) the slow subsystem and (7.3) the fast subsystem of system (7.1). In the present paper we use a method for the qualitative asymptotic analysis of differential equations with singular perturbations. The method relies on the theory of integral manifolds, which essentially replaces the original system by another system on an integral manifold with dimension equal to that of the slow subsystem. In the zero-epsilon approximation ($\epsilon = 0$), this method leads to a modification of the quasi-steady-state approximation. Recall, that a smooth surface S in $R^m \times R^n \times R$ is called an integral manifold of the system (7.1) if any trajectory of the system that has at least one point in common with S lies entirely in S. Formally, if $(x(t_0), y(t_0), t_0) \in S$, then the trajectory $(x(t,\varepsilon), y(t,\varepsilon), t)$ lies entirely in S. An integral manifold of an autonomous system

$$\dot{x} = f(x, y, \varepsilon),$$
$$\varepsilon\dot{y} = g(x, y, \varepsilon)$$

has the form $S_1 \times (-\infty, \infty)$, where S_1 is a surface in the phase space $R^m \times R^n$. The only integral manifolds of system (7.1) of relevance here are those of dimension m (the dimension of the slow variables) that can be represented as the graphs of vector-valued functions

$$y = h(x, t, \varepsilon).$$

We also stipulate that $h(x, t, 0) = h^{(0)}(x, t)$, where $h^{(0)}(x, t)$ is a function whose graph is a sheet of the slow surface, and we assume that $h(x, t, \varepsilon)$ is a sufficiently smooth function of ε. In autonomous systems the integral manifolds will be graphs of functions

$$y = h(x, \varepsilon).$$

Such integral manifolds are called manifolds of slow motions – the origin of this term lies in nonlinear mechanics. An integral manifold may be regarded as a surface on which the phase velocity has a local minimum, that is, a surface characterized by the most persistent phase changes (motions). Integral manifolds of slow motions constitute a refinement of the sheets of the slow surface, obtained by taking the small parameter ε into consideration.

The motion along an integral manifold is governed by the equation

$$\dot{x} = f(x, h(x,t,\varepsilon), t, \varepsilon).$$

If $x(t,\varepsilon)$ is a solution of this equation, then the pair $\Big(x(t,\varepsilon), y(t,\varepsilon)\Big)$, where $y(t,\varepsilon) = h(x(t,\varepsilon),t,\varepsilon)$, is a solution of the original system (7.1), since it defines a trajectory on the integral manifold.

Consider the *associated* subsystem, that is,

$$\frac{dy}{d\tau} = g(x,y,t,0), \quad \tau = t/\varepsilon,$$

treating x and t as parameters. We shall assume that some of the steady states $y^0 = y^0(x,t)$ of this subsystem are asymptotically stable and that a trajectory starting at any point of the domain approaches one of these states as closely as desired as $t \to \infty$. This assumption will hold, for example, if the matrix

$$(\partial g/\partial y)(x, y^0(x,t), t, 0)$$

is stable for part of the stationary states and the domain can be represented as the union of the domains of attraction of the asymptotically stable steady states.

Let I_i be the interval $I_i := \{\varepsilon \in R : 0 < \varepsilon < \varepsilon_i\}$, where $0 < \varepsilon_i \ll 1$, $i = 0, 1, \dots$.

$(\mathbf{A_1})$. $f : R^m \times R^n \times R \times \overline{I_0} \to R^m$, $g : R^m \times R^n \times R \times \overline{I_0} \to R^n$ are sufficiently smooth and uniformly bounded together with their derivatives.

$(\mathbf{A_2})$. There is some region $G \in R^m$ and a map $h : G \times R \to R^m$ of the same smoothness as g such that

$$g(x, h(x,t), t, 0) \equiv 0, \qquad \forall (x,t) \in G \times R.$$

$(\mathbf{A_3})$. The spectrum of the Jacobian matrix $g_y(x, h(x,t), t, 0)$ is uniformly separated from the imaginary axis for all $(x,t) \in G \times R$.

Then the following result is valid (see e.g. [43, 49]):

Proposition 1.1. *Under the assumptions* $(\mathbf{A_1})$–$(\mathbf{A_3})$ *there is a sufficiently small positive* ε_1, $\varepsilon_1 \le \varepsilon_0$, *such that, for* $\varepsilon \in \overline{I_1}$, *system* (7.1) *has a smooth integral manifold* $\mathcal{M}_\varepsilon$ *with the representation*

$$\mathcal{M}_\varepsilon := \{(x,y,t) \in R^{n+m+1} : y = \psi(x,t,\varepsilon),\ (x,t) \in G \times R\}.$$

Remark. The global boundedness assumption in $(\mathbf{A_1})$ with respect to (x,y) can be relaxed by modifying f and g outside some bounded region of $R^n \times R^m$.

7.1.2 Asymptotic representation of integral manifolds

When the method of integral manifolds is being used to solve a specific problem, a central question is the calculation of the function $h(x,t,\varepsilon)$ in terms of the manifold described. Exact calculation is generally impossible, and various approximations are necessary. One possibility is the asymptotic expansion of $h(x,t,\varepsilon)$ in integer powers of the small parameter:

$$h(x,t,\varepsilon) = h_0(x,t) + \varepsilon h_1(x,t) + \ldots + \varepsilon^k h_k(x,t) + \ldots .$$

Substituting this formal expansion in equation (7.3) i.e.,

$$\varepsilon\frac{\partial h}{\partial t} + \varepsilon\frac{\partial h}{\partial x} f(x, h(x,t,\varepsilon), t, \varepsilon) = g(x,h,\varepsilon),$$

we obtain the relationship

$$\varepsilon\sum_{k\geq 0}\varepsilon^k\frac{\partial h_k}{\partial t} + \varepsilon\sum_{k\geq 0}\varepsilon^k\frac{\partial h_k}{\partial x} f(x, \sum_{k\geq 0}\varepsilon^k h_k, t, \varepsilon)$$
$$= g(x, \sum_{k\geq 0}\varepsilon^k h_k, t, \varepsilon). \tag{7.4}$$

We use the formal asymptotic representations

$$f(x, \sum_{k\geq 0}\varepsilon^k h_k, t, \varepsilon) = \sum_{k\geq 0}\varepsilon^k f^{(k)}(x, h_0, \ldots, h_{k-1}, t),$$

and

$$g(x, \sum_{k\geq 0}\varepsilon^k h_k, t, \varepsilon) = B(x,t)\sum_{k\geq 1}\varepsilon^k h_k + \sum_{k\geq 1}\varepsilon^k g^{(k)}(x, h_0, ..., h_{k-1}, t),$$

where the matrix $B(x,t) \equiv (\partial g/\partial y)(x, h_0, t, 0)$, and where

$$g(x, h^{(0)}(x,t), t, 0) = 0.$$

Substituting these formal expansions into (7.4) and equating powers of ε, we obtain

$$\frac{\partial h_{k-1}}{\partial t} + \sum_{0\leq p\leq k-1}\frac{\partial h_p}{\partial x} f^{(k-1-p)} = Bh_k + g^{(k)}.$$

Since B is invertible

$$h_k = B^{-1}\left[g^{(k)} - \frac{\partial h_{k-1}}{\partial t} - \sum_{0\leq p\leq k-1}\frac{\partial h_p}{\partial x} f^{(k-1-p)}\right]. \tag{7.5}$$

Note that asymptotic expansions of slow integral manifolds were first used in [38, 40, 41] and, for systems with several small parameters in [36].

7.1.3 Stability of slow integral manifolds

In applications it is often assumed that the spectrum of the Jacobian matrix

$$g_y(x, h(x,t), t, 0)$$

is located in the left half plane. Under this additional hypothesis the manifold $\mathcal{M}_\varepsilon$ is exponentially attracting for $\varepsilon \in I_1$. In this case, the solution $x = x(t,\varepsilon)$, $y = y(t,\varepsilon)$ of the original system that satisfied the initial condition $x(t_0,\varepsilon) = x^0$, $y(t_0,\varepsilon) = y^0$ can be represented as

$$\begin{aligned} x(t,\varepsilon) &= v(t,\varepsilon) + \varepsilon\varphi_1(t,\varepsilon), \\ y(t,\varepsilon) &= \bar{y}(t,\varepsilon) + \varphi_2(t,\varepsilon). \end{aligned} \tag{7.6}$$

There exists a point v^0 which is the initial value for a solution $v(t,\varepsilon)$ of the equation $\dot{v} = f(v, h(v,t,\varepsilon), t, \varepsilon)$; the functions $\varphi_1(t,\varepsilon)$, $\varphi_2(t,\varepsilon)$ are corrections that determine the degree to which trajectories passing near the manifold tend asymptotically to the corresponding trajectories on the manifold as t increases. They satisfy the following inequalities:

$$|\varphi_i(t,\varepsilon)| \leq N|y^0 - h(x^0, t_0, \varepsilon)| \exp[-\beta(t-t_0)/\varepsilon], \quad i = 1, 2, \tag{7.7}$$

for $t \geq t_0$.

From (7.6) and (7.7) we obtain the following *reduction principle* for a stable integral manifold defined by a function $y = h(x,t,\varepsilon)$: a solution $x = x(t,\varepsilon)$, $y = h(x(t,\varepsilon),t,\varepsilon)$ of the original system (7.1) is stable (asymptotically stable, unstable) if and only if the corresponding solution of the system of equations $\dot{v} = F(v,t,\varepsilon) = f(x, h(x,t,\varepsilon), t, \varepsilon)$ on the integral manifold is stable (asymptotically stable, unstable). The Lyapunov reduction principle was extended to ordinary differential systems with Lipschitz right–hand sides by Pliss [28], and to singularly perturbed systems in [42, 43]. Thanks to the reduction principle and the representation (7.6), the qualitative behavior of trajectories of the original system near the integral manifold may be investigated by analyzing the equation on the manifold.

7.1.4 Critical cases

The case in which the assumption $(\mathbf{A_3})$ is violated is called critical. We distinguish the following subcases:

1. The Jacobian matrix $g_y(x,y,t,0)$ is singular on some subspace of $R^m \times R^n \times R$. In that case, system (7.1) is referred to as a singular singularly perturbed system. This subcase has been treated in [10, 11, 16, 27, 31, 46]. In this case it is possible to introduce new variables in such a way that the transformed differential system has a structurally hyperbolic fast subsystem of lower dimension. This means that the original differential system has a slow integral manifold of higher dimension. This case and an application to a high-gain control problem [33] are investigated in the next section.

2. The Jacobian matrix $g_y(x,y,t,0)$ has eigenvalues on the imaginary axis with nonvanishing imaginary parts. A similar case has been investigated in [35, 32, 38, 43]. If this part of the eigenvalues is pure imaginary but, after taking into account the perturbations of higher order, they move to the complex left half-plane, then the system under consideration has stable slow integral manifolds. Some problems of mechanics of gyroscopes and manipulators with high-frequency and weakly damped transient regimes are discussed in Section 3.

3. The Jacobian matrix $g_y(x,y,t,0)$ is singular on the set $\mathcal{M}_0 := \{(x,y,t) \in R^m \times R^n \times R : y = h(x,t), (x,t) \in G \times R\}$. In that case, $y = h(x,t)$ is generically an isolated root of $g = 0$ but not a simple one. This case will be studied in Section 4.

   Further interesting situations associated with the phenomena of exchange of stability are considered in Chapters *4* and *8* in this book.

7.2 Singular Singularly Perturbed Systems

Consider the system

$$\varepsilon \dot{z} = Z(z,t,\varepsilon), \qquad z \in R^{m+n}, \quad t \in R, \tag{7.8}$$

where $0 \leq \varepsilon \ll 1$, and the vector-function Z is sufficiently smooth. Suppose that the limit system $Z(z,t,0) = 0$ $(\varepsilon = 0)$ has a family of solutions

$$z = \psi(v,t), \qquad v \in R^m, \quad t \in R, \tag{7.9}$$

with a sufficiently smooth vector-function ψ. We try to find a slow integral manifold

$$z = P(v,t,\varepsilon), \tag{7.10}$$

with a flow described by the equation

$$\dot{v} = Q(v,t,\varepsilon). \tag{7.11}$$

We shall restrict our consideration to smooth integral surfaces situated in the ε-neighborhood of the slow surface $z = \psi(v,t)$, i. e.

$$P(v,t,0) = \psi(v,t),$$

a motion described by differential equations of form (7.11) with a smooth right hand side.

The equation (7.8) describes motions with speeds of order $O(\varepsilon^{-1})$, while (7.11) describes motions with speeds of order $O(1)$. Thus, the integral manifold (7.10) is a manifold of slow motions or a slow integral manifold.

Consider the singularly perturbed system

$$\dot{x} = X(x,y,t,\varepsilon), \qquad x \in R^m, \quad t \in R, \tag{7.12}$$

$$\varepsilon \dot{y} = Y(x, y, t, \varepsilon), \qquad y \in R^n. \tag{7.13}$$

If the equation $Y(x, y, t, 0) = 0$ has the isolated solution

$$y = \varphi(x, t), \tag{7.14}$$

and $\det B(x, t) \neq 0$ for $B(x, t) = Y_y(x, \varphi, t, 0)$, then, under some additional conditions on the eigenvalues of $B(x, t)$, the system (7.12), (7.13) has, in an ε–neighborhood of the slow surface (7.14), the slow integral surface

$$y = h(x, t, \varepsilon), \tag{7.15}$$

the motion on which is described by the equation

$$\dot{x} = X(x, h(x, t, \varepsilon), t, \varepsilon). \tag{7.16}$$

For the system (7.12), (7.13) the role of slow variable is played by the vector x, the integral manifold (7.10) is described by equation (7.15), and the equation (7.11) takes the form (7.16).

7.2.1 Existence of slow integral manifolds

Suppose the following hypotheses hold: the rank of matrix $\psi_v(v, t)$ is equal to m; the rank of the matrix $A(v, t) = Z_z(\psi(v, t), t, 0)$ is equal to n; the matrix $A(v, t)$ has an m–fold zero eigenvalue and n other eigenvalues $\lambda_i(v, t)$ which satisfy the inequality

$$Re\lambda_i(v, t) \leq -2\alpha < 0, \qquad t \in R, \qquad v \in R^m. \tag{7.17}$$

Differentiation of $Z(\psi(v, t), t, 0) = 0$ with respect to v gives

$$A(v, t)\psi_v(v, t) = 0.$$

The last equality means that the $(m + n) \times (m + n)$–matrix $A(v, t)$ possesses m linearly independent eigenvectors, which are columns of $\psi_v(v, t)$, corresponding to multiple zero eigenvalues.

Let D_1^T be an $(m + n) \times n$–matrix, the columns of which give the basis of the A^T kernel, and D_2^T be such an $(m + n) \times m$–matrix so that (D_1^T, D_2^T) is a non-singular matrix. Then

$$A^T(D_1^T \ D_2^T) = (0 \ B^T),$$

or

$$DA = \begin{pmatrix} 0 \\ B \end{pmatrix}, \text{ for } D = \begin{pmatrix} D_1 \\ D_2 \end{pmatrix}.$$

Thus, the result of multiplying the non-singular matrix D on the left by A provides the zero $m \times (m + n)$ –block and the non-singular $n \times (m + n)$–block B.

The rank of B is equal to n. Consequently, without loss of generality, the system (7.8) may be considered to be of the form

$$\varepsilon \dot{x} = f_1(x, y_2, t, \varepsilon), \qquad x \in R^m, \tag{7.18}$$

$$\varepsilon\dot{y}_2 = f_2(x, y_2, t, \varepsilon), \qquad y_2 \in R^n, \tag{7.19}$$

and the following conditions apply.

(B₁). The equation $f_2(x, y_2, t, 0) = 0$ has a smooth isolated root $y_2 = \varphi(x, t)$ with $x \in R^m, t \in R$, and $f_2(x, \varphi(x, t), t, 0) = 0$.

(B₂). The Jacobian matrix

$$A(x,t) = \begin{pmatrix} f_{1x} & f_{1y_2} \\ f_{2x} & f_{2y_2} \end{pmatrix}\Bigg|_{y_2=\varphi(x,t),\varepsilon=0}$$

on the surface $y_2 = \varphi(x, t)$ has an m–fold zero eigenvalue and m–dimensional kernel, and the matrix $B(x, t) = f_{2y_2}(x, \varphi(x, t), t, 0)$ has n eigenvalues satisfying (7.17) when $v = x$.

(B₃). In the domain

$$\Omega = \{(x, y_2, t, \varepsilon) \mid x \in R^m, \ \ ||y_2 - \varphi(x, t)|| \le p, \ \ t \in R, \ \ 0 \le \varepsilon \le \varepsilon_0\},$$

the functions f_1, f_2 and the matrix A are continuously differentiable $(k + 2)$ times $(k \ge 0)$.

Using the change of variable $y_2 = y_1 + \varphi(x, t)$ in (7.18), (7.19), we obtain the following equations for x and y_1

$$\varepsilon\dot{x} = C(x, t)y_1 + F_1(x, y_1, t) + \varepsilon X(x, y_1, t, \varepsilon), \tag{7.20}$$

$$\varepsilon\dot{y}_1 = B(x, t)y_1 + F_2(x, y_1, t) + \varepsilon Y(x, y_1, t, \varepsilon), \tag{7.21}$$

where

$$C(x, t) = f_{1y_2}(x, \varphi(x, t), t, 0), \qquad B(x, t) = f_{2y_2}(x, \varphi(x, t), t, 0),$$

$$F_1(x, y_1, y) = f_1(x, y_1 + \varphi(x, t), t, 0) - C(x, t),$$

$$F_2(x, y_1, t) = f_2(x, y_1 + \varphi(x, t), t, 0) - B(x, t),$$

$$\varepsilon X(x, y_1, t, \varepsilon) = f_1(x, y_1 + \varphi(x, t), t, \varepsilon) - f_1(x, y_1 + \varphi(x, t), t, 0),$$

$$\varepsilon Y(x, y_1, t.\varepsilon) = f_2(x, y_1 + \varphi(x, t), t, \varepsilon) - f_2(x, y_1 + \varphi(x, t), t, 0).$$

Note that the vector-functions F_i $(i = 1, 2)$ satisfy the relations $||F_i(x, y_1, t)|| = O(||y_1||^2)$. Thus, $\varepsilon^{-1}F_i(x, \varepsilon y, t)$ are continuous in Ω.

We now have the following theorem.

Theorem 7.1. *Let the assumptions* **(B₁)**–**(B₃)** *be satisfied. Then there exists* ε_1, $0 < \varepsilon_1 < \varepsilon_0$, *such that for any* $\varepsilon \in (0, \varepsilon_1)$ *the system (7.20), (7.21) possesses a unique slow integral manifold* $y_1 = \varepsilon p(x, t, \varepsilon)$. *On this manifold the flow of the system is governed by the equation*

$$\dot{x} = X_1(x, t, \varepsilon),$$

where $X_1(x,t,\varepsilon) = C(x,t)p(x,t,\varepsilon)+X(x,\varepsilon p,t,\varepsilon)+\varepsilon^{-1}F_1(x,\varepsilon p,t)$, *and the function* $p(x,t,\varepsilon)$ *is* k *times continuously differentiable with respect to* x *and* t.

Note that the change of variable $y_1 = \varepsilon y$ converts the system (7.20), (7.21) to the form (7.12), (7.13) and the role of the slow variable is now played by x.

7.2.2 Explicit and implicit slow integral manifolds

To describe slow integral manifolds for systems like (7.12), (7.13) the explicit representation (7.15) is usual. In this case the approximation to $h(x,t,\varepsilon)$ may be obtained as an asymptotic expansion in powers of ε.

In general, it is impossible to find the function $h(x,t,0)$ exactly from the equation

$$Y(x,y,t,0) = 0.$$

In this case the slow integral manifold can be obtained in an implicit form. The flow on the slow integral manifold as a zero approximation is governed by the following differential-algebraic system:

$$\dot{x} = X(x,y,t,0), \tag{7.22}$$

$$0 = Y(x,y,t,0). \tag{7.23}$$

To obtain the first approximation, it is necessary to differentiate $Y(x,y,t,\varepsilon)$ and by virtue of (7.12), (7.13)

$$\varepsilon\frac{d}{dt}Y = Y_yY + \varepsilon Y_t + \varepsilon Y_xX.$$

As a first approximation, the flow on the slow integral manifolds is governed by the differential-algebraic system

$$\dot{x} = X(x,y,t,\varepsilon), \tag{7.24}$$

$$Y_yY + \varepsilon Y_t + \varepsilon Y_xX = 0, \tag{7.25}$$

where terms of order $o(\varepsilon)$ can be neglected.

To obtain the second order approximation, it is necessary to differentiate $Y(x,y,t,\varepsilon)$ twice and use (7.12), (7.13). Unfortunately, the corresponding relationships are too cumbersome. Because of this, we consider the case of autonomous systems. Then the second order approximation takes the form of (7.24) and

$$\begin{aligned} &Y + \varepsilon(Y_y)^{-1}Y_xX \\ &+\varepsilon^2Y_y^{-2}\{Y_xX_x + Y_{xx}X - Y_{xy}(Y_y)^{-1}Y_xX - Y_xX_y(Y_y)^{-1}Y_x \\ &-Y_{yx}X(Y_y)^{-1}Y_x + Y_{yy}(Y_y)^{-1}Y_xX(Y_y)^{-1}Y_xX\} = 0. \end{aligned} \tag{7.26}$$

In (7.24), (7.26) all terms multiplied by ε^3 can be neglected.

To obtain the k–th order approximation, it is necessary to differentiate $Y(x,y,t,\varepsilon)$ k times with respect to t and use (7.12), (7.13).

To check these formulae it is sufficient to note that in the calculation of the asymptotic expansions of $h(x,t,\varepsilon)$

$$h = h_0 + O(\varepsilon), \quad h = h_0 + \varepsilon h_1 + O(\varepsilon^2),$$

$$h = h_0 + \varepsilon h_1 + \varepsilon^2 h_2 + O(\varepsilon^3),$$

the use of (7.22) – (7.26) gives the same result as (7.5).

To illustrate this, consider the example

$$\dot{x} = y, \qquad \varepsilon \dot{y} = x^2 + y^2 - a, \qquad a > 0.$$

The first approximation of the slow integral manifold is

$$y^2 + x^2 - a + \varepsilon x = 0.$$

It is easy to check that the second order approximation

$$y^2 + (x + \varepsilon/2)^2 = a - \varepsilon^2/4$$

gives the exact equation for this manifold.

7.2.3 Parametric representation of integral manifolds

The implicit form of integral manifolds has evident disadvantages, but for numerous problems it is impossible to find a solution of $Y(x,y,t,0) = 0$ in the explicit form $y = \varphi(x,t)$. However, sometimes the solution of $Y(x,y,t,0) = 0$ can be found as a parametric function

$$x = \chi_0(v,t), \qquad y = \varphi_0(v,t),$$

where $v \in R^m$, and the following identity holds

$$Y(\chi_0(v,t), \varphi_0(v,t), t, 0) \equiv 0, t \in R, \quad v \in R^m.$$

In this case the slow integral manifold may be found in parametric form

$$x = \chi(v,t,\varepsilon), \qquad y = \varphi(v,t,\varepsilon),$$

where $t \in R$, $v \in R^m$, $\chi(v,t,0) = \chi_0$, $\varphi(v,t,0) = \varphi_0$. The flow on the manifold is governed by the equation

$$\dot{v} = F(v,t,\varepsilon), \tag{7.27}$$

and the function $F(v,t,\varepsilon)$ will be determined below. The functions χ, φ, F can be found as asymptotic expansions

$$\begin{aligned}\chi(v,t,\varepsilon) &= \chi_0(v,t) + \varepsilon\chi_1(v,t) + \ldots + \varepsilon^k\chi_k(v,t) + \ldots,\\ \varphi(v,t,\varepsilon) &= \varphi_0(v,t) + \varepsilon\varphi_1(v,t) + \ldots + \varepsilon^k\varphi_k(v,t) + \ldots,\\ F(v,t,\varepsilon) &= F_0(v,t) + \varepsilon F_1(v,t) + \ldots + \varepsilon^k F_k(v,t) + \ldots,\end{aligned} \tag{7.28}$$

in agreement with (7.27), from the equations

$$\frac{\partial\chi}{\partial t}+\frac{\partial\chi}{\partial v}F=X(\chi,\varphi,t,\varepsilon), \tag{7.29}$$

$$\varepsilon\frac{\partial\varphi}{\partial t}+\varepsilon\frac{\partial\varphi}{\partial v}F=Y(\chi,\varphi,t,\varepsilon). \tag{7.30}$$

Equating coefficients of powers of the small parameter ε we obtain

$$\frac{\partial\chi_0}{\partial t}+\frac{\partial\chi_0}{\partial v}F_0=X(\chi_0,\varphi_0,t,0), \qquad Y(\chi_0,\varphi_0,t,0)=0,$$

$$\frac{\partial\chi_1}{\partial t}+\frac{\partial\chi_1}{\partial v}F_0+\frac{\partial\chi_0}{\partial v}F_1=X_x(\chi_0,\varphi_0,t,0)\chi_1$$
$$+X_y(\chi_0,\varphi_0,t,0)\varphi_1+X_1,$$

$$\frac{\partial\varphi_0}{\partial t}+\frac{\partial\varphi_0}{\partial v}F_0=Y_x(\chi_0,\varphi_0,t,0)\chi_1+Y_y(\chi_0,\varphi_0,t,0)\varphi_1+Y_1,$$
$$X_1=X_\varepsilon(\chi_0,\varphi_0,t,0), \qquad Y_1=Y_\varepsilon(\chi_0,\varphi_0,t,0).$$

The relationships (7.29), (7.30) contain unknown functions χ, φ, F. In a concrete problem it is possible to consider one of these functions or any m scalar components of χ,φ and F as known functions, and all others may be found from (7.29), (7.30). Moreover, it is possible at any step of the calculation of coefficients in (7.28) to choose any m components of these coefficients as given functions. In the case that F is a given function, equations (7.28), (7.29) are used to calculate the coefficients of asymptotic expansions of χ and φ. If it is possible to predetermine the function χ, then these equations allow the calculation of F and φ.

Note that in the case of the explicit form $y=h(x,t,\varepsilon)$,

$$v=x, \quad \chi=v, \quad \varphi=h(v,t,\varepsilon), \quad F=X(v,h(v,t,\varepsilon),t,\varepsilon),$$

(7.30) takes the form

$$\varepsilon\frac{\partial h}{\partial t}+\varepsilon\frac{\partial h}{\partial v}X(v,h,t,\varepsilon)=Y(v,h,t,\varepsilon), \qquad h=h(v,t,\varepsilon).$$

If $dim\ x=dim\ y$ and the role of v is that of y, then $\varphi=v$ and

$$\frac{\partial\chi}{\partial t}+\frac{\partial\chi}{\partial v}F=X(\chi,v,t,\varepsilon), \quad \varepsilon F=Y(\chi,v,t,\varepsilon). \tag{7.31}$$

The equation for χ follows immediately

$$\varepsilon\frac{\partial\chi}{\partial t}+\frac{\partial\chi}{\partial v}Y(\chi,v,t,\varepsilon)=\varepsilon X(\chi,v,t,\varepsilon),$$

whence, under the assumption $\det(\frac{\partial\chi_0}{\partial v})\neq 0$, it is possible to calculate χ as an asymptotic expansion. Note that $Y(\chi_0,\varphi_0,t,0)=0$ means that equation (7.27) is regularly perturbed, because the last equation in (7.31) implies, in this case, $F=O(1)$ as $\varepsilon\to 0$.

Returning back to the system (7.8), we use the parametric form to describe the slow integral manifolds and the reduced differential equation

$$z = P(v, t, \varepsilon), \qquad \dot{v} = Q(v, t, \varepsilon).$$

The functions P and Q will be found as asymptotic expansions

$$P(v, t, \varepsilon) = P_0(v, t) + \varepsilon P_1(v, t) + \ldots + \varepsilon^k P_k(v, t) + \ldots,$$

$$Q(v, t, \varepsilon) = Q_0(v, t) + \varepsilon Q_1(v, t) + \ldots + \varepsilon^k Q_k(v, t) + \ldots.$$

Differentiating P with respect to t, and using (7.8), (7.11), gives

$$\varepsilon \frac{\partial P}{\partial t} + \varepsilon \frac{\partial P}{\partial v} Q = Z(P, t, \varepsilon). \tag{7.32}$$

Write the Taylor series expansion of $Z(P, t, \varepsilon)$ about $\varepsilon = 0$ as

$$\begin{aligned} Z(P, t, \varepsilon) = Z(P_0, t, 0) + \varepsilon Z_1(P_0, P_1, t) + \ldots \\ + \varepsilon^k Z_k(P_0, P_1, \ldots, P_k, t) \ldots, \end{aligned}$$

and represent $Z_k (K \geq 1)$ in the form

$$Z_k(P_0, \ldots, P_k, t) = Z_2(P_0, t, 0) P_k + R_k(P_0, P_1, \ldots, P_{k-1}, t).$$

In particular, $Z_1(P_0, P_1, t) = Z_2(P_0, t, 0) P_1 + Z_\varepsilon(P_0, t, 0)$. Equating powers of ε, we obtain from (7.32) with $\varepsilon = 0$

$$Z(P_0, t, 0) = 0.$$

In keeping with (7.9), let

$$P_0(v, t) = \psi(V, t).$$

Using the following notation

$$A(v, t) = Z_z(\psi(v, t), t, 0),$$

we obtain at order ε

$$\frac{\partial \psi}{\partial t} + \frac{\partial \psi}{\partial v} Q_0 = A P_1 + R_1. \tag{7.33}$$

Equation (7.33) contains two unknown functions P_1 and Q_0. Where P_1 is concerned, equation (7.33) may be considered as a nonhomogeneous linear algebraic system with a singular matrix, $det A(v, t) \equiv 0$, $v \in R^m$, $t \in R$. Thereby, Q_0 is required to ensure the solvability of the system. It is apparent that we have some freedom in choosing the form of Q_0 and P_1 . To determinate these functions uniquely, multiply equation (7.33) on the left by the matrix D, introduced in Subsection 2.1, and obtain the pair of equations

$$D_1 \frac{\partial \psi}{\partial t} + D_1 \frac{\partial \psi}{\partial v} Q_0 = D_1 R_1, \tag{7.34}$$

and

$$D_2\frac{\partial\psi}{\partial t} + D_2\frac{\partial\psi}{\partial v}Q_0 = BP_1 + D_2R_1. \qquad (7.35)$$

If it is assumed additionally that the matrix

$$D_1 = \partial\psi/\partial v$$

is invertible, so that (7.34) gives

$$Q_0 = (D_1\psi_v)^{-1}D_1(R_1 - \psi_t),$$

and this permits us to determine P_1 uniquely from (7.35)

$$P_1 = B^{-1}D_2(\psi_t + \psi_v Q_0 - R_1).$$

The determination of the pairs of later coefficients P_k, Q_{k-1} is carried out in the same way. Equating coefficients of ε^k , we obtain the equation

$$\frac{\partial P_{k-1}}{\partial t} + \frac{\partial\varphi}{\partial v}Q_{k-1} + \sum_{i=1}^{k-1}\frac{\partial P_i}{\partial v}Q_{k-i-1} = AP_k + R_k,$$

which is decomposed into the two equations

$$\Big(D_1\frac{\partial\psi}{\partial v}\Big)Q_{k-1} + D_1\Big(\frac{\partial P_{k-1}}{\partial t} + \sum_{i-1}^{k-1}\frac{\partial P_i}{\partial v}Q_{k-i-1}\Big) = D_1R_k,$$

$$D_2\Big[\frac{\partial P_{k-1}}{\partial t} + \frac{\partial\psi}{\partial v}Q_{k-1} + \sum_{i=1}^{k-1}\frac{\partial P_i}{\partial v}Q_{k-i-1}\Big] = BP_k + D_2R_k,$$

by multiplying on the left by D. The last equations give

$$Q_{k-1} = \Big(D_1\psi_v\Big)^{-1}D_1\Big[R_k - \sum_{i=1}^{k-1}\frac{\partial P_i}{\partial v}Q_{k-i-1} - \frac{\partial P_{k-1}}{\partial t}\Big],$$

$$P_k = B^{-1}D_2\Big[\frac{\partial P_{k-1}}{\partial t} + \psi_v Q_{k-1} + \sum_{i=1}^{k-1}\frac{\partial P_i}{\partial v}Q_{k-i-1} - R_k\Big].$$

Example

As a very simple example consider the system

$$\varepsilon\dot{x} = f(x,y,t,\varepsilon) + \varepsilon f_1(x,y,t,\varepsilon),$$

$$\varepsilon\dot{y} = kf(x,y,t,\varepsilon) + \varepsilon f_2(x,y,t,\varepsilon).$$

Introducing a new variable $x_1 = y - kx$, we obtain the following differential equation for the slow variable

$$\dot{x}_1 = f_2(x,y,t,\varepsilon) - kf_1(x,y,t,\varepsilon).$$

To obtain the full solution, it is possible to use either of the two equations for x or y as a fast equation.

7.2.4 High-gain control

Consider the control system

$$\dot{x} = f(x) + B(x)u, \quad x(0) = x_0,$$

where $x \in R^n$, $u \in R^r$ and $t \geq 0$. The vector function f and the matrix function B are taken to be sufficiently smooth and bounded. The control vector u is to be selected in such a way as to transfer the vector x from $x = x_0$ to a sufficiently small neighborhood of a smooth m–dimensional surface $S(x) = 0$. A commonly employed feedback control is

$$u = -\frac{1}{\varepsilon}KS(x),$$

where K is a constant $r \times m$–matrix and ε is a small positive parameter, see [48].

Suppose that we can choose the matrix K in such a way that the matrix $-N(x,t) = -GBK$ is stable[1] and its inverse matrix is bounded, and introduce the additional variable $y = S(x)$, then x and y satisfy the system

$$\varepsilon\dot{x} = \varepsilon f(x) - B(x)Ky, \quad x(0) = x_0,$$

$$\varepsilon\dot{y} = \varepsilon G(x)f(x) - G(x)B(x)Ky, \quad y(0) = y_0 = S(x_0),$$

where $G(x) = \partial S/\partial x$. The reduced ($\varepsilon = 0$) algebraic problem possesses an n–parameter family of solutions $x = v$, $y = 0$. The role of A is played by the singular matrix

$$\begin{pmatrix} 0 & -B_1K \\ 0 & -N \end{pmatrix}.$$

The latter singular singularly perturbed differential system possesses an n–dimensional slow integral manifold

$$x = v, \qquad y = \varepsilon N^{-1}(v,t)G(v)f(v,t) + O(\varepsilon^2).$$

The flow on the manifold is governed by

$$\dot{v} = [I - B_1(v,t)KN^{-1}(v,t)G(v)]f(v,t) + O(\varepsilon^2).$$

Introduce the new variables

$$x = v + B_1(v,t)KN^{-1}(v,t)z; \qquad y = z + \varepsilon N^{-1}(x,t)G(x)f(x,t).$$

Then for v, z we obtain the equations

$$\dot{v} = (I - B_1KN^{-1}G)f + O(\varepsilon), \quad \varepsilon\dot{z} = -(N + O(\varepsilon))z.$$

It is now clear that the representations

$$x = v + O(e^{-\nu t/\varepsilon}), \quad y = \varepsilon\varphi(v,t,\varepsilon) + O(e^{-\nu t/\varepsilon}), \quad \varphi = N^{-1}Gf + O(\varepsilon)$$

[1] A stable matrix is one whose eigenvalues all have strictly negative real parts.

are valid for some $\nu > 0$ for all $t > 0$. Thus, under the given control law

$$u = -\frac{1}{\varepsilon} KS(x),$$

the trajectory very quickly attains the ε–neighborhood of $S(x) = 0$.

It is easy to see that the modified control

$$u = -\frac{1}{\varepsilon} K \left[S(x) - \varepsilon N^{-1}(x,t) G(x) f(x,t) \right],$$

with the stable matrix $-N(x,t) = -GBK$, is preferable, because it guides the trajectory of x in a time Δt to the $e^{-\nu \Delta t \varepsilon^{-1}}$–neighborhood of $S(x) = 0$. Under this control for the variable x we obtain the equation

$$\varepsilon \dot{x} = \varepsilon \left[I - B(x,t) K (GBK)^{-1} G(x) \right] f(x,t) - B_1(x,t) KS(x),$$

and for the variable $y = S(x)$ we obtain the equation

$$\varepsilon \dot{y} = -N(x,t) y,$$

$$y = O\left(e^{-\nu \varepsilon^{-1} t} \right), \quad \nu > 0, \ \ t > 0, \ \ \varepsilon \to 0.$$

7.2.5 Asymptotics with fractional exponents of the small parameter

Firstly consider cases when some of the assumptions $(\mathbf{A_1})$–$(\mathbf{A_3})$ break down. Thus, for the system

$$\varepsilon \dot{x}_1 = \varepsilon f_1(x_1, x_2, x_3, t, \varepsilon), \quad \varepsilon \dot{x}_2 = \varepsilon f_2(x_1, x_2, x_3, t, \varepsilon),$$

$$\varepsilon \dot{x}_3 = D(x_1, t) x_2 + \varepsilon f_3(x_1, x_2, x_3, t, \varepsilon),$$

$$x_i \in R^{n_i}, \quad i = 1, 2, 3,$$

where

$$A = A(x_1, t) = \begin{pmatrix} 0 & 0 & 0 \\ 0 & 0 & 0 \\ 0 & D & 0 \end{pmatrix},$$

the multiplicity $n_1 + n_2 + n_3 = n$ of zero eigenvalues of $A(v,t)$ being greater than the number of corresponding eigenvectors $n - k$ $(k > 0)$ under the condition $D \neq 0$. Using a new variable $\varepsilon^{1/2} x_2$ rather than x_2, we obtain the system

$$\dot{x} = f_1(x_1, \sqrt{\varepsilon} x_2, x_3, t, \varepsilon), \quad \sqrt{\varepsilon} \dot{x}_2 = f_2(x_1, \sqrt{\varepsilon} x_2, x_3, t, \varepsilon),$$

$$\sqrt{\varepsilon} \dot{x}_3 = D(x_1, t) x_2 + \sqrt{\varepsilon} f_3(x_1, \sqrt{\varepsilon} x_2, x_3, t, \varepsilon). \tag{7.36}$$

In this case the slow integral manifold can be found using asymptotic expansions in powers of $\sqrt{\varepsilon}$, i.e., we need to use fractional powers of ε.

Let $x_3 = \varphi(x_1, t)$ be an isolated root of

$$f_2(x_1, 0, x_3, t, 0) = 0.$$

The last differential system has the form (7.12), (7.13) with a small parameter $\sqrt{\varepsilon}$. The role of B is played by the matrix

$$B = \begin{pmatrix} O & C(x_1,t) \\ D(x_1,t) & O \end{pmatrix},$$

where $C(x_1, t) = f_{2x_2}(x_1, 0, \varphi(x_1, t), t, 0)$. If $detB \neq 0$, then the following situations are typical.

The eigenvalues of B have nonzero real parts and (7.36) possesses a conditionally stable slow integral manifold (the usual situation for optimal control problems [15, 37, 34, 33]).

The eigenvalues of B are pure imaginary. Such systems appear in modelling gyroscopic systems and double spin satellites [38, 40, 41, 43].

Asymptotic expansions with fractional exponents of a small parameter will be used in Section 4.

7.3 Systems with Slow Dissipation

Let some of the eigenvalues have a pure imaginary part which moves to the complex left half plane at higher orders of the perturbations. In this case the system has a stable slow integral manifold. Some important problems of the mechanics of gyroscopes, satellites and manipulators with high-frequency and weakly damped transient regimes were studied in [35, 32, 38, 43].

7.3.1 Gyroscopic systems

The general equations of gyroscopic systems on a fixed base may be represented in the form [43]:

$$\frac{dx}{dt} = y,$$

$$\varepsilon\frac{d}{dt}(Ay) = -(G + \varepsilon B)y + \varepsilon R + \varepsilon Q, \; R = \frac{1}{2}\left[\frac{\partial(Ay)}{\partial x}\right]^T y. \tag{7.37}$$

Here $x \in R^n$, $A(t, x)$ is a symmetric positive definite matrix, $G(x, t)$ is a skew-symmetric matrix of gyroscopic forces, and $B(x, t)$ is a symmetric positive definite matrix of dissipative forces, $Q(x, t)$ is a vector of generalized forces and $\varepsilon = H^{-1}$ is a small positive parameter.

The precessional equations take the form

$$(G + \varepsilon B)\frac{dx}{dt} = \varepsilon Q. \tag{7.38}$$

Equations (7.38) are obtained from (7.37) by neglecting some of the terms multiplied by the small parameter. All roots of the characteristic equation

$$\det(G - \lambda I) = 0$$

are situated on the imaginary axis, so the main assumption of the Tikhonov theorem [44] is violated. To justify the permissibility of the precessional equations we use the integral manifold method.

7.3.2 Precessional and nutational motions

Return to the system (7.37). The initial variables x and y are connected with the new slow variable v and the new fast variable z by the following relations

$$x = v + \varepsilon H(v, z, t, \varepsilon), \tag{7.39}$$

$$y = z + \varepsilon h(x, t, \varepsilon) = z + \varepsilon h(v + \varepsilon H(v, z, t, \varepsilon), t, \varepsilon). \tag{7.40}$$

Differential equations for v and z are

$$\frac{dv}{dt} = \varepsilon h(v, t, \varepsilon), \tag{7.41}$$

$$\varepsilon \frac{d}{dt}(Az) = -(G + \varepsilon B)z + \varepsilon R(v, z, t) + \varepsilon^2 R_1(v, z, t, \varepsilon), \tag{7.42}$$

where

$$A = A(v + \varepsilon H, t), \quad B = B(v + \varepsilon H, t), \quad G = G(v + \varepsilon H, t),$$

$$h = h(v + \varepsilon H, t, \varepsilon), \quad H = H(v, z, t, \varepsilon), \quad R_1 = P(v + \varepsilon H, z, t, \varepsilon),$$

and

$$P(x, y, t, \varepsilon) = \frac{1}{2}\left(\frac{\partial A}{\partial x} y\right)^T + \frac{1}{2}\left(\frac{\partial A}{\partial x} h\right)^T y - \frac{\partial (Ah)}{\partial x} y.$$

The following asymptotic formulae hold:

$$h(x, t, \varepsilon) = h_1(x, t) + \varepsilon h_2(x, t) + O(\varepsilon^2),$$

$$H(v, z, t, \varepsilon) = H_1(v, z, t) + \varepsilon H_2(v, z, t) + O(\varepsilon^2),$$

$$h_1 = G^{-1} Q, \quad h_2 = -G^{-1}\left[B h_1 + \frac{\partial (A h_1)}{\partial t}\right],$$

$$H_1 = -G^{-1} A z, \quad H_2 = -\left[\frac{\partial G^{-1}}{\partial t} A - G^{-1} B\right] G^{-1} A z + O(||z||^2).$$

In the expressions for h_1, h_2, matrices A, B, G and the function Q depend on x and t, and in the expressions for H_1, H_2 these functions and matrices depend on v and t.

The equation (7.41) describes slow precessional motions of a gyroscopic system, and the equation (7.42) describes fast nutational motions. The formula (7.39) shows that the vector of generalized coordinates x is a superposition of precessional and nutational motions.

The original equations are split into two subsystems (7.41), (7.42) by the use of (7.39), (7.40), and the approximate representations for h and H are obtained.

Let us consider an initial value problem for (7.37) with initial conditions $x(t_0) = x_0$, $y(t_0) = \varepsilon y_0$. Then $z(t_0) = \varepsilon z_0 = \varepsilon(y_0 - h(x_0, t_0, \varepsilon))$, and an initial value, $v(t_0) = v_0$, for the equation of precessional motions (7.41) is found from the equation

$$x_0 = v_0 + \varepsilon H(v_0, \varepsilon z_0, t_0, \varepsilon)$$

as an asymptotic expansion

$$v_0 = x_0 + \varepsilon^2 G^{-1}(x_0, t_0)[y_0 - G^{-1}(x_0, t_0)Q(x_0, t_0)] + O(\varepsilon^3).$$

Returning to the problem of the justification of precessional theory, we see that the r.h.s. of the precessional equations (7.38) represented by the form

$$\dot{v} = \varepsilon[G(v, t) + \varepsilon B(v, t)]^{-1}Q(v, t),$$

coincides with the r.h.s. of (7.41) with an accuracy of order $O(\varepsilon)$ in the case of a nonautonomous system, and with an accuracy of order $O(\varepsilon^2)$ in the case of an autonomous one. Taking into account that for the system under consideration nutation oscillations are quenched, it may be inferred that the use of the precession equations is justified. We present without proof some results from [35] concerning investigations of integral manifolds in gyroscopic type systems. Consider the following system of ordinary differential equations

$$\begin{aligned}
\frac{da}{dt} &= Aa + P_1 y + P_2 z + Q_1(t, a, x, y, z, \varepsilon);\\
\frac{dx}{dt} &= Bx + P_3 y + P_4 z + Q_2(t, a, x, y, z, \varepsilon);\\
\varepsilon\frac{dy}{dt} &= C(\varepsilon)y + \varepsilon\delta_1 P_5 z + \varepsilon Q_3(t, a, x, y, z, \varepsilon);\\
\varepsilon\frac{dz}{dt} &= Dz + \delta_2 P_6 y + Q_4(t, a, x, y, z, \varepsilon),
\end{aligned} \tag{7.43}$$

where $Q_1, a \in R^k; Q_2, x \in R^l; Q_3, y \in R^m; Q_4, z \in R^n; A, B, C(\varepsilon), D, P_i (i = 1, ..., 6)$ are matrices of corresponding dimensions, ε is a small positive parameter, $C(\varepsilon) = C_0 + \varepsilon C_1, \delta_1\delta_2 = 0$. Let $\|\cdot\|$ denote a norm in a finite-dimensional space. Suppose the following assumptions hold.

(C₁). The functions $Q_i (i = 1, ..., 4)$ are continuous in

$$\Omega = \{-\infty < t < \infty, a \in R^k, \|x\| \le r_1, \|y\| \le r_2, \|z\| \le r_3, 0 \le \varepsilon \le \varepsilon_0\},$$

satisfy $\|Q_i(t, a, 0, 0, 0, \varepsilon)\| < M$ and a Lipschitz condition with the constant λ with respect to a, x, y, z, where M and λ are sufficiently small positive numbers.

(C_2). The eigenvalues λ_i of $A, B, C(\varepsilon), D$ satisfy:

1. A: $Re\lambda_i = 0$, $(i = 1, ..., k)$;

2. $B: Re\lambda_i < -\alpha_1 < 0$, $(i = 1, ..., l)$;

3. $C(\varepsilon)$: $Re\lambda_i(\varepsilon) < -\varepsilon\alpha_2(\alpha_2 > 0)$, $(i = 1, ..., m)$;

4. D: $Re\lambda_i < -\alpha_3 < 0$, $(i = 1, ..., n)$;

and

$$\|e^{At}\| \leq Ke^{\alpha|t|/2}; \|e^{Bt}\| \leq Ke^{-\alpha t}; \|e^{C(\varepsilon)t/\varepsilon}\| \leq Ke^{-\alpha t}; \|e^{D/(\varepsilon)t}\| \leq Ke^{-\alpha t/\varepsilon}$$

for some $K, \alpha > 0$ and any $t > 0$.

As usual, by an integral manifold of (7.43) is meant some set S in $R^1 \times R^k \times R^l \times R^m \times R^n$ consisting of trajectories of this system. The integral manifold $x = f(t, a, \varepsilon), y = g(t, a, \varepsilon), z = h(t, a, \varepsilon)$ is called an (L, Δ) - manifold if norms of the functions f, g, h under all t, a, ε from Ω are bounded by L and these functions satisfy the Lipschitz condition with respect to a with a constant Δ.

Theorem 7.2. *Let conditions* $(\mathbf{C_1})$ *and* $(\mathbf{C_2})$ *be satisfied. There exist* $L_0 > 0, \Delta_0 > 0$, *such that for all* $L < L_0, \Delta < \Delta_0$ *with sufficiently small* M, λ, ε *the system (7.43) possesses a unique* (L, Δ) *- integral manifold, given by equations* $x = f(t, a, \varepsilon), y = g(t, a, \varepsilon), z = h(t, a, \varepsilon)$. *The behaviour of (7.43) on this manifold is described by the equation*

$$\frac{da}{dt} = Aa + P_1 g(t, a, \varepsilon) + P_2 h(t, a, \varepsilon) + Q_1(t, a, f(t, a, \varepsilon), g(t, a, \varepsilon), h(t, a, \varepsilon), \varepsilon).$$

Let $a(t), x(t), y(t), z(t)$ be solutions of (7.43) satisfying the initial conditions $a(t_0) = a_0, x(t_0) = t_0, y(t_0) = y_0, z(t_0) = z_0$. The following statement characterises the stability of an integral manifold S.

Theorem 7.3. *If the system (7.43) satisfies* $(\mathbf{C_1})$ *and* $(\mathbf{C_2})$, *then there exist positive numbers* N, γ, *such that for* $t > t_0$ *the following inequalities hold*

$$\|x(t) - f(t, a(t), \varepsilon)\| \leq N\eta_0 e^{-\gamma(t-t_0)},$$

$$\|y(t) - g(t, a(t), \varepsilon)\| \leq N\eta_0 e^{-\gamma(t-t_0)},$$

$$\|z(t) - h(t, a(t), \varepsilon)\| \leq N\eta_0 e^{\frac{-\gamma}{\varepsilon}(t-t_0)},$$

where $\eta_0 = \|x_0 - f(t_0, a_0, \varepsilon)\| + \|y_0 - g(t_0, a_0, \varepsilon)\| + \|z_0 - h(t_0, a_0, \varepsilon)\|$, *and* f, g, h *are the functions defined in Theorem 7.2.*

We can establish asymptotic properties of f, g, h, by assuming that C_0 is a regular matrix, A is similar to a diagonal matrix and the functions $Q_i (i = 1, ..., 4)$ have bounded partial derivatives with respect to all variables up to $(r + 1)$ order.

Lemma 7.4. *If the conditions of Theorem 7.2 and the inequalities*

$$\|Q_i(t, a, 0, 0, 0, \varepsilon)\| \leq \varepsilon^{r+1} M, (i = 2, ..., 4)$$

hold, then the norms of the functions f, g, h *are bounded by the value* $\varepsilon^{r+1}L$.

Introduce new variables in (7.43):

$$\begin{aligned} x &= x_1 + f_0(t,a) + \varepsilon f_1(t,a) + \ldots + \varepsilon^r f_r(t,a), \\ y &= y_1 + g_0(t,a) + \varepsilon g_1(t,a) + \ldots + \varepsilon^r g_r(t,a), \\ z &= z_1 + h_0(t,a) + \varepsilon h_1(t,a) + \ldots + \varepsilon^r h_r(t,a). \end{aligned} \tag{7.44}$$

Substituting expressions (7.44) into (7.43) and equating x_1, y_1, z_1 to zero, we obtain the formal equalities for $f_i, g_i, h_i, (i = 1, ..., r)$. These equalities with an accuracy of terms multiplied by ε^{r+1} may be represented in the form

$$\begin{aligned} &\sum_{i=0}^{r} \varepsilon^i \frac{\partial f_i}{\partial t} + \sum_{i=0}^{r} \varepsilon^i \frac{\partial f_i}{\partial a} \Big(Aa + P_1 \sum_{j=0}^{r} \varepsilon^j g_j + P_2 \sum_{j=0}^{r} \varepsilon^j h_j \\ &+ Q_1(t, a, \sum_{j=0}^{r} \varepsilon^j f_j, \sum_{j=0}^{r} \varepsilon^j g_j, \sum_{j=0}^{r} \varepsilon^j h_j, \varepsilon) \Big) \equiv B \sum_{i=0}^{r} \varepsilon^i f_i \\ &+ P_3 \sum_{i=0}^{r} \varepsilon^i g_i + P_4 \sum_{i=0}^{r} \varepsilon^i h_i + Q_2(t, a, \sum_{i=0}^{r} \varepsilon^i f_i, \sum_{i=0}^{r} \varepsilon^i g_i, \sum_{i=0}^{r} \varepsilon^i h_i, \varepsilon), \end{aligned} \tag{7.45}$$

$$\begin{aligned} &\varepsilon \Big(\sum_{i=0}^{r} \varepsilon^i \frac{\partial g_i}{\partial t} \sum_{i=0}^{r} \varepsilon^i \frac{\partial g_i}{\partial a} \Big(Aa + P_1 \sum_{j=0}^{r} \varepsilon^j g_j + P_2 \sum_{j=0}^{r} \varepsilon^j h_j \\ &+ Q_1(t, a, \sum_{j=0}^{r} \varepsilon^j f_j, \sum_{j=0}^{r} \varepsilon^j g_j, \sum_{j=0}^{r} \varepsilon^j h_j, \varepsilon) \Big) \Big) \equiv C(\varepsilon) \sum_{i=0}^{r} \varepsilon^i g_i + \varepsilon \delta_1 P_5 \sum_{j=0}^{r} \varepsilon^j h_j \\ &+ \varepsilon Q_3(t, a, \sum_{j=0}^{r} \varepsilon^j f_j, \sum_{j=0}^{r} \varepsilon^j g_j, \sum_{j=0}^{r} \varepsilon^j h_j, \varepsilon) \Big) \Big) \\ &\varepsilon \Big(\sum_{i=0}^{r} \varepsilon^i \frac{\partial h_i}{\partial t} \sum_{i=0}^{r} \varepsilon^i \frac{\partial h_i}{\partial a} \Big(Aa + P_1 \sum_{j=0}^{r} \varepsilon^j g_j + P_2 \sum_{j=0}^{r} \varepsilon^j h_j \\ &+ Q_1(t, a, \sum_{j=0}^{r} \varepsilon^j f_j, \sum_{j=0}^{r} \varepsilon^j g_j, \sum_{j=0}^{r} \varepsilon^j h_j, \varepsilon) \Big) \Big) \equiv D \sum_{i=0}^{r} \varepsilon^i h_i + \delta_2 P_6 \sum_{j=0}^{r} \varepsilon^j g_j \\ &+ Q_4(t, a, \sum_{j=0}^{r} \varepsilon^j f_j, \sum_{j=0}^{r} \varepsilon^j g_j, \sum_{j=0}^{r} \varepsilon^j h_j, \varepsilon) \Big) \Big). \end{aligned} \tag{7.46}$$

Expanding all functions in (7.45)–(7.46) into formal asymptotic series in powers of the small parameter and equating coefficients of the same powers of ε, we obtain equations which allow us to find $f_i, g_i, h_i, (i = 1, ..., r)$.

The differential system for the variable a and the variables x_1, y_1, z_1, introduced by (7.44), possesses an integral manifold described by

$$f_{r+1}(t, a, \varepsilon), g_{r+1}(t, a, \varepsilon), h_{r+1}(t, a, \varepsilon),$$

and the norms of these functions are bounded by the value $\varepsilon^{r+1}L$. Returning to the variables a, x, y, z we see that the system (7.43) possesses the integral manifold $x = f(t, a, \varepsilon), y = g(t, a, \varepsilon), z = h(t, a, \varepsilon)$, with

$$\begin{aligned} f(t, a, \varepsilon) &= f_0(t, a) + \varepsilon f_1(t, a) + \ldots + \varepsilon^r f_r(t, a) + \varepsilon^{r+1}\tilde{f}_{r+1}(t, a, \varepsilon), \\ g(t, a, \varepsilon) &= \varepsilon g_1(t, a) + \ldots + \varepsilon^r g_r(t, a) + \varepsilon^{r+1}\tilde{g}_{r+1}(t, a, \varepsilon), \\ h(t, a, \varepsilon) &= h_0(t, a) + \varepsilon h_1(t, a) + \ldots + \varepsilon^r h_r(t, a) + \varepsilon^{r+1}\tilde{h}_{r+1}(t, a, \varepsilon), \end{aligned} \tag{7.47}$$

where

$$\varepsilon^{r+1}\tilde{f}_{r+1} = f_{r+1}(t, a, \varepsilon), \varepsilon^{r+1}\tilde{g}_{r+1} = g_{r+1}(t, a, \varepsilon), \varepsilon^{r+1}\tilde{h}_{r+1} = h_{r+1}(t, a, \varepsilon).$$

Thus, the following statement holds

Theorem 7.5. *If* $(\mathbf{C_1})$–$(\mathbf{C_2})$ *hold, and the functions* Q_i $(i = 1, ..., 4)$ *have bounded partial derivatives with respect to all variables to* $(r+1)$*st order,* A *is similar to a diagonal matrix and* $\det C_0 \neq 0$,, *then there exist numbers* $M_1 (M_1 < M_0), \lambda_1 (\lambda_1 < \lambda_0), \varepsilon_2 (\varepsilon_2 < \varepsilon_1)$, *such that for all* $M < M_1, \lambda < \lambda_1, 0 < \varepsilon < \varepsilon_2$ *the functions* f, g, h, *in Theorem 7.2, have bounded partial derivatives with respect to* a *up to* r*th order and can be represented in the form (7.47). The coefficients in the expansions (7.47) are uniquely determined by (7.45)–(7.46).*

More general results concerning gyroscopic systems may be found in [5, 43].

7.3.3 Vertical gyro with radial corrections

The equations of small oscillations of a gyroscopic system about the equilibrium position have the form

$$A\ddot{x} + (HG + B)\dot{x} + Cx = 0,$$

where A is a symmetric positive-definite matrix of inertia, G is a skew-symmetric matrix of gyroscopic forces, B is a symmetric positive-definite matrix of dissipative forces, C is the matrix of potential and nonconservative forces, and H is the gyroscope angular momentum. We let $\varepsilon = H^{-1}$ be a small positive parameter. If G is a non-singular matrix, the characteristic roots of this linear autonomous system break down into groups of roots of order $O(\varepsilon)$ and order $O(1/\varepsilon)$. In those cases in which the roots of order $O(1/\varepsilon)$ lie in the left half-plane, we can set up a slow invariant manifold for which the reduction principle is valid. The equations of motion on this manifold describe only precessional oscillations.

The corresponding precessional equations which can be considered as approximate equations on the integral manifolds are

$$(HG + B)\dot{x} + Cx = 0.$$

Investigation of a vertical gyroscope with radial corrections leads to the equations

$$J\ddot{\alpha} - H\dot{\beta} + d\dot{\alpha} - k\beta = 0, \ J\ddot{\beta} + H\dot{\alpha} + d\dot{\beta} + k\alpha = 0.$$

In these equations J is the equatorial mass moment of inertia of the gyroscope, H is its angular momentum, d is the coefficient of friction. Forces $-H\dot{\beta}$ and $H\dot{\alpha}$ are gyroscopic, while $-k\beta$ and $k\alpha$ are nonconservative forces, and in the theory of gyroscopic systems are referred to as forces of radial corrections [43].

For this system the precessional equations take the form

$$H\dot{\beta} + k\beta - d\dot{\alpha} = 0, \ H\dot{\alpha} + k\alpha + d\dot{\beta} = 0.$$

It is easy to see that, even in the case $d = 0$, the trivial solution of the precessional equations is asymptotically stable, whereas the trivial solution of the original equations is unstable. This means that in the case $d = 0$ the use of precessional equations is inappropriate. More detailed analysis shows that the value of the damping factor for which asymptotic stability will prevail is

$$d > kJ/H.$$

We note that the angular momentum H of the gyroscope is large compared to kJ. Therefore, the lower limit for the damping factor is small.

7.3.4 Heavy gyroscope

Using either the general equations of motion or Lagrange equations, differential equations governing the motions of the axis of the heavy gyroscope in cardanic suspension can be derived [43] as

$$A(\beta)\ddot{\alpha} + H\cos\beta \cdot \beta + E\sin 2\beta \cdot \dot{\alpha} \cdot \dot{\beta} = -m_1\dot{\alpha},$$

$$B_0\ddot{\beta} - H\cos\beta \cdot \dot{\alpha} - \frac{1}{2}E\sin 2\beta \cdot \dot{\alpha}^2 = -m_2\dot{\beta} - Pl\cos\beta,$$

where

$$A(\beta) = (A + A_1)\cos^2\beta + C_1\sin^2\beta + A_2,$$

$$B_0 = A + B_1, \quad E = C_1 - A - A_1.$$

Introduce the new time variable τ by $t = T\tau$ and the dimensionless parameters

$$\frac{A(\beta)}{B_0} = a(\beta), \quad \frac{E}{B_0} = e, \quad \frac{PlT}{B_0} = \nu,$$

$$\frac{m_1 T}{B_0} = b_1, \quad \frac{m_2 T}{B_0} = b_2, \quad B_0T/H = \varepsilon,$$

and note that T can be chosen in such a way that $a(\beta)$, e, ν, b_1, b_2 are values of order $O(1)$, $\varepsilon \ll 1$.

Introducing the new dependent variables $\alpha_1 = \dot{\alpha}T$, $\beta_1 = \dot{\beta}T$ leads to equations of the form (7.37) with

$$x = \begin{pmatrix} \alpha \\ \beta \end{pmatrix}, \quad y = \begin{pmatrix} \alpha_1 \\ \beta_1 \end{pmatrix}, \quad A(x,t) = \begin{pmatrix} a(\beta) & 0 \\ 0 & 1 \end{pmatrix},$$

$$G(x,t) = \begin{pmatrix} 0 & \cos\beta \\ -\cos\beta & 0 \end{pmatrix}, \quad B(x,t) = \begin{pmatrix} b_1 & 0 \\ 0 & b_2 \end{pmatrix},$$

$$Q(x,t) = \begin{pmatrix} 0 \\ -\nu\cos\beta \end{pmatrix}.$$

The representations for the slow variables are

$$\alpha = v_1 + \varepsilon\frac{z_2}{\cos v_2} + O(\varepsilon^2||z||), \quad v = \begin{pmatrix} v_1 \\ v_2 \end{pmatrix},$$

$$\beta = v_2 - \varepsilon\frac{a(v_2)z_1}{\cos v_2} + O(\varepsilon^2||z||), \quad z = \begin{pmatrix} z_1 \\ z_2 \end{pmatrix},$$

where v_1, v_2, z_1, z_2 satisfy the equations

$$\begin{aligned}
\dot{v}_1 &= \varepsilon\nu - \varepsilon^3\left(\nu^2 e\cdot\sin v_2 + \frac{\nu b_1 b_2}{\cos^2 v_2}\right) + O(\varepsilon^4),\\
\dot{v}_2 &= -\varepsilon\frac{\nu b_1}{\cos^2 v_2} + O(\varepsilon^4),\\
\varepsilon\dot{z}_1 &= -\varepsilon\frac{b_1}{a(v_2)}z_1 - \frac{\cos v_2}{a(v_2)}z_2 - \varepsilon z_1 z_2 tg v_2 + O(\varepsilon^2||z||),\\
\varepsilon\dot{z}_2 &= \cos v_2\cdot z_1 - \varepsilon b_2 z_2 + \varepsilon\gamma_0 z_1^2 tg v_2 + O(\varepsilon^2||z||),\\
\gamma_0 &= a(v_2) + \frac{1}{2}e\cdot\cos^2 v_2 = \frac{C_1 + A_2}{B_0}.
\end{aligned}$$

These relationships show that the motion of a heavy gyroscope is very close to regular precession on bounded time intervals, but over times of order $O(1/\varepsilon^2)$ the angle β tends to the value $\pi/2$, which is to say that the gyroscopic frames tend to the same plane.

The influence of random perturbations on gyroscopic systems will be studied in the Appendix to this Chapter.

7.3.5 A one rigid-link flexible-joint manipulator

Consider a simple model of a rigid-link flexible joint manipulator [7, 39], where J_m is the motor inertia, J_1 is the link inertia, M is the link mass, k is the stiffness. The model is described by the equations:

$$J_1\ddot{q}_1 + Mg\sin q_1 + k(q_1 - q_m) = 0,$$

$$J_m\ddot{q}_m - k(q_1 - q_m) = u.$$

Here q_1 is the link angle, q_m is the rotor angle, and u is the torque input. The differential system may be considered as a singularly perturbed one with $\varepsilon = \mu^2 = 1/k$, and may be expressed as

$$\mu \dot{q}_1 = w_1,$$
$$\mu \dot{w}_1 = -\varepsilon^2 \frac{Mg}{J_1} \sin q_1 - \frac{q_1 - q_m}{J_1},$$
$$\mu \dot{q}_m = w_m,$$
$$\mu \dot{w}_m = (q_1 - q_m + \mu^2 u)/J_m.$$

It is clear that we obtain a singular singularly perturbed system. If we rewrite the original system in the form

$$J_1 \ddot{q}_1 + J_m 1 \ddot{q}_m + Mg \sin q_1 = u,$$
$$\ddot{q}_1 - \ddot{q}_m + Mg \sin q_1 + k(1/J_1 + 1/J_m)(q_1 - q_m) = u,$$

then the use of new variables

$$x_1 = (J_1 q_1 + J_m q_m)/(J_1 + J_m), \quad x_2 = \dot{x}_1,$$
$$y = q_1 - q_m,$$

yields the system

$$\dot{x}_1 = x_2,$$
$$\dot{x}_2 = -\frac{Mg}{J_1 + J_m} \sin(x^1 + \frac{J_m}{J_1 + J_m} y) + \frac{u}{J_1 + J_m},$$
$$\varepsilon \ddot{y} = -(1/J_1 + 1/J_m) y - \varepsilon \frac{Mg}{J_1} \sin(x^1 + \frac{J_m}{J_1 + J_m} y) - \varepsilon \frac{u}{J_m}.$$

Now, for the investigation of robotic-like systems with weak energy dissipation the following results from [32] are useful.

Consider the differential system

$$\dot{x} = Ax + B\dot{y} + Qy + X(x, y, \dot{y}, \varepsilon),$$
$$\varepsilon \ddot{y} = \varepsilon C \dot{y} + Py + \varepsilon Rx + \varepsilon Y(x, y, \dot{y}, \varepsilon), \tag{7.48}$$

with a small positive parameter ε. Here x, X are m-vectors; y, Y are n-vectors; A, B, Q, C, P, R are constant matrices of appropriate dimensions. Suppose the following assumptions hold:

($\mathbf{D_1}$). The functions X and Y are defined and continuous in

$$\Omega = \{x \in R^m, \|y\| < \rho, 0 \le \varepsilon \le \varepsilon_0\},$$

bounded by M and satisfy a Lipschitz condition with constant λ with respect to $x, y, \dot{y}$, where M and λ are sufficiently small positive numbers.

($\mathbf{D_2}$). There exist numbers $K, \alpha, \beta (K > 0, \alpha > \beta \ge 0)$, such that the following inequalities are satisfied

$$\|e^{At}\| \le K e^{\beta |t|} \quad (-\infty < t < \infty),$$

$$\|e^{D(\varepsilon)t}\| \le Ke^{-\alpha\varepsilon t} \quad (0 \le t < \infty),$$

$$D(\varepsilon) = \begin{pmatrix} 0 & \varepsilon I \\ P & \varepsilon C \end{pmatrix},$$

where I is the identity matrix. Setting $z = (y, \dot{y})^T$, rewrite (7.48) as

$$\begin{aligned} \dot{x} &= Ax + P_1 z + X_1(x, z, \varepsilon), \\ \varepsilon\dot{z} &= D(\varepsilon)z + \varepsilon Z(x, z, \varepsilon), \end{aligned} \tag{7.49}$$

where $P_1 = (QB)$, $X_1(x, z, \varepsilon) = X(x, y, \dot{y}, \varepsilon)$, and

$$Z(x, z, \varepsilon) = \begin{pmatrix} 0 \\ Y(x, y, \dot{y}, \varepsilon) + R(x) \end{pmatrix}.$$

The system (7.49) possesses a local integral manifold $z = H(x, \varepsilon)$, with the flow on this manifold described by

$$\dot{x} = Ax + P_1 H(x, \varepsilon) + X_1(x, H(x, \varepsilon), \varepsilon).$$

If P is a regular matrix, then the functions describing the integral manifold of (7.48) can be found as asymptotic expansions

$$y = F(x, \varepsilon) = \varepsilon f_1(x) + \ldots + \varepsilon^k f_k(x) + \varepsilon^{k+1} f_{k+1}(x, \varepsilon),$$

$$\dot{y} = G(x, \varepsilon) = \varepsilon g_1(x) + \ldots + \varepsilon^k g_k(x) + \varepsilon^{k+1} g_{k+1}(x, \varepsilon),$$

where functions f_i, g_i are calculated using the following identities:

$$0 = Pf_1 + Rx + Y(x, 0, 0, 0),$$

$$g_1 = \frac{\partial f_1}{\partial x}(Ax + X(x, 0, 0, 0)),$$

$$\tfrac{\partial g_1}{\partial x}(Ax + X(x, 0, 0, 0)) = Cg_1 + Pf_2 + \tfrac{\partial Y}{\partial y}(x, 0, 0, 0) + \tfrac{\partial Y}{\partial \dot{y}}(x, 0, 0, 0),$$

$$g_2 = \tfrac{\partial f_2}{\partial x}(Ax + X(x, 0, 0, 0))$$
$$+ \tfrac{\partial f_1}{\partial x}\left(Bg_1 + Qf_1 + \left[\tfrac{\partial x}{\partial y}(x, 0, 0, 0)\right] f_1 + \left[\tfrac{\partial x}{\partial \dot{y}}(x, 0, 0, 0)\right] g_1\right),$$

and so on. The flow on this manifold is given by the equation

$$\dot{x} = Ax + BG(x, \varepsilon) + QF(x, \varepsilon) + X(x, F(x, \varepsilon), G(x, \varepsilon), \varepsilon).$$

Note that the reducibility principle holds for this integral manifold.

In [32] these results were used to obtain a sufficient condition for the orientation stability of satellites with double spin in the case of small friction and large damping stiffness.

These results may be applied to the analysis of the manipulator model under consideration. Following [15, 39], the control function u is written as a sum $u = u_f + u_s$, where $u_f = u_f(y, \dot{y}), u_s = u_s(x^1, x^2)$.

Setting

$$u_f = 2b\dot{y}, b > 0;$$

$$u_s = Mg\sin x^1 + \varepsilon\frac{Mg}{J_1 + J_m}\cos x^1 (\frac{J_m}{J_1 + J_m} Mg\sin x^1 - 2bx^2) + \varepsilon(J_1 + J_m)v_s,$$

we obtain, to an accuracy of order $O(\varepsilon^2)$, the equations

$$\dot{x}^1 = x^2,$$

$$\dot{x}^2 = \varepsilon v_s,$$

on the slow integral manifold with a new control function v_s.

The corresponding fast variable z which describes the behaviour of the system under consideration near this integral manifold, with an accuracy of the same order, satisfies the following fast equation

$$\varepsilon\ddot{z} = -(1/J_1 + 1/J_m)z - \varepsilon\frac{2b\dot{z}}{J_m}.$$

This means that transition regimes have a slowly damped high frequency oscillation.

7.4 Branching of Slow Integral Manifolds

In this section we formulate the problem and recall the method of gauge functions by considering a degenerate two-dimensional autonomous singularly perturbed differential system. The case of a quasi-homogeneous degenerate system and the case of an autonomous homogeneous singularly perturbed system are treated. Several examples are given, and one example describes a partially cheap optimal control problem.

We consider singularly perturbed differential systems whose degenerate equations have an isolated but not simple solution. In this case, the standard theory to establish a slow integral manifold near this solution does not work. Applying scaling transformations, and using the technique of gauge functions, we reduce the original singularly perturbed problem to a regularized one such that the existence of slow integral manifolds can be established by means of the standard theory. We illustrate the method by several examples.

7.4.1 Formulation of the problem. Preliminaries

We consider system (7.1) under the assumptions $(\mathbf{A_1})$ and $(\mathbf{A_2})$. Instead of hypothesis $(\mathbf{A_3})$ we suppose

$$det\ g_y(x, h(x,t), t, 0) \equiv 0 \quad \forall (x,t) \in G \times R, \tag{7.50}$$

that is, $y = h(x,t)$ is not a simple root of the degenerate equation

$$g(x, y, t, 0) = 0. \tag{7.51}$$

Under this assumption we cannot apply Proposition 1.1 to system (7.1) in order to establish the existence of a slow integral manifold near $\mathcal{M}_0$ for small ε. Our goal is to derive conditions which imply that, for sufficiently small ε, system (7.1) has at least one integral manifold $\mathcal{M}_\varepsilon$ with the representation

$$y = \psi_i(x,t,\varepsilon) = h(x,t) + \varepsilon^{q_i} h_{1,i}(x,t) + \varepsilon^{2q_i} h_{2,i}(x,t) + \dots .$$

where $q_i, 0 < q_i < 1$, is a rational number.
The key idea to solve this problem consists in looking for scalings and transformations of the type

$$\varepsilon = \mu^r, \; y = \tilde{y}(\mu, z, x, t), \; t = \tilde{t}(\mu, \tau),$$

such that system (7.1) can be reduced to a system

$$\begin{aligned} \frac{dx}{d\tau} &= f(x,z,\tau,\mu), \\ \mu \frac{dz}{d\tau} &= g(x,z,\tau,\mu), \end{aligned}$$

to which Proposition 1.1 can be applied. In this process the method of gauge functions plays an important role.

We illustrate our approach by considering a simple example, and at the same time we recall the method of undetermined gauges.

Example 4.1. Let us consider the system

$$\begin{aligned} \frac{dx}{dt} &= y, \\ \varepsilon \frac{dy}{dt} &= -y^2 - y^3 + \varepsilon\phi^2(x,t), \end{aligned} \tag{7.52}$$

where ϕ is a smooth positive function. The degenerate equation for (7.52) reads

$$0 = -y^2 - y^3$$

and has the isolated, but multiple, root $y = 0$. To find a transformation reducing the system (7.52) to a system to which Proposition 1.1 can be applied, we look for an approximation of the roots of the equation

$$0 = -y^2 - y^3 + \varepsilon\phi^2(x,t) \tag{7.53}$$

by means of the method of undetermined gauges (see e.g. [22]). For this purpose we postulate a solution of (7.53) in the form

$$y \cong \delta_1(\varepsilon) y_1(x,t) + \delta_2(\varepsilon) y_2(x,t) + \dots . \tag{7.54}$$

The functions $\delta_i(\varepsilon)$, called gauges, must be determined along with the functions $y_i(x,t)$. We suppose that the gauge functions $\delta_i(\varepsilon)$ are monotone in the interval I_0

and satisfy $\delta_i(\varepsilon) \to 0$ and $\delta_{i+1}(\varepsilon)/\delta_i(\varepsilon) \to 0$ as $\varepsilon \to 0$ for all i.
Substituting

$$y \cong \delta_1(\varepsilon) y_1$$

into (7.53) leads to the equation

$$0 \cong -y_1^2\delta_1^2(\varepsilon) + y_1^3\delta_1^3(\varepsilon) + \varepsilon\phi^2(x,t). \tag{7.55}$$

As $\delta_1^3(\varepsilon) \ll \delta_1^2(\varepsilon)$ for sufficiently small ε we simplify (7.55) to

$$0 \cong -y_1^2\delta_1^2 + \varepsilon\phi^2(x,t). \tag{7.56}$$

Now we have to compare the order functions $\delta_1(\varepsilon)$ and ε. Supposing that $\delta_1^2(\varepsilon)$ is the leading term in (7.56) we get $y_1 = 0$, if we assume that $\varepsilon\phi^2(x,t)$ is leading we are not able to determine y_1. If we suppose that $\delta_1^2(\varepsilon)$ and $\varepsilon\phi^2(x,t)$ are of equal significance for small ε, then we can set

$$\delta_1(\varepsilon) := \sqrt{\varepsilon}. \tag{7.57}$$

We note that this is not the only possible choice for $\delta_1(\varepsilon)$ (see [22]). Putting (7.57) into (7.56) we obtain

$$y_1(x,t) = \pm\phi(x,t).$$

Similarly we can determine higher order gauges and coefficients.
Now we use the relations (7.54) and (7.57) to scale the the variable y by $\varepsilon = \mu^2$, $y = \mu z$. Substituting these relations into (7.52) we get

$$\begin{aligned} \frac{dx}{dt} &= \mu z, \\ \mu\frac{dz}{dt} &= -z^2 + \phi^2(x,t) - \mu z^3. \end{aligned} \tag{7.58}$$

Taking into account that the degenerate equation for (7.58) has the two isolated simple solutions $z = \pm\phi(x,t)$ we may apply Proposition 1.1 to system (7.58) with respect to these roots, and get that system (7.52) has two integral manifolds with the representation

$$y = \pm\phi(x,t)\sqrt{\varepsilon} + O(\varepsilon).$$

In the following sections we study the existence and approximation of slow integral manifolds of system (7.1) in some degenerate cases.

7.4.2 Quasi-homogeneous degenerate equations

We study system (7.1) under the assumption $(\mathbf{A_1})$. We replace the assumptions $(\mathbf{A_2})$ and $(\mathbf{A_3})$ by the the following hypotheses.

$(\mathbf{E_1})$. The function $g(x,y,t,0)$ can be represented in the form

$$g(x,y,t,0) \equiv g_1(x,y,t) + g_2(x,y,t), \tag{7.59}$$

where the functions g_1 and g_2 have the following properties

1. g_1 is homogeneous in y of degree $r \geq 2$, i.e., for $\forall \lambda \in R$ we have

$$g_1(x, \lambda y, t) = \lambda^r g_1(x, y, t), \quad \forall (x, y, t) \in R^n \times R^m \times R. \tag{7.60}$$

2. The relationship

$$g_2(x, y, t) = O(|y|^{r+1}), \text{ as } y \to 0, \tag{7.61}$$

holds uniformly in $(x, t) \in R^n \times R$.

Hypothesis $(\mathbf{E_1})$ implies that $y = h(x, t) \equiv 0$ is a solution of the degenerate equation (7.51) satisfying (7.50).

$(\mathbf{E_2})$.

$$g_\varepsilon(x, 0, t, 0) \neq 0, \quad \forall \ (x, t) \in R^n \times R.$$

By means of the scaling

$$\varepsilon = \mu^r, y = \mu z, \tag{7.62}$$

we get from (7.59)–(7.61)

$$g(x, \mu z, t, \mu^r) = \mu^r \Big(g_1(x, z, t) + g_\varepsilon(x, 0, t, 0) + \mu \bar{g}(x, z, t, \mu) \Big), \tag{7.63}$$

where $\bar{g}(x, z, t, \mu)$ is smooth. Substituting (7.62) into (7.1) and taking into account (7.63) we obtain

$$\begin{aligned} \dot{x} &= f(x, \mu z, t, \mu^r), \\ \mu \dot{z} &= g_1(x, z, t) + g_\varepsilon(x, 0, t, 0) + \mu \bar{g}(x, z, t, \mu). \end{aligned} \tag{7.64}$$

The degenerate equation of (7.64) reads

$$g_1(x, z, t) + g_\varepsilon(x, 0, t, 0) = 0. \tag{7.65}$$

We further assume:

$(\mathbf{E_3})$. There is a smooth function $\bar{h} : R^m \times R \to R$ such that

1. $z = \bar{h}(x, t)$ is a root of (7.65),

2. the spectrum of the Jacobian matrix $\frac{\partial g_1}{\partial z}(x, \bar{h}(x, t), t)$ is separated from the imaginary axis for $(x, t) \in G \times R$.

Applying Proposition 1.1 to system (7.64) we get

Theorem 4.1. *Under the hypotheses* $(\mathbf{A_1}), (\mathbf{E_1}), (\mathbf{E_2})$, *and* $(\mathbf{E_3})$ *there is a sufficiently small positive* $\varepsilon_2, \varepsilon_2 \leq \varepsilon_1$, *such that for* $\varepsilon \in I_2$ *system (7.1) has the integral manifold*

$$\mathcal{M}_\varepsilon := \{(x, y, t) \in R^{n+m+1} : y = \bar{\psi}(x, t, \varepsilon), (x, t, \varepsilon) \in G \times T \times I_2\}$$

with the asymptotic representation

$$y = \bar{\psi}(x,t,\varepsilon) = \varepsilon^{1/r}\bar{h}(x,t) + \varepsilon^{2/r}\bar{h}_1(x,t) + \dots .$$

Remark 4.1. From Theorem *4.1* it follows that the integral manifold $\mathcal{M}_\varepsilon$ converges to the root $y = 0$ of the degenerate equation *(7.51)* as ε tends to 0. If equation *(7.64)* has more than one simple solution then several integral manifolds branch from the non–simple solution $y = 0$.

To illustrate Theorem 4.1 we consider Example 4.1 from the previous subsection. In that case we have $g_1(x,y,t) \equiv -y^2$, $g_2(x,y,t) \equiv -y^3$, $g_\varepsilon(x,y,t) \equiv \phi^2(x,t)$ such that the degenerate system (7.64) reads

$$y^2 - \phi^2(x,t) = 0.$$

Here, we have two slow integral manifolds of system (7.52) branching from the multiple solution $y = 0$.

7.4.3 Homogeneous systems

Consider the autonomous system

$$\begin{aligned} \frac{dx}{dt} &= f(x,y,\varepsilon), \\ \varepsilon\,\frac{dy}{dt} &= g(x,y,\varepsilon), \end{aligned} \tag{7.66}$$

under the assumption

(**F**). f and g are homogeneous polynomials in x and y of degree $r, r \geq 2$, with coefficients smoothly depending on ε.

It follows from hypothesis (**F**) that for $\forall \lambda \in R$ and $\forall\ (x,y,\varepsilon) \in R^m \times R^n \times I_0$

$$\begin{aligned} f(\lambda x, \lambda y, \varepsilon) &= \lambda^r f(x,y,\varepsilon), \\ g(\lambda x, \lambda y, \varepsilon) &= \lambda^r g(x,y,\varepsilon). \end{aligned}$$

Thus, $y = 0$ is a non-simple root of the degenerate equation $0 = g(x,y,0)$. Furthermore, if we replace x by λx, y by λy and t by $\lambda^{1-r}t$ in (7.66), then the system (7.66) is invariant under this transformation. Thus, if $(x(t), y(t))$ is a solution of (7.66) then $(\lambda x(\lambda^{r-1}t), \lambda y(\lambda^{r-1}t))$ is also a solution of (7.66). This property implies that any slow invariant manifold $y = \psi(x,\varepsilon)$ of (7.66) has the form

$$y = L(\varepsilon)x, \tag{7.67}$$

where $L(\varepsilon)$ is a $(n \times m)$–matrix. Thus, under our conditions, any slow invariant manifold of system (7.66) is a linear manifold.
Exploiting the invariance of $y = L(\varepsilon)x$ with respect to the system (7.66) we get the relation

$$\varepsilon\ L(\varepsilon)\ f(x, L(\varepsilon)x, \varepsilon) \equiv g(x, L(\varepsilon)x, \varepsilon), \quad \forall\ x \in R^m. \tag{7.68}$$

We consider (7.68) to be an equation to determine the entries in the matrix $L(\varepsilon)$. Since that equation can have more than one solution we call (7.68) a bifurcation equation. Thus, we have the following result:

Theorem 4.2. *Under the assumption* (**F**), *any slow invariant manifold of* (7.66) *is a linear manifold* (7.67), *where the matrix* $L(\varepsilon)$ *is determined by the bifurcation equation* (7.68).

To illustrate Theorem 4.2 we consider the following example.

Example 4.2.

$$\frac{dx}{dt} = 3x^3, \qquad \varepsilon\frac{dy}{dt} = y^3 + \varepsilon^3 x^3. \tag{7.69}$$

The corresponding degenerate equation is $y^3 = 0$.
According to Theorem 4.2, any slow invariant manifold of (7.69) has the form $y = L(\varepsilon)x$, where L is a scalar function. By (7.68) the corresponding bifurcation equation reads

$$L^3 - 3\varepsilon L + \varepsilon^3 = 0.$$

This equation possesses three solutions. For small ε we find, by means of the method of undetermined gauges, the representations

$$\begin{aligned} L_1(\varepsilon) &= \frac{1}{3}\varepsilon^2 + \frac{1}{81}\varepsilon^5 + o(\varepsilon^5), \\ L_2(\varepsilon) &= -\varepsilon^{1/2}\sqrt{3} - \frac{1}{6}\varepsilon^2 + o(\varepsilon^2), \\ L_3(\varepsilon) &= \varepsilon^{1/2}\sqrt{3} - \frac{1}{6}\varepsilon^2 + o(\varepsilon^2). \end{aligned}$$

Thus, the differential system (7.69) under consideration has three slow invariant manifolds $y = L_k(\varepsilon)x, k = 1, 2, 3$.

7.4.4 Quasi-polynomial degenerate equations

Consider the system

$$\begin{aligned} \frac{dx}{dt} &= f(x, y, t, \varepsilon), \\ \varepsilon\,\frac{dy}{dt} &= g(x, y, t, \varepsilon), \end{aligned} \tag{7.70}$$

with $x \in R^m, y \in R, t \in R, \varepsilon \in I_0$. In what follows we assume

($\mathbf{G_1}$). f and g satisfy assumption ($\mathbf{A_1}$). Additionally we suppose that g is a polynomial with respect to y and ε.

By assumption ($\mathbf{G_1}$), g can be represented in the form

$$g(x, y, t, \varepsilon) \equiv \sum_{i=k_0}^{n_0} a_{0i}(x, t)y^i$$

$$+\varepsilon \sum_{i=k_1}^{n_1} a_{1i}(x, t)y^i + \ldots + \varepsilon^m \sum_{i=k_m}^{n_m} a_{mi}(x, t)y^i.$$

For what follows we further suppose

(G_2). $k_0 \geq 2$, $a_{jk_j}(x,t) \neq 0$ for $j = 0, ..., m$, and $\forall (x,t)$. It follows from hypothesis (**G_2**) that $y = 0$ is a multiple root of the degenerate equation of (7.70)

$$g(x,y,t,0) \equiv \sum_{i=k_0}^{n_0} a_{0i}(x,t)y^i = 0.$$

As in the previous sections we scale the parameter ε and the variable y by

$$\varepsilon = \mu^q, y = \mu^p z \tag{7.71}$$

and look for conditions on the coefficients $a_{jk_j}(x,t)$ such that the equation

$$\varepsilon \frac{dy}{dt} = g(x,y,t,\varepsilon)$$

can be transformed into an equation of the type

$$\mu \frac{dz}{dt} = \tilde{g}(x,z,t,\mu), \tag{7.72}$$

whose corresponding degenerate equation

$$0 = \tilde{g}(x,z,t) \tag{7.73}$$

has a simple root $z = \tilde{h}(x,t)$ to which Proposition 1.1 can be applied.

Substitute (7.71) into the right hand side of (7.70) and rewrite the latter in the form

$$\mu^{p+q} \frac{dz}{dt} = a_{0k_0}(x,t)\mu^{k_0 p} z^{k_0 p}$$
$$+ \sum_{j=1}^{m} a_{jk_j}(x,t)\mu^{jq+k_j p} z^{k_j p} + o(\mu^{mq+k_m p}). \tag{7.74}$$

Let r be the leading order of the right hand side of (7.74). Then equation (7.74) can be reduced to the form (7.72) if

$$p + q = r + 1. \tag{7.75}$$

To eliminate $z = 0$ as a multiple root of (7.73) we have to look for a scaling (7.71) such that the first term on the right hand side of (7.74) determines the leading order, that is

$$r = k_0 p,$$

and that there exist at least two terms of the leading order on the right hand side of (7.74). If we require that the jth term on the right hand side of (7.74) has the same order as the first term, then we get the relation

$$jq = (k_0 - k_j)p, \tag{7.76}$$

that can be considered as a determining equation for q.

From (7.75)–(7.76) we obtain

$$p = \frac{1}{j(1-k_0)+k_0-k_j}, \quad 1 \le j \le m. \tag{7.77}$$

Since $k_j, j = 0, ..., m$, are positive integers, where $k_0 \ge 2$, and since p is also a positive integer, it is easy to check that only for $j = 1$ and for $k_1 = 0$ does (7.77) yield a positive integer, namely $p = 1$. Thus, in order to be able to reduce (7.74) to an equation of the type (7.72) we have to require $a_{1,0}(x,t) \neq 0$. This implies that g can be represented in the form

$$g(x,y,t,\varepsilon) = a_{0,k_0}(x,t)y^{k_0} + \varepsilon a_{1,k_1}(x,t)o(y^{k_0}) + o(\varepsilon).$$

But this representation is the same as treated in Theorem 2.1. Therefore, we have the following result

Theorem 4.3. *Suppose hypotheses* $(\mathbf{G_1})$ *and* $(\mathbf{G_2})$ *to be valid. Then the condition* $a_{1,k_1}(x,t) \neq 0, \quad \forall (x,t)$, *is necessary and sufficient for the existence of a slow integral manifold of system* (7.1).

Remark 4.2. If *(7.70)* is an autonomous system with a structure such that after some scaling of y, ε and t it can be represented in the form

$$\begin{aligned} \frac{dx}{dt} &= f(x,z,\nu), \\ \nu^k \frac{dz}{dt} &= g(x,z,\nu), \end{aligned}$$

with $k \ge 2$, then the existence of a slow integral manifold can be established under relaxed conditions.

We illustrate Remark 4.2 by the following example.

Example 4.3. We consider the two-dimensional system

$$\begin{aligned} \frac{dx}{dt} &= y, \\ \varepsilon \frac{dy}{dt} &= \alpha(x)y^3 + \varepsilon\beta(x)y + \varepsilon^2\gamma(x), \end{aligned} \tag{7.78}$$

where all coefficients are sufficiently smooth, and α and β satisfy $\alpha(x)\beta(x) < 0, \quad \forall\, x$.

Using the scaling

$$y = \mu^p,\ \varepsilon = \mu^q,$$

we obtain from (7.78)

$$\begin{aligned} \frac{dx}{dt} &= \mu^p z, \\ \mu^{p+q} \frac{dz}{dt} &= \alpha(x)\mu^{3p}z^3 + \beta(x)\mu^{q+p}z + \gamma(x)\mu^{2q}. \end{aligned} \tag{7.79}$$

As can be verified, only the choice $q = 2p$ provides two terms on the right hand side of (7.79) with leading order $3p$. Thus, we get

$$\frac{dx}{dt} = \mu^p z,$$
$$\mu^{3p}\frac{dz}{dt} = \mu^{3p}(\alpha(x)z^3 + \beta(x)z + \mu\gamma(x)).$$

If we cancel the factor μ^{3p} in the last equation, we do not obtain a singularly perturbed equation. But after introducing the scaled time $\tau = t\mu^p$ and setting $p = 1$ we get

$$\begin{aligned} \frac{dx}{d\tau} &= z, \\ \mu\frac{dz}{d\tau} &= \alpha(x)z^3 + \beta(x)z + \mu\gamma(x). \end{aligned} \tag{7.80}$$

The degenerate equation of (7.80) is

$$\alpha(x)z^3 + \beta(x)z = 0$$

and has the three simple roots $z = 0$, $z = \pm\sqrt{-\frac{\beta(x)}{\alpha(x)}}$.

Thus, the original system (7.78) has three slow invariant manifolds

$$y = O(\mu),$$
$$y = \pm\sqrt{-\frac{\beta(x)}{\alpha(x)}}\mu + O(\mu^2).$$

The case that the variable y in (7.70) is a n-vector can be treated similarly. As an illustration we consider the "cheap control" problem [26].

7.4.5 Cheap control

Consider the linear-quadratic optimal control problem

$$\dot{x} = A(t,\varepsilon)x + B(t,\varepsilon)u;$$

$$J = \frac{1}{2}x^T(1)Fx(1) + \frac{1}{2}\int_0^1 [x^T(t)Q(t,\varepsilon)x(t) + \varepsilon^2 u^T(t)R(t,\varepsilon)u(t)]dt,$$

where $Q = Q^T \geq 0$, $F = F^T \geq 0$, $R = R^T > 0$, $t \in [0,1]$, ε is a small positive parameter. Usually, such a problem is called a "cheap control" problem because there is a small parameter multiplied by a control function in the cost functional. The solution of this problem is given by the formula

$$u = -\varepsilon^{-2}R^{-1}B^T Kx,$$

where K is the solution of matrix Riccati equation

$$\varepsilon^2(K + A^TK + KA + Q) = KSK, \quad S = BR^{-1}B^T; \quad K(1) = F.$$

Under the condition $\varepsilon = 0$, the degenerate equation has the multiple solution $K = 0$ and the branching of slow integral manifolds takes place for this Riccati equation. For simplicity, we confine consideration to the problem of minimizing the functional

$$J = \frac{1}{2}\int_0^1 [q(t)y^2(t) + \varepsilon^2u^2(t)]dt + \frac{1}{2}y^2(1),$$

under the restriction

$$y^{(n)} + a_1(t)y^{(n-1)} + \ldots + a_n(t)y = u(t),$$

where y and u are scalars. Let

$$K = (\mu^{i+j-1}K_{ij})_{i,j=1,\ldots,n}, \quad \mu = \varepsilon^{1/n}.$$

Then the matrix $X = (K_{ij})_{i,j=1,\ldots,n}$ satisfies the following singularly perturbed matrix Riccati equation

$$\mu\dot{X} + (A_0^T + \mu A_1^T(t,\mu))X + X(A_0 + \mu A_1(t,\mu)) + Q = XSX, \tag{7.81}$$

where

$$A_0 = \begin{pmatrix} 0 & 1 & 0\ldots 0 \\ & \cdots & \\ 0 & 0 & 0\ldots 1 \\ 0 & 0 & 0\ldots 0 \end{pmatrix}.$$

The corresponding Lurie (degenerate) equation

$$A_0^TX + XA_0 - XSX + Q = 0 \tag{7.82}$$

has several real solutions and every such solution corresponds to a slow integral manifold (i.e., a uniformly bounded solution for all t) of the Riccati equation (7.81). All these slow integral manifolds coincide, under the condition $\mu = 0$, and the branching of slow integral manifolds takes place. Note that the Lurie equation (7.82) has the unique positive solution $X = C_0$, such that the matrix $A - SC_0$ is stable. The corresponding slow integral manifold $X = C(t,\mu)$ may be found as an asymptotic expansion with respect to the small parameter μ.

Let $C(t,\mu) = (C_{i,j}(t,\mu))_{i,j=1,\ldots,n}$. Then for suboptimal control we obtain the expression

$$u = -\mu^{-n}(\mu^{n-1}C_{nn}y^{(n-1)} + \ldots + \mu C_{n1}y).$$

In this case the error of the cost functional is of order $O(e^{-1/\mu})$.

Example 4.4.

We investigate the optimal control problem

$$\dot{x}_1 = u_1,$$
$$\dot{x}_2 = x_1 + x_2 + u_2,$$

with the cost functional

$$J = \frac{1}{2}\int_0^T [x_1^2(t) + \varepsilon x_2^2(t) + u_1^2(t) + \varepsilon^2 u_2^2(t)]dt \to \min .$$

This problem is called a "partially cheap control" problem because one of the control terms in the cost functional is multiplied by a small parameter [26]. It is well known [12, 29] that optimal control in this problem is given by the formula

$$\begin{pmatrix} u_1 \\ u_2 \end{pmatrix} = -\begin{pmatrix} 1 & 0 \\ 0 & \varepsilon \end{pmatrix}^{-1} K \begin{pmatrix} x_1 \\ x_2 \end{pmatrix},$$

where the matrix K is a nonnegative solution of the matrix Riccati equation

$$\frac{dK}{dt} = -K\begin{pmatrix} 0 & 0 \\ 1 & 1 \end{pmatrix} - \begin{pmatrix} 0 & 1 \\ 0 & 1 \end{pmatrix} K + K\begin{pmatrix} 1 & 0 \\ 0 & \varepsilon^2 \end{pmatrix}^{-1} K - \begin{pmatrix} 1 & 0 \\ 0 & \varepsilon \end{pmatrix}$$

satisfying the condition $K(T) = 0$. If we put

$$K = \begin{pmatrix} k_1 & \varepsilon k_2 \\ \varepsilon k_2 & \varepsilon k_3 \end{pmatrix},$$

we obtain the differential system

$$\begin{aligned} \frac{dk_1}{dt} &= k_1^2 + k_2^2 - 2\varepsilon k_2 - 1, \\ \varepsilon\,\frac{dk_2}{dt} &= \varepsilon k_1 k_2 + k_2 k_3 - \varepsilon(k_2 + k_3), \\ \varepsilon\,\frac{k_3}{dt} &= k_3^2 + \varepsilon^2 k_2^2 - 2\varepsilon k_3 - \varepsilon. \end{aligned} \tag{7.83}$$

The corresponding degenerate system

$$k_2 k_3 = 0, \ k_3^2 = 0$$

has the solution $k_2 = k_3 = 0$, but this solution is not simple. In order to get simple roots we apply the scaling

$$k_3 = \mu\kappa_3, \ \varepsilon = \mu^2. \tag{7.84}$$

Substituting (7.84) into (7.83) we obtain

$$\begin{aligned} \frac{dk_1}{dt} &= k_1^2 + k_2^2 - 2\mu^2 k_2 - 1, \\ \mu\,\frac{dk_2}{dt} &= \mu k_1 k_2 + k_2\kappa_3 - \mu(k_2 + \mu\kappa_3), \\ \mu\,\frac{d\kappa_3}{dt} &= \kappa_3^2 - 1 + \mu^2 k_2^2 - 2\mu\kappa_3. \end{aligned}$$

The corresponding degenerate system

$$k_2\kappa_3 = 0, \ \kappa_3^2 = 1,$$

has the simple nonnegative solution $k_2 = 0, \kappa_3 = 1$. Applying Proposition 1.1 we get that the original system (7.83) has the invariant manifold

$$k_2 = O(\mu), \ k_3 = \mu + O(\mu^2). \tag{7.85}$$

Substituting (7.85) into (7.83) leads to the initial value problem

$$\frac{dk_1}{dt} = k_1^2 - 1 + O(\mu^2), \ k_1(T) = 0.$$

In this way, the elements of the matrix K can be determined approximately.

Note that some results of this section may be found in [30].

7.5 Appendix: Optimal Estimation in Gyroscopic Systems

by Gorelova, E. Ya. and Sobolev, V. A.

7.5.1 Introduction

The effect of random inputs on the movement of systems of solid bodies was investigated by many authors, see, for example, [1]. This Appendix deals with the analysis of the equations of gyroscopic systems under the influence of random forces. The possibility of the replacement of the equations of motion by the corresponding precessional equations is investigated. This approach is widespread in mechanics and gives suitable results in numerous cases. But there are a great number of examples when the substitution of the original equations by the precessional ones leads to inaccurate or qualitatively incorrect results. In this respect, there have been a few works studying either the reasoning behind such a procedure, or the conditions under which it gives an appropriate result [18, 19].

This problem was solved by the method of integral manifolds [38]. The essence of this method is in the separation of the class of slow motions of the original system. The dimension of the system is reduced, but the system obtained, while of lower dimension, inherits the main features of its qualitative behaviour. In this Appendix the equations of motion of the gyroscopic system of the form suggested by Merkin [19] are analysed. It is shown that the method of integral manifolds can be applied to systems of this type.

Note that the equations of the flow along the integral manifold to the specified accuracy coincide with the corresponding precessional equations. In most applications the restrictions under which this slow integral manifold is stable are fulfilled. This means that any solution of the original equations, starting in the vicinity of the integral manifold, may be represented as a sum of some solution of the precessional equations and a small rapidly vanishing term. In this sense conversion to the precessional equations is permissible.

The main result of the Appendix is concerned with the possibility of conversion to the precessional equations in the presence of random terms. It is shown that the use of precessional equations as the basis for equations of the filtering error in the problem of optimal estimation may provide inadmissible errors.

7.5.2 The equations of a Kalman filter for gyroscopic systems

We derive the equations of the Kalman filter for gyroscopic systems. Consider the equations of motion of gyroscopic system in the non-stationary case under the action of random forces in the form in [19]

$$\ddot{x} + [HG_0(t) + G_1(t)]\dot{x} + N(t)x = B(t)\dot{\omega}(t). \tag{7.86}$$

Here x is the $n-$dimensional vector of the system state, $G_0(t)$ is a skew-symmetric matrix of gyroscopic forces, and possessing a bounded inverse for all $t \geq 0$, $G_1(t)$

is a symmetric matrix of damping forces, $N(t)$ is the matrix of potential and non-conservative forces, H is a large parameter proportional to the angular velocity of the proper rotation of the gyroscope and which is much larger than the values of all the other system parameters for many gyroscopic systems.

Let the observation take place in the presence of Gaussian white noise described by the equation

$$z = C(t)x + \dot{\xi}(t), \tag{7.87}$$

where z is $m-$dimensional vector, $C(t)$ is $m \times n$ matrix. Let $\dot{\omega}(t)$ and $\dot{\xi}(t)$ be independent Gaussian white noise with zero expected values and correlation matrices $Q(t)\delta(t-s)$ and $R(t)\delta(t-s)$, respectively, where $Q(t)$ and $R(t)$ are symmetric positive semidefinite $m \times m-$matrices.

Introducing $\varepsilon = 1/H$, we rewrite (7.86) as a system

$$\begin{cases} \dot{x} &= y \\ \varepsilon\dot{y} &= -[G_0(t) + \varepsilon G_1(t)]y - \varepsilon N(t)x + \varepsilon B(t)\dot{\omega}. \end{cases} \tag{7.88}$$

For simplicity of presentation we assume that $x_0 = x(0)$ and $y_0 = y(0)$ are known vectors.

We are required to obtain an estimate $(\hat{x}(t), \hat{y}(t))^T$ of the state $(x(t), y(t))^T$, (^{T} stands for transposition), of system (7.88) in accordance with the vector-function $z(t)$ available for measurement at $t > 0$. The vector-function $x(t)$ is not available for measurement. The system which determines the vector $(\hat{x}(t), \hat{y}(t))^T$ is usually called the filter. We examine filters which are non-stationary linear systems of the form

$$\dot{\rho} = F(t)\rho + G(t)z,$$

where $\rho(t)$ is a $2n$ dimensional vector, $F(t)$ is a $2n \times 2n$ matrix, $G(t)$ is a $2n \times m$ matrix. It is known ([12, 29]) that the filter which provides an unbiased estimate

$$e(t) = (x(t), y(t))^T - (\hat{x}(t), \hat{y}(t))^T$$

for the system

$$\dot{x} = A(t)x + B(T)\dot{w},$$

with the observation (7.87), is defined by the differential equation

$$\frac{d\rho}{dt} = [A(t) - G(t)C(t)]\rho + G(t)z(t), \tag{7.89}$$

and satisfies the initial condition

$$\rho(0) = E[(x(0), y(0))^T].$$

Here $E[\cdot]$ is an expected value. Those filters which satisfy equation (7.89) contain the matrix $G(t)$ as a parameter, and it should be chosen to minimize the variance of the error $e(t)$. To ensure that the estimate is unbiased, we require that

$$E[(x(t), y(t))^T] = E[\rho(t)],$$

at all $t > 0$, whence $E[e(t)] = 0$.

Consequently, the correlation matrix $P(t)$ of the error $e(t)$ has the form

$$P(t) = E[e(t)e^T(t)].$$

It is clear that $P(t)$ is a symmetric matrix satisfying the initial condition

$$P(0) = E[e(0)e^T(0)] = P_0,$$

and the differential equation

$$\frac{dP}{dt} = [A(t) - G(t)C(t)]P + P[A(t) - G(t)C(t)]^T + B(t)Q(t)B^T(t) + G(t)R(t)G^T(t).$$

Note that matrix $G(t)$ is still unknown. Following [12] the filter is optimal if

$$G(t) = P(t)C^T(t)R^{-1}(t). \tag{7.90}$$

Taking (7.90) into consideration we obtain the equation for the correlation matrix of errors in the form of the Riccati equation

$$\frac{dP}{dt} = A(t)P + PA^T(t) - PC^TR^{-1}CP + BQB^T, \tag{7.91}$$

$$P(0) = P_0. \tag{7.92}$$

It was shown in [12] that, if P_0 is a positive definite matrix, equation (7.91) can be solved uniquely for the matrix $P(t)$, which exists for all $t \geq 0$. Then the equation for the optimal filter, on using (7.89) and (7.90), takes the form

$$\frac{d\rho}{dt} = [A(t) - P(t)C^T(t)R^{-1}(t)C(t)]\rho + P(t)C^T(t)R^{-1}(t)z(t),$$

$$\rho(0) = E[(x(0), y(0))^T],$$

where $P(t)$ is the solution of the differential Riccati equation (7.91) satisfying the initial conditions (7.92).

Let $m_1(t,\varepsilon)$ and $m_2(t,\varepsilon)$ be the mathematical expectations of the vectors $x(t)$ and $y(t)$ of system (7.88), i. e.,

$$m_1(t,\varepsilon) = E[x(t)], m_2(t,\varepsilon) = E[y(t)].$$

Then the vector $m(t,\varepsilon) = (m_1(t,\varepsilon), m_2(t,\varepsilon))^T$ satisfies the differential equation

$$\dot{m} = A(t)m + PC^T(t)R^{-1}(t)(z - C(t)m). \tag{7.93}$$

We apply the above results to system (7.88). $A(t)$ is the matrix of linear terms of the system (7.88) and is defined by

$$A(t) = \begin{pmatrix} 0 & I \\ -N(t) & -\frac{1}{\varepsilon}G_0(t) - G_1(t) \end{pmatrix}.$$

Let $B_1(t)$ and $C_1(t)$ denote the block matrices

$$B_1(t) = \begin{pmatrix} 0 \\ -B(t) \end{pmatrix}, \quad C_1(t) = \begin{pmatrix} C(t) & 0 \end{pmatrix}.$$

Then the Riccati equation for the correlation matrix $P(t, \varepsilon)$ of system (7.88) is

$$\frac{dP}{dt} = A(t)P + PA^T(t) - PC_1^T R^{-1} C_1 P + B_1 Q B_1^T. \tag{7.94}$$

We designate the $n \times n$ blocks of the matrix $P(t, \varepsilon)$ as follows:

$$P(t, \varepsilon) = \begin{pmatrix} P_1(t, \varepsilon) & P_2(t, \varepsilon) \\ P_2^T(t, \varepsilon) & P_3(t, \varepsilon) \end{pmatrix}.$$

Then equation (7.94) implies the system

$$\dot{P}_1 = P_2^T + P_2 - P_1 S P_1, \tag{7.95}$$

$$\varepsilon \dot{P}_2 = \varepsilon P_3 - \varepsilon P_1 N^T - P_2(G_0 + \varepsilon G_1)^T - \varepsilon P_1 S P_2, \tag{7.96}$$

$$\begin{aligned} \varepsilon \dot{P}_3 = & -\varepsilon(N P_2 + P_2^T N^T) - P_3(G_0 + \varepsilon G_1)^T \\ & -(G_0 + \varepsilon G_1) P_3 - \varepsilon P_2^T S P_2 + \varepsilon L, \end{aligned} \tag{7.97}$$

where $S = C^T R^{-1} C, L = BQB^T$.

Equation (7.93) may also be rewritten as a system:

$$\begin{cases} \dot{m}_1 &= m_2 + P_1 C R^{-1}(z - Cm_1), \\ \varepsilon \dot{m}_2 &= -(G_0 + \varepsilon G_1) m_2 - \varepsilon N m_1 + \varepsilon P_2 C R^{-1}(z - Cm_1), \end{cases}$$

where $m_1(t, \varepsilon)$ and $m_2(t, \varepsilon)$ satisfy the initial conditions

$$m_1(0, \varepsilon) = x_0, \; m_2(0, \varepsilon) = y_0.$$

We now use some results of integral manifold theory from this Chapter. The existence of an attracting integral manifold permits us to reduce the singularly perturbed system to a system of lower dimension.

7.5.3 Precessional equations in the deterministic case

In this subsection, we employ the above results (see Section *3. Systems with Slow Dissipation*, in this chapter) to analyze the equations of a gyroscopic system

$$\ddot{x} + (HG_0 + G_1)\dot{x} + Nx = 0,$$

in the deterministic case. The notation coincides with that introduced above. Having denoted $\varepsilon = 1/H$, we obtain

$$\varepsilon \ddot{x} + (G_0 + \varepsilon G_1)\dot{x} + \varepsilon N x = 0. \tag{7.98}$$

It is a commonly held view that equations (7.98) may be replaced by the corresponding precessional equations

$$(G_0 + \varepsilon G_1)\dot{x} + \varepsilon N x = 0. \tag{7.99}$$

Note that the dimension of (7.99) is half the dimension of (7.98). We shall apply the results of the theory of integral manifolds to prove the possibility of such a replacement. With that aim in view, we transform Equation (7.99) into the first order system

$$\dot{x} = y, \ \varepsilon \dot{y} = -(G_0 + \varepsilon G_1)y - \varepsilon N x. \tag{7.100}$$

In terms given in the preceding subsection, we have the equation $g(t, x, y, 0) = 0$ in the form $G_0 y = 0$. Hence, $y = h_0(t, x) = 0$, and the flow on the integral manifold is described by an equation

$$y = h(x, \varepsilon). \tag{7.101}$$

The function $h(x, \varepsilon)$ may be found as an asymptotic series

$$h(x, \varepsilon) = \sum_{i \geq 1} \varepsilon^i h_i(x) \tag{7.102}$$

from the equation

$$\varepsilon \frac{\partial h(x, \varepsilon)}{\partial x} = -(G_0 + \varepsilon G_1)h(x, \varepsilon) - \varepsilon N x. \tag{7.103}$$

Now, the usual technique of asymptotic analysis is applied. The expansion (7.102) is put into (7.103). Having equated the coefficients of powers of the small parameter ε, we compute the approximate solution of (7.103) in the form

$$h(x, \varepsilon) = -(G_0 + \varepsilon G_1)^{-1} \varepsilon N x + O(\varepsilon^2).$$

Thus, Equation (7.100) turns into

$$\dot{x} = -(G_0 + \varepsilon G_1)^{-1} \varepsilon N x + O(\varepsilon^3). \tag{7.104}$$

We compare equations (7.99) and (7.104). Evidently, they coincide to the accuracy of $O(\varepsilon^3)$. Consequently, the solutions of the system (7.100) and the solutions of the precessional equations (7.99) differ in the rapidly vanishing terms only, which correspond to the so-called nutational oscillations in the gyroscopic system. So it is quite correct to examine the precessional equation instead of the full equations of the gyroscopic system in the deterministic case.

Notice that the dimension of the slow integral manifold coincides with the dimension of vector of slow variables.

7.5.4 Optimal filtering in the precessional equations of gyroscopic systems

Let us now examine optimal filtering in gyroscopic systems described by the precessional equations.

We do not discuss here the physical aspects of obtaining the precessional equations. We remark only that such equations may be derived by neglecting the second derivative terms in (7.86). Consider precessional equations corresponding to (7.86) in the form

$$\dot{x} = -(G_0 + \varepsilon G_1)^{-1}\varepsilon N x + \varepsilon(G_0 + \varepsilon G_1)^{-1} B\dot{w}.$$

Denote the correlation matrix of the vector $x(t)$ by $\Phi(t)$. Then, according to (7.94), this matrix must satisfy the equation

$$\begin{aligned}(G_0 + \varepsilon G_1)\dot{\Phi} &= -\varepsilon N\Phi - \varepsilon(G_0 + \varepsilon G_1)\Phi N^T((G_0 + \varepsilon G_1)^{-1})^T \\ &- (G_0 + \varepsilon G_1)\Phi C^T R^1 C\Phi + \varepsilon^2 BQB^T((G_0 + \varepsilon G_1)^T)^{-1}. \qquad (7.105)\end{aligned}$$

Notice that at $\varepsilon = 0$ equation (7.105) has much in common with equation (7.95). Still this similarity is not sufficient to consider the precessional equations (7.86) to be acceptable as the basis for Kalman filtering.

We examine this topic in detail. System (7.95)–(7.97) has a stable integral manifold of slow motions [43]. The flow along this manifold is governed by the regularly (not singularly) perturbed equations of this system. At first sight only equation (7.95) is regular, and (7.105), being quite similar to it, may replace the full system (7.95)–(7.97). But, in fact, there are more regular equations in the system (7.95)–(7.97). We require that the matrix $G_0(t)$ has no zero eigenvalues for all $t \in R$. But the linear operator

$$LY = YG - GY$$

has a nontrivial kernel, since differences $(\lambda_i(t) - \lambda_j(t))$, $i, j = 1, \ldots, n$, form its spectrum. That is why there are many regular scalar equations in (7.97), since this operator has many zero eigenvalues. Thus, the dimension of the slow integral manifold of (7.95)–(7.97) is greater than the dimension of the matrix $\Phi(t)$, and the use of equation (7.92) for filtering can give unacceptable results.

This situation has much in common with that in the gyroscopic systems with a degenerate matrix of gyroscopic forces, where one should use the so-called "full" precessional equations to obtain acceptable results instead of the system given by the traditional precessional equations.

An additional advantage of the approach used here is that it allows us to consider equation (7.92), and regular equations from (7.97), instead of the full system (7.95)–(7.97).

Next we consider one example illustrating this result: The plane gyroscopic pendulum.

The gyroscopic pendulum is the simplest apparatus for indicating the proper vertical line direction in a moving ship or aeroplane.

Consider the equations of the plane gyroscopic pendulum with the horizontal axis of a gimbal. This pendulum is provided with a gyroscope which can turn near the axis of its housing. The turning of the gyroscope housing is limited by a spring. We investigate the movement of a plane gyroscopic pendulum under the rolling of a ship. Assume that the system is supplied with an apparatus for radial correction.

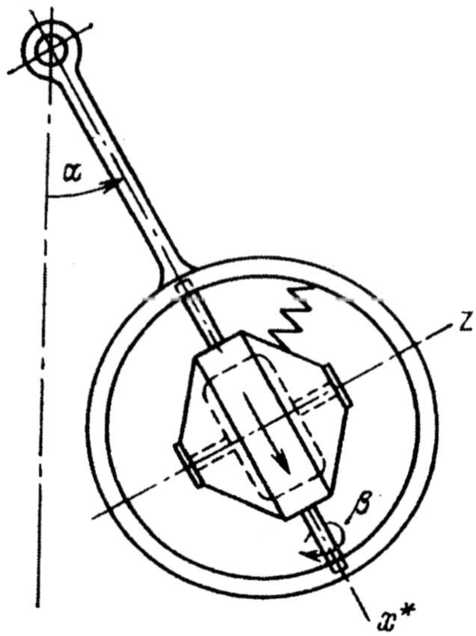

Figure 7.1. *The plane gyroscopic pendulum*

The latter imposes the moment proportional to the rotation angle of the gyroscope housing round the axis of the pendulum oscillation. Then the equations of motion of the plane gyroscopic pendulum are of the form

$$
\begin{aligned}
I_1\ddot{\alpha} + H\dot{\beta} + lp\alpha + M\beta + n\dot{\alpha} + b\dot{w} &= 0, \\
I_2\ddot{\beta} - H\dot{\alpha} + E\dot{\beta} + \kappa\beta &= 0.
\end{aligned} \tag{7.106}
$$

Here α is the angle of the pendulum rotation around its axis; β is the angle of gyroscope rotation around its housing axis; I_1 and I_2 are the corresponding moments of inertia; H is a moment of momentum of the gyroscope; lp is the static moment of the pendulum; M is the steepness of the moment of the radial correction; κ is the rigidity of the spring connecting the gyroscope housing with the pendulum; E and n are the coefficients of the viscous friction; $\dot{w}$ is a stationary random process corresponding to the angle of roll of the ship. Let $\dot{w}$ be a Gaussian white noise process with zero mean value and correlation function $q\delta(t-s)$.

Let the variable $z = \beta + \dot{\xi}$ be observed. At first, we consider the precessional equations for (7.106) neglecting the inertial terms $I_1\ddot{\alpha}$ and $I_2\ddot{\beta}$:

$$
\begin{aligned}
H\dot{\beta} + lp\alpha + n\dot{\alpha} + M\beta + b\dot{w} &= 0, \\
-H\dot{\alpha} + E\dot{\beta} + k\beta &= 0.
\end{aligned} \tag{7.107}
$$

Having divided both parts of the equations (7.107) by H and set $1/H = \varepsilon$,

$(\alpha\, \beta)^T = \omega$ we obtain:

$$\dot{\omega} = \varepsilon \begin{pmatrix} -\varepsilon Elp & -\varepsilon EM + \kappa \\ -lp & -M - \varepsilon n\kappa \end{pmatrix} \omega - \varepsilon \begin{pmatrix} \varepsilon Eb \\ b \end{pmatrix} + O(\varepsilon^3).$$

Then the equations of the Kalman filter for the correlation matrix P of the errors in the angles take the form

$$\dot{P} = \varepsilon \begin{pmatrix} \varepsilon Elp & -\varepsilon EM + \kappa \\ -lp & -M - \varepsilon n\kappa \end{pmatrix} P + \varepsilon P \begin{pmatrix} -\varepsilon Elp & -lp \\ -\varepsilon EM + \kappa & -M - \varepsilon n\kappa \end{pmatrix}$$

$$- P^T SP + \varepsilon^2 q \begin{pmatrix} \varepsilon^2 E^2 b^2 & b^2 E \\ \varepsilon E b^2 & b^2 \end{pmatrix} + O(\varepsilon^3), \tag{7.108}$$

where

$$S = \begin{pmatrix} 0 & 0 \\ 0 & 1/r \end{pmatrix}.$$

We seek a solution of (7.108) as a series:

$$P(\varepsilon) = D_0 + \varepsilon D_1 + O(\varepsilon^2).$$

From (7.108) we obtain that $D_0 = 0$, and D_1 satisfies the equation

$$\dot{D}_1 = \begin{pmatrix} 0 & \kappa \\ -lp & -M \end{pmatrix} D_1 + D_1 \begin{pmatrix} 0 & -lp \\ \kappa & -M \end{pmatrix} - D_1 S D_1 + \begin{pmatrix} 0 & 0 \\ 0 & qb^2 \end{pmatrix}.$$

It should be noted, that this mechanical system (plane gyroscopic pendulum) was examined in [29] by means of the precessional theory of gyroscopes, provided that $n = E = 0$. Under such assumptions, Equation (7.108) does not contain $O(\varepsilon^3)$ terms and, in coordinate form, is as follows:

$$\begin{aligned} \dot{d}_1 &= 2\frac{\kappa}{H} d_2 - \frac{d_2^2}{r}, \\ \dot{d}_2 &= -\frac{lp}{H} d_1 - \frac{M}{H} d_2 + \frac{\kappa}{H} d_3 - \frac{d_2 d_3}{r}, \\ \dot{d}_3 &= -2\frac{lp}{H} d_2 - 2\frac{M}{H} d_3 - \frac{d_3^2}{r} + \frac{qb^2}{H^2}. \end{aligned}$$

Here d_1, d_2 and d_3 denote the elements of the symmetric correlation matrix D. But we cannot compare these equations with those obtained on the basis of the theory of integral manifolds, since, for $n = E = 0$, the equations of motion of the plane gyroscopic pendulum may have no attracting integral manifold.

Next we consider the full equations (7.106) in the form

$$\begin{aligned} \varepsilon\ddot{\alpha} + \frac{\beta}{I_1} + \varepsilon\frac{n}{I_1}\dot{\alpha} + \varepsilon\frac{lp}{I_1}\alpha + \varepsilon\frac{M}{I_1}\beta + \varepsilon\frac{b}{I_1}\dot{w} &= 0, \\ \varepsilon\ddot{\beta} - \frac{1}{I_2}\alpha + \varepsilon\frac{E}{I_2}\beta + \varepsilon\frac{\kappa}{I_2}\beta &= 0, \end{aligned}$$

or, in the more convenient form,

$$\varepsilon\begin{pmatrix}\ddot\alpha\\ \ddot\beta\end{pmatrix}+\begin{pmatrix}0 & 1/I_1\\ -1/I_2 & 0\end{pmatrix}\begin{pmatrix}\dot\alpha\\ \dot\beta\end{pmatrix}+\varepsilon\begin{pmatrix}n/I_1 & 0\\ 0 & E/I_2\end{pmatrix}\begin{pmatrix}\alpha\\ \beta\end{pmatrix}$$

$$+\varepsilon\begin{pmatrix}lp/I_1 & M/I_1\\ 0 & \kappa/I_2\end{pmatrix}\begin{pmatrix}\alpha\\ \beta\end{pmatrix}=-\varepsilon\begin{pmatrix}b/I_1\\ 0\end{pmatrix}\dot w. \tag{7.109}$$

We use the following notation:

$$G_0=\begin{pmatrix}0 & 1/I_1\\ -1/I_2 & 0\end{pmatrix},\quad G_1=\begin{pmatrix}n/I_1 & 0\\ 0 & E/I_2\end{pmatrix},$$

$$N=\begin{pmatrix}lp/I_1 & M/I_1\\ 0 & \kappa/I_2\end{pmatrix},\quad B_2=\begin{pmatrix}b/I_1\\ 0\end{pmatrix}.$$

Then the equations for the Kalman filter, according to (7.109), may be written as follows:

$$\dot P_1=P_2^T+P_2-P_1SP_2, \tag{7.110}$$

$$\varepsilon\dot P_2=\varepsilon P_3-\varepsilon P_1N^T-P_2(G_0+\varepsilon G_1)^T-\varepsilon P_1SP_2, \tag{7.111}$$

$$\varepsilon\dot P_3=-\varepsilon(NP_2+P_2^TN^T)-P_3(G_0+\varepsilon G_1)^T$$
$$-(G_0+\varepsilon G_1)P_3-\varepsilon P_2^TSP_2+\varepsilon B_2QB_2^T. \tag{7.112}$$

The matrices $B_2QB_2^T$ and $CR^{-1}C=S$ can be computed easily in the form

$$B_2QB_2^T=\begin{pmatrix}b^2q/I_1^2 & 0\\ 0 & 0\end{pmatrix},\quad S=\begin{pmatrix}0 & 0\\ 0 & 1/r\end{pmatrix}.$$

We designate the elements of the 2×2 matrices P_1, P_2, P_3 as follows:

$$P_1=\begin{pmatrix}p_1 & p_2\\ p_2 & p_3\end{pmatrix},\ P_2=\begin{pmatrix}p_4 & p_7\\ p_5 & p_8\end{pmatrix},\ P_3=\begin{pmatrix}p_6 & p_9\\ p_9 & p_{10}\end{pmatrix}.$$

Then equation (7.112) may be transformed into a system of three scalar equations:

$$\varepsilon\dot p_6=-\frac{2}{I_1}p_9-2\varepsilon\left(\frac{lp}{I_1}p_4+\frac{M}{I_1}\right)p_5-2\varepsilon\frac{n}{I_1}p_6+\varepsilon\frac{b^2}{I_1^2}q-\varepsilon\frac{p_5^2}{r},$$

$$\varepsilon\dot p_9=\frac{p_6}{I_2}-\frac{p_{10}}{I_1}-\varepsilon\left(\frac{\kappa}{I_2}p_5+\frac{M}{I_1}p_8+\frac{lp}{I_1}p_7+\left(\frac{E}{I_2}+\frac{n}{I_1}\right)p_9\right)-\varepsilon\frac{p_5p_8}{r}, \tag{7.113}$$

$$\varepsilon\dot p_{10}=2\frac{1}{I_2}p_9-2\varepsilon\frac{\kappa}{I_2}p_8-2\varepsilon\frac{E}{I_2}p_{10}-\varepsilon\frac{p_8^2}{r}.$$

In (7.113) we introduce the change of variables

$$(p_6\ p_9\ p_{10})^T=T(\omega_6\ \omega_9\ \omega_{10})^T, \tag{7.114}$$

where T is the matrix

$$T = \begin{pmatrix} I_2 & I_2 & I_2 \\ 0 & \sqrt{I_1 I_2} & -\sqrt{I_1 I_2} \\ I_1 & -I_1 & -I_1 \end{pmatrix},$$

with the inverse

$$T^{-1} = \begin{pmatrix} \frac{1}{2I_2} & 0 & \frac{1}{2I_1} \\ \frac{1}{4I_2} & \frac{1}{2\sqrt{I_1 I_2}} & -\frac{1}{4I_1} \\ \frac{1}{4I_2} & -\frac{1}{2\sqrt{I_1 I_2}} & -\frac{1}{4I_1} \end{pmatrix}.$$

This matrix T transforms the matrix of linear terms of the system (7.113) to the skew-symmetric matrix

$$\begin{pmatrix} 0 & 0 & 0 \\ 0 & 0 & \frac{2}{\sqrt{I_1 I_2}} \\ 0 & -\frac{2}{\sqrt{I_1 I_2}} & 0 \end{pmatrix}.$$

It may be easily checked that, after this transformation of variables, system (7.113) becomes

$$\dot{\omega}_6 = -\left(\frac{n}{I_1} + \frac{E}{I_2}\right)\omega_6 + \left(-\frac{n}{I_1} - \frac{E}{I_2}\right)\omega_9 + \left(-\frac{n}{I_1} + \frac{E}{I_2}\right)\omega_{10}$$
$$- \left(\frac{lp}{I_1 I_2}p_4 + \frac{M}{I_1 I_2}p_5\right) - \frac{\kappa}{I_1 I_2}p_8 - \frac{p_5^2}{2I_2 r} - \frac{p_8^2}{2I_1 r} + \frac{b^2 q}{I_1^2 I_2}, \tag{7.115}$$

$$\varepsilon\dot{\omega}_9 = \varepsilon\left(-\frac{n}{2I_1} + \frac{E}{2I_2}\right)\omega_6 - \varepsilon\left(\frac{n}{I_1} + \frac{E}{I_2}\right)\omega_{10} + \frac{2}{\sqrt{I_1 I_2}}\omega_{10} - \varepsilon\frac{1}{2I_2}\left(\frac{lp}{I_1}p_4 + \frac{M}{I_1}p_5\right)$$
$$- \frac{\varepsilon}{\sqrt{I_1 I_2}}\left(\frac{\kappa}{I_2}p_5 + \frac{M}{I_1}p_8 + \frac{lp}{2I_1}p_7 + \frac{1}{2r}p_5 p_8\right) \tag{7.116}$$
$$+ \varepsilon\frac{\kappa}{2I_1 I_2}p_8 - \varepsilon\frac{1}{4I_2 r}p_5^2 + \varepsilon\frac{1}{4I_1 r}p_8^2 + \varepsilon\frac{b^2}{4I_1^2 I_2}q,$$

$$\varepsilon\dot{\omega}_{10} = -\varepsilon\left(-\frac{n}{2I_1} + \frac{E}{2I_2}\right)\omega_6 - \varepsilon\left(\frac{n}{I_1} + \frac{E}{I_2}\right)\omega_9 - \frac{2}{\sqrt{I_1 I_2}}\omega_9 + \varepsilon\frac{1}{2I_2}\left(\frac{lp}{I_1}p_4 + \frac{M}{I_1}p_5\right)$$
$$- \frac{\varepsilon}{\sqrt{I_1 I_2}}\left(\frac{\kappa}{I_2}p_5 + \frac{M}{I_1}p_8 + \frac{lp}{2I_1}p_7 + \frac{1}{r}p_5 p_8\right) \tag{7.117}$$
$$+ \varepsilon\frac{\kappa}{2I_1 I_2}p_8 - \varepsilon\frac{1}{4I_2 r}p_5^2 + \varepsilon\frac{1}{4I_1 r}p_8^2 + \varepsilon\frac{b^2}{4I_1^2 I_2}q.$$

Now we consider system (7.110), (7.111), (7.115)–(7.117). There are four regular equations, hence four slow variables. This system has a four-dimensional slow integral manifold, and it is stable. We search for this manifold as an asymptotic series

$$P_2 = P_2^{(0)}(P_1, \omega_6) + \varepsilon P_2^{(1)}(P_1, \omega_6) + O(\varepsilon^2),$$
$$\omega_j = \omega_j^{(0)}(P_1, \omega_6) + \varepsilon\omega_j^{(1)}(P_1, \omega_6) + O(\varepsilon^2), j = 9, 10. \tag{7.118}$$

We substitute these expansions into the singularly perturbed equations (7.110), (7.111), (7.115)–(7.117), and denote by $\partial P_2/\partial P_1 \dot{P}_1$ the matrix

$$\frac{\partial P_2}{\partial P_1}\dot{P}_1 = \begin{pmatrix} \sum_{i=1}^3 \frac{\partial p_4}{\partial p_i}\dot{p}_i & \sum_{i=1}^3 \frac{\partial p_5}{\partial p_i}\dot{p}_i \\ \sum_{i=1}^3 \frac{\partial p_7}{\partial p_i}\dot{p}_i & \sum_{i=1}^3 \frac{\partial p_8}{\partial p_i}\dot{p}_i \end{pmatrix},$$

where the notation $\partial \omega_k/\partial P_1 \dot{P}_1$, $k = 1, 2$, is interpreted as

$$\frac{\partial \omega_k}{\partial P_1}\dot{P}_1 = \sum_{i=1}^{3} \frac{\partial \omega_k}{\partial p_i}\dot{p}_i.$$

Note that, after the change of variables (7.114), the matrix P_3 becomes

$$P_3 = \begin{pmatrix} I_2 & 0 \\ 0 & I_1 \end{pmatrix}\omega_6 + \begin{pmatrix} I_2 & \sqrt{I_1 I_2} \\ \sqrt{I_1 I_2} & -I_1 \end{pmatrix}\omega_9 + \begin{pmatrix} I_2 & -\sqrt{I_1 I_2} \\ -\sqrt{I_1 I_2} & -I_1 \end{pmatrix}\omega_{10}.$$

Hence the equations from which the slow integral manifold (7.118) is calculated are:

$$\varepsilon \frac{\partial P_2}{\partial P_1}\dot{P}_1 + \varepsilon \frac{\partial P_2}{\partial \omega_6}\dot{\omega}_6 = \varepsilon P_3 - \varepsilon P_1 N^T - P_2 G_0^T - \varepsilon P_2 G_1^T - \varepsilon P_1 S P_2, \tag{7.119}$$

$$\varepsilon \frac{\partial \omega_9}{\partial P_1}\dot{P}_1 + \varepsilon \frac{\partial \omega_9}{\partial \omega_6}\dot{\omega}_6 = \frac{2}{\sqrt{I_1 I_2}}\omega_{10} + O(\varepsilon), \tag{7.120}$$

$$\varepsilon \frac{\partial \omega_{10}}{\partial P_1}\dot{P}_1 + \varepsilon \frac{\partial \omega_{10}}{\partial \omega_6}\dot{\omega}_6 = -\frac{2}{\sqrt{I_1 I_2}}\omega_9 + O(\varepsilon). \tag{7.121}$$

Here the expressions for $\dot{P}_1$ and $\dot{\omega}_6$ should be substituted into (7.119)–(7.121) from (7.109) and(7.114). From (7.116) and (7.117) it immediately follows that

$$\omega_9^{(0)} = \omega_{10}^{(0)} = 0.$$

The terms $\omega_9^{(1)}$ and $\omega_{10}^{(1)}$ satisfy the equations

$$\begin{aligned} \left(-\frac{n}{2I_1} + \frac{E}{2I_2}\right)\omega_6 + \frac{2}{\sqrt{I_1 I_2}}\omega_{10}^{(1)} + \frac{b^2}{4I_1^2 I_2}q = 0, \\ \left(-\frac{n}{2I_1} + \frac{E}{2I_2}\right)\omega_6 - \frac{2}{\sqrt{I_1 I_2}}\omega_9^{(1)} + \frac{b^2}{4I_1^2 I_2}q = 0. \end{aligned} \tag{7.122}$$

Now, we turn to equation (7.119). Evidently, $P_2^{(0)} = 0$ and

$$P_2^{(1)} = \begin{pmatrix} p_2\kappa & I_1 I_2 \omega_6 - p_1 lp - p_2 M \\ p_3\kappa - I_1 I_2 \omega_6 & -p_2 lp - p_3 M \end{pmatrix}.$$

The next approximation $P_2{}^{(2)}$ is

$$P_2{}^{(2)} = \begin{pmatrix} p_4^{(2)} & p_7^{(2)} \\ p_5^{(2)} & p_8^{(2)} \end{pmatrix},$$

where

$$
\begin{aligned}
p_4^{(2)} &= -0.5(nI_2^2 + EI_1I_2 + I_1I_2^2p_1)\omega_6 - E(p_1lp + p_2M) \\
&\quad + I_2b^2\frac{q}{2I_1} + \frac{1}{r}(I_2lpp_2^2 + I_2Mp_2p_3 + I_2\kappa p_1p_3), \\
p_7^{(2)} &= -p_2\kappa n + \frac{I_1^2I_2}{r}p_2\omega_6, \\
p_5^{(2)} &= (p_2lp + p_3M)\left(\frac{I_2(p_1+p_3)}{r} - E\right), \\
p_8^{(2)} &= I_1^2I_2\left(-\frac{3}{2}\frac{n}{I_1}\omega_6 - \frac{E}{2I_2}\omega_6\right) + \frac{3}{4}b^2q \\
&\quad - p_3\kappa n + \frac{I_1}{r}\left(\kappa p_3^2 + p_2^2lp + p_2p_3M\right).
\end{aligned}
\tag{7.123}
$$

The approximations derived above permit us to follow how equation (7.108), derived on the basis of precessional equations, differs from the equations which describe the flow along the attracting integral manifold of the system (7.109)–(7.111).

Consider the system describing the flow on the slow integral manifold of (7.109)–(7.111). According to the results of the theory of integral manifolds, this flow is determined by the regular equations of this system, namely, by the equations (7.109) and (7.114). To derive the equations of this flow one should substitute the asymptotic expansions (7.117) into the right-hand sides of these equations:

$$
\begin{aligned}
\dot{p}_1 &= 2\frac{\kappa}{H}p_2 - \frac{p_2^2}{r} + \frac{I_2b^2q}{I_1H^2} \\
&\quad - (nI_2^2 + EI_1I_2^2p_1)\omega_6/H^2 + O(1/H^3), \\
\dot{p}_2 &= -\frac{lp}{H}p_1 - \frac{M}{H}p_2 + \frac{\kappa}{H}p_3 - \frac{p_2p_3}{r} + O(1/H^3), \\
\dot{p}_3 &= -2\frac{lp}{H}p_2 - 2\frac{M}{H}p_3 - \frac{d_3^2}{r} + \frac{qb^2}{H^2} \\
&\quad + \frac{2}{H^2}\left(I_1^2I_2\left(-\frac{3n}{2I_1}\omega_6 - \frac{E}{2I_2}\omega_6 + I_1I_2n\omega_6\right) + \frac{3}{4}b^2q\right) + O(1/H^3).
\end{aligned}
$$

The calculations may be carried to any desired accuracy. We compare the results obtained in this Example for the full equations of motion, and those got on the basis of the precessional equations. The 4×4 correlation matrix $P(t, \varepsilon)$, corresponding to the full equations, is calculated from (7.110)–(7.112). This system has a 4 dimensional stable integral manifold of slow motions. So, every solution of this system, starting in the vicinity of this manifold, tends to it as $t \to \infty$. Consequently, to investigate the trajectories of (7.110)–(7.112) it is enough to follow the trajectories lying on the manifold. Equations (7.110) and (7.115) describe this motion. We compared the solutions of (7.110), (7.115) and of (7.108) numerically. Both systems were solved numerically with the same initial conditions (Dormand-Prince method), and the solutions obtained differed in the $O(\varepsilon^3)$ terms. By this is meant that the use of the 4 dimensional stable integral manifold of slow motions gives an accurate account of the behaviour of the original system, whereas the use of precessional equations, instead of the original ones, for calculating the filtering error may lead to an intolerable error if the motion is performed under the action of random forces of Gaussian white noise type.

7.6 Acknowledgements

The author is grateful to Michael Mortell and Robert O'Malley for helpful discussions. This work was supported in part the Boole Centre for Research in Informatics, University College Cork, by the Russian Foundation for Basic Research under grant no. 04–01–96515, and by the Presidium of the Russian Academy of Science Program no. 19.

Bibliography

[1] M. Arato, A. N. Kolmogorov, and J. G. Sinai, *On the estimation of the parameters of complex stationary Gaussian Markov process*, Dokl. Akad. Nauk SSSR, 146(4) (1962), pp. 747–750.

[2] Ja. S. Baris, *The integral variety of an irregularly perturbed differential system* (in Russian, MR 38:3535), Ukrain. Mat. Ž., 20 (1968), pp. 439–448.

[3] Ja. S. Baris and V. I. Fodčuk, *Investigation of the bounded solutions of nonlinear irregularly perturbed systems by the method of integral manifolds* (in Russian, MR 41:596), Ukrain. Mat. Ž., 22 (1970) pp. 3–11.

[4] E. Benoit, ed., *Dynamic Bifurcations*, Springer Lect. Notes in Math., 1493, 1991.

[5] S. V. Bogatyrev and V. A. Sobolev, *Separation of rapid and slow motions in problems of the dynamics of systems of rigid bodies and gyroscopes*, J. Appl. Math. Mech., 52(1) (1988), pp. 34-41.

[6] N. Fenichel, *Geometric singular perturbation theory for ordinary differential equations*, J. Diff. Eq., 31 (1979), pp. 53–98.

[7] F. Ghorbel and M. W. Spong, *Integral manifolds of singularly perturbed systems with application to rigid-link flexible-joint multibody systems*, Int. J. of Non-Linear Mechanics, 35 (2000), pp. 133-155.

[8] V. M. Gol'dshtein and V. A. Sobolev, *Integral manifolds in chemical kinetics and combustion*, in Singularity Theory and Some Problems of Functional Analysis, AMS Translations, Series 2, 153 (1992), pp. 73–92.

[9] J. Guckenheimer, *Towards a global theory of singularly perturbed dynamical systems*, Progr. Nonlinear Differential Equations Appl., 19 (1996) pp. 213–225.

[10] Z. Gu, N. N. Nefedov, and R. E. O'Malley, *On singular singularly perturbed initial values problems*, SIAM J. Appl. Math., 49 (1989), pp. 1–25.

[11] L. V. Kalachev and R. E. O'Malley, *The regularization of linear differential-algebraic equations*, SIAM J. Math. Anal., 27(1) (1996), pp. 258–273.

[12] R. E. KALMAN AND R. S. BUCY, *New results in linear filtering and prediction theory* J. Basic Eng., 83D (1961), pp. 95–108.

[13] T. J. KAPER *An introduction to geometric methods and dynamical systems theory for singular perturbation problems*, in Analyzing Multiscale Phenomena using Singular Perturbation Methods, Proc. Sympos. Appl. Math., 56, R. E. O'Malley and J. Cronin, eds., 1999, pp. 85–131.

[14] H. W. KNOBLOCH AND B. AULBACK, *Singular perturbations and integral manifolds*, J. Math. Phys. Sci., 18(5) (1984), pp. 415–424.

[15] P. V. Kokotović, K. H. Khalil, and J. O'Reilly, *Singular Perturbation Methods in Control: Analysis and Design*, Academic Press, Inc., London, 1986.

[16] L. I. KONONENKO AND V. A. SOBOLEV, *Asymptotic expansion of slow integral manifolds*, Sib. Math. J., 35 (1994), pp. 1119–1132.

[17] M. KRUPA AND P. SZMOLYAN, *Extending geometric singular perturbation theory to nonhyperbolic points—fold and canard points in two dimensions*, SIAM J. Math. Anal., 33(2) (2001), pp. 286–314.

[18] K. MAGNUS, *Kreisel. Theorie und Anwendungen.*, Springer-Verlag, Berlin-New York, 1971 (in German).

[19] D. R. MERKIN, *Gyroscopic Systems*, Nauka, Moscow, 1974 (in Russian).

[20] YU. A. MITROPOL'SKII AND O. B. LYKOVA, *Integral Manifolds in Nonlinear Mechanics* (in Russian, MR 51:1025), Nauka, Moscow, 1975.

[21] E. F. MISHCHENKO AND N. KH. ROZOV, *Differential Equations with Small Parameters and Relaxation Oscillations,* Plenum Press, New York, 1980.

[22] J. A. MURDOCK, *Perturbations. Theory and Methods*, John Wiley & Sons Inc., 1991.

[23] D. S. NAIDU, *Singular perturbations and time scales in control theory and applications: an overview.*, Dyn. Contin. Discrete Impuls. Syst. Ser. B Appl. Algorithms, 9(2) (2002), pp. 233–278.

[24] K. NIPP, *An algorithmic approach for solving singularly perturbed initial value problems*, Dynamics Reported, 1 (1988), pp. 173–263.

[25] R. E. O'MALLEY, *Singular Perturbation Methods for Ordinary Differential Equations*, Appl. Math. Sci., 89, Springer–Verlag, New-York, 1991.

[26] R. E. O'MALLEY AND A. JAMESON, *Singular perturbations and singular arcs, I, II*, IEEE Trans. Autom. Control, AC-20 (1975), 218–226; AC-22 (1977), pp. 328–337.

[27] R. E. O'MALLEY AND L. V. KALACHEV *Regularization of nonlinear differential-algebraic equations*, SIAM J. Math. Anal., 25(2) (1994), pp. 615–629.

[28] V. A. Pliss, *A reduction principle in the theory of stability of motion*, (in Russian, MR 32:7861) Izv. Akad. Nauk SSSR Ser. Mat., 28 (1964), pp. 1297–1324.

[29] Ya. N. Roitenberg, *Automatic Control* (in Russian, MR 93m:93001), Nauka, Moscow, 1992.

[30] K. R. Schneider and V. A. Sobolev, *Existence and approximation of slow integral manifolds in some degenerate cases*, Weierstraß–Institut für Angewandte Analysis und Stochastik, Preprint 782, Berlin, 2002.

[31] K. R. Schneider and T. Wilhelm, *Model reduction by extended quasi-steady state assumption*, J. Math. Biology, 40 (2000), pp. 443–450.

[32] V. A. Sobolev, *Asymptotic behavior of the integral manifolds of a class of systems of differential equations with a small parameter multiplying the derivatives* (in Russian, MR 83d:34090), Approximate methods for investigating differential equations and their applications, Kuĭbyshev. Gos. Univ., 4 (1978), pp. 85–90,.

[33] ——, *Geometrical theory of singularly perturbed control systems*, Proc. 11th Congress of IFAC, Tallinn, 6 (1990), pp. 163–168.

[34] ——, *Integral manifolds and decomposition of singularly perturbed systems*, System and Control Lett., 5 (1984), pp. 169–179.

[35] ——, *Integral manifolds of singularly perturbed systems in one critical case* (in Russian, MR 58:6522), Differential equations, Kuĭbyshev. Gos. Univ., (1976), pp. 63–71.

[36] ——, *Investigation of differential equations with small parameters multiplying derivatives by the method of integral manifolds* (in Russian, MR 80k:34075), Proceedings of a Seminar on Differential Equations, Kuĭbyshev. Gos. Univ., 3 (1977), pp. 78–85.

[37] ——, *Singular perturbations in linearly quadratic optimal control problems*, Autom. Rem. Control, 52 (1991), pp. 180–189.

[38] V. A. Sobolev and V. V. Strygin, *Permissibility of changing over to precession equations of gyroscopic systems*, Mechanics of Solids, 5 (1978), pp. 7–13.

[39] M. W. Spong, K. Khorasani, and P. V. Kokotovic, *An integral manifold approach to feedback control of flexible joint robots*, IEEE Journal of Robotics and Automation, 3(4) (1987), pp. 291–301.

[40] V. V. Strygin and V. A. Sobolev, *Asymptotic methods in the problem of stabilization of rotating bodies by using passive dampers*, Mechanics of Solids, 5 (1977), pp. 19–25.

[41] ——, *Effect of geometric and kinetic parameters and energy dissipation on orientation stability of satellites with double spin*, Cosmic Research, 14(3) (1976), pp. 331–335.

[42] ——, *Reduction principles in the theory of differential equations with a manifold of stationary states* (in Russian, MR 80k:34069), Proceedings of a Seminar on Differential Equations, Kuĭbyshev. Gos. Univ., 3 (1977), pp. 92–103.

[43] ——, *Separation of Motions by the Integral Manifolds Method* (in Russian, MR 89k:34071), Nauka, Moscow, 1988.

[44] A. N. Tikhonov, *Systems of differential equations with small parameters multiplying the derivatives*, Matem. sb., 31(3) (1952), pp. 575–586.

[45] H. C. Tsengand and D. D. Siljak, *A learning scheme for dynamic neural networks: equilibrium manifold and connective stability*, Neural Networks, 8(6) (1995), pp. 853–864 .

[46] A. B. Vasil'eva and V. F. Butuzov, *Singularly Perturbed Equations in the Critical Case*, Tech. Report MRC–TSR 2039, University of Wisconsin, Madison, WI, 1980.

[47] A. B. Vasil'eva, V. F. Butuzov, and L. V. Kalachev, *The Boundary Function Method for Singular Perturbation Problems*, SIAM Studies in Appl. Math., 14, 1995.

[48] K.–K. D. Young, P. V. Kokotovic, and V. I. Utkin, *A singular perturbation analysis of high-gain feedback systems*, IEEE Trans. Automat. Control, AC-22(6) (1977), pp. 931–938.

[49] K. V. Zadiraka, *On a non-local integral manifold of a singularly perturbed differential system* (in Russian, MR 34:4633), Ukrain. Mat. Ž., 17(1) (1965), pp. 47–63.

[50] ——, *On the integral manifold of a system of differential equations containing a small parameter*, (in Russian, MR 19:858a), Dokl. Akad. Nauk SSSR, 115 (1957), pp. 646–649.

Chapter 8

Black Swans and Canards in Laser and Combustion Models

E. Shchepakina and V. Sobolev

The paper is devoted to the investigation of the relationship between slow integral manifolds of singularly perturbed differential equations and critical phenomena in chemical kinetics. We consider different problems using the techniques of canards and black swans. The language of singular perturbations seems to apply to all critical phenomena even in the most disparate chemical systems.

In a majority of papers devoted to canards the term "canard" is associated with periodic trajectories. In our work a canard is a trajectory of a singularly perturbed system of differential equations if it, at first, follows a stable integral manifold, and then an unstable one. In both cases the distances travelled are more than infinitesimally small. A canard may be considered as the result of gluing stable (attractive) and unstable (repelling) slow integral manifolds at one point of the breakdown surface, due to the availability of an additional scalar parameter in the differential system. If we take an additional function of a vector variable parameterizing the breakdown surface, we can glue the stable (attractive) and unstable (repelling) slow integral manifolds at all points of the breakdown curve at the same time. As a result we obtain the continuous stable/unstable (attractive/repelling) integral surface or black swan. Such surfaces are considered as a multidimensional analogue of the notion of a canard. It is possible to consider the gluing function as a special kind of partial feedback control. This guarantees the safety of chemical regimes, even with perturbations, during a chemical process.

We shall use canards as *separating solutions* corresponding to the critical regimes of chemical reactions. This approach was proposed for the first time in [16] and was then applied in [15, 57]. Later this approach was extended to black swans.

The main object of our consideration is the following singularly perturbed system

$$\begin{aligned}\dot{x} &= f(x, y, z, \varepsilon),\\ \dot{y} &= g(x, y, z, \varepsilon),\\ \varepsilon\dot{z} &= p(x, y, z, \alpha, \varepsilon),\end{aligned} \tag{8.1}$$

where ε is a small positive parameter, α is a scalar parameter, x and z are scalar variables, y is a vector of dimension n, and the dot refers to differentiation with respect to time t. Note that we detach the variable x for the following reason: it will be used as a new independent variable when the original variable t is excluded. For nonautonomous systems the variable t plays the role of the variable x and $f \equiv 1$ in this case. The case of a vector variable z will also be considered.

Recall that the slow surface S (or S_α) of the system (8.1) is the surface described by the equation

$$p(x, y, z, \alpha, 0) = 0. \tag{8.2}$$

Let $z = \phi(x, y, \alpha)$ be an isolated solution of equation (8.2). We call the subset S_α^s (S_α^u) of S, defined by

$$\frac{\partial p}{\partial z}(x, y, \phi(x, y), \alpha, 0) < 0, \quad (> 0),$$

the stable (unstable) subset of S_α.

The subset of S_α defined by

$$\frac{\partial p}{\partial z}(x, y, \phi(x, y), \alpha, 0) = 0$$

is called the *breakdown surface*. Its dimension is equal to $\dim y$.

In an ε–neighborhood of S_α^s (S_α^u) there exists a stable (unstable) slow integral manifold. The slow integral manifold is defined as a invariant surface of slow motions.

The availability of the additional scalar parameter α provides the possibility of gluing the stable and unstable integral manifolds at one point of the breakdown surface. The canard trajectory passes through this point.

It should be noted that in the early papers devoted to canards in the case $\dim y = 0$, the existence of a unique canard corresponding to a unique value of the parameter $\alpha = \alpha^*$ was stated (more precisely, the "canard" value of the parameter α^* exists on an interval of order $O(e^{-1/\varepsilon})$). This property is known as the *short life of canards*. But, in the case $\dim y = 1$, another picture is beginning to emerge. It was shown that a one-parameter family of canards exists [55]. If we take the parameter α as a function of y we can glue the stable and unstable integral manifolds along all points of the breakdown curve at the same time. This approach is obviously associated with Krasnosel'skii's method of functionalization of a parameter [23].

We consider several simple examples.

Example 1 ($\dim y = 0$).

As the simplest system with a canard we propose

$$\dot{x} = 1, \quad \varepsilon\dot{z} = 2xz + \alpha.$$

It is clear that for $\alpha = 0$, the trajectory $z = 0$ is a canard.
Example 2 (dim $y = 0$).
For plane systems

$$\dot{x} = f(x, z, \varepsilon),$$

$$\varepsilon\dot{z} = p(x, z, \alpha, \varepsilon),$$

the stable and unstable parts of the slow curve are separated by points at which $\partial p/\partial z = 0$. Such points are called irregular [33]. Usually, the investigation of such systems in the neighborhood of an irregular point is carried out on the assumption that the inequality

$$\left(\frac{\partial p}{\partial x}\right)^2 + \left(\frac{\partial p}{\partial z}\right)^2 > 0$$

holds at these points.

However, there is a class of problems where this condition is not fulfilled for some value of α. For example, in the system

$$\frac{dx}{dt} = 1, \qquad \varepsilon\frac{dz}{dt} = z^2 - x^2 + \alpha,$$

with $\alpha = \pm\varepsilon$, the lines $z = \pm x$ pass along the slow curve $z^2 - x^2 + \alpha = 0$ over an infinitely long distance, see Fig. 8.1 (a). Note that the canard is only the $z = x$ trajectory. In this example [16], the point $x = 0$, $z = 0$ is the point of self-intersection of the slow curve at $\alpha = 0$. Such problems were examined in [9, 15, 16]. The same systems appear in thermal explosion models in the case of autocatalytic reactions. In this case the canards are the natural mathematical objects which allow us to model critical phenomena and discover critical parameter values as asymptotic expansions involving powers of the small parameter ε.

As mentioned above, statements of the type "The life of a canard is very short" can often be found in papers. However, it is not difficult to give examples of canards living for centuries.

Example 3. Consider the system [22]

$$\dot{x} = z,$$

$$\varepsilon\dot{z} = x^2 + z^2 - \alpha^2.$$

The circle $(x + \varepsilon/2)^2 + z^2 = \alpha^2 - \varepsilon^2/4$ is a canard. The upper semicircle is unstable and the lower one is stable, see Fig. 8.1 (b). This canard exists provided $\alpha^2 > \varepsilon^2/4$.

Let us consider an extension of Example 3. The system

$$\dot{x} = z,$$

$$\varepsilon\dot{z} = bz^2 + f(x, \varepsilon),$$

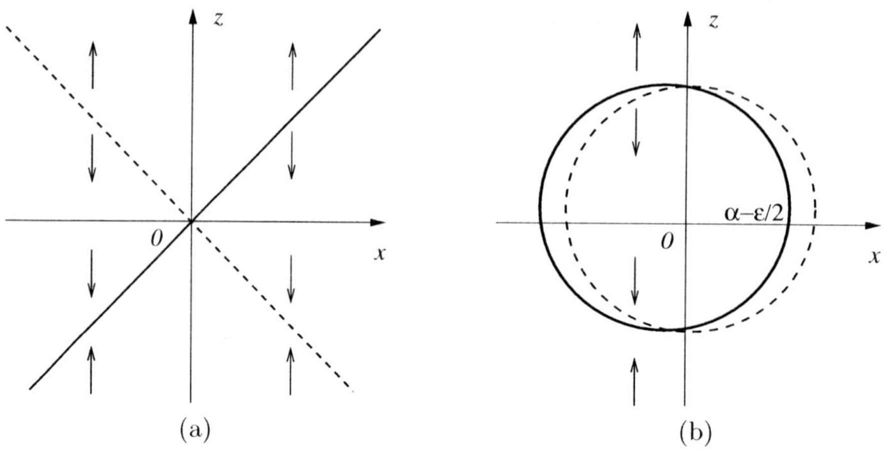

Figure 8.1.

has singular points at $z = 0$, $f(x_s, \varepsilon) = 0$. Using the Jacobian matrix

$$\begin{pmatrix} 0 & 1 \\ \frac{1}{\varepsilon} f'(x_s, \varepsilon) & 0 \end{pmatrix},$$

the singular point of the linearized system is a saddle (center) if $f'(x_s, \varepsilon) > 0$ ($f'(x_s, \varepsilon) < 0$). Here $f' = df/dx$.

For the original nonlinear system under consideration the singular points are of the same type. In the case of a saddle it is a known fact. As to a center, we can apply the Lyapunov theorem concerned with a center singular point due to the existence of an analytic first integral of the original system [28]. This integral can be obtained from the following equation

$$\frac{\varepsilon}{2}(z^2)' = bz^2 + f(x, \varepsilon). \tag{8.3}$$

The solution of (8.3) is given by

$$z^2 = Ce^{2bx/\varepsilon} + \frac{2}{\varepsilon}\int_0^x e^{2b(x-\tau)/\varepsilon} f(\tau, \varepsilon)d\tau.$$

Thus, the analytical first integral is

$$\left[z^2 - \frac{2}{\varepsilon}e^{2bx/\varepsilon}\int_0^x e^{-2b\tau/\varepsilon} f(\tau, \varepsilon)d\tau\right] e^{-2bx/\varepsilon} = C.$$

If the function $f(x, \varepsilon)$ is a polynomial

$$f(x, \varepsilon) = \sum_{n=0}^{N} p_n x^n,$$

then the particular solution of (8.3) satisfies the relation

$$z^2 = \sum_{n=0}^{N} q_n x^n.$$

By substituting these expressions and

$$\left(z^2\right)' = \phi(x, \varepsilon) = \sum_{n=1}^{N} n q_n x^{n-1} = \sum_{n=0}^{N-1} (n+1) q_{n+1} x^n$$

into (8.3), we get

$$\frac{\varepsilon}{2} \sum_{n=0}^{N-1} (n+1) q_{n+1} x^n = b \sum_{n=0}^{N} q_n x^n + \sum_{n=0}^{N} p_n x^n.$$

From this we have the following formulae

$$bq_N + p_N = 0,$$

$$\frac{\varepsilon}{2}(n+1) q_{n+1} = bq_n + p_n, \quad n = N-1, \ldots, 0,$$

so that

$$q_N = -p_N/b,$$

$$q_n = \left(-p_n + \frac{\varepsilon}{2}(n+1) q_{n+1}\right)/b, \quad n = N-1, \ldots, 0.$$

The invariant manifold of (8.3) is defined by $z^2 = \phi(x, \varepsilon)$ and it is attractive when $bz < 0$ (repelling when $bz > 0$).

Example 4. For $N = 1$ in the above example, we have

$$f(x, \varepsilon) = p_1 x + p_0, \quad p_0 = \alpha, \quad p_1 = a,$$

$$q_1 = -a/b, \quad q_0 = -\left(\alpha + \frac{\varepsilon}{2b} a\right)/b,$$

and the slow integral manifold is described by the equation

$$bz^2 + ax + \alpha + \frac{\varepsilon}{2b} a = 0.$$

The corresponding curve is a parabola with attractive (repelling) part for $b < 0$, $z > 0$ $(z < 0)$, see Fig. 8.2 (a).

Example 5. For $N = 2$ we get

$$f(x, \varepsilon) = p_2 x^2 + p_1 x + p_0, \quad p_1 = 0, \quad p_2 = a, \quad p_0 = \alpha,$$

$$q_2 = -a/b, \quad q_1 = -\varepsilon a/b^2, \quad q_0 = -\left(\alpha + \frac{\varepsilon^2}{2b^2} a)\right)/b.$$

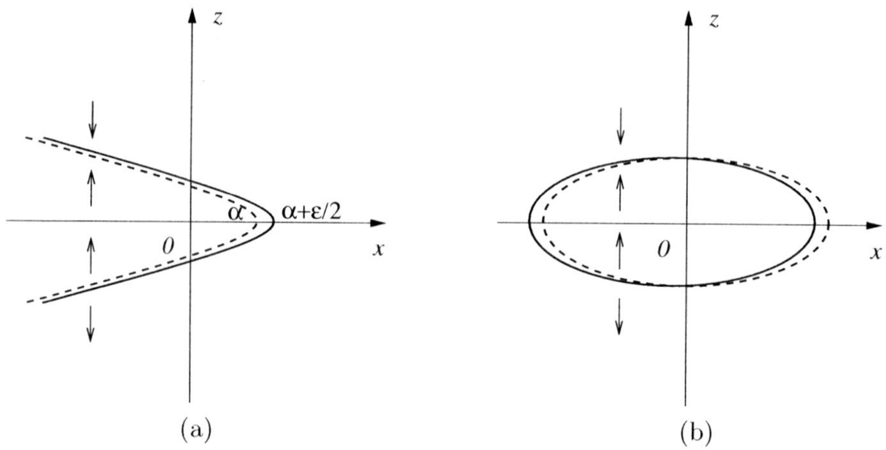

Figure 8.2. *The case $b < 0$*

The slow integral manifold equation takes the form

$$bz^2 + ax^2 + \frac{\varepsilon a}{b}x + \alpha + \frac{\varepsilon^2 a}{2b^2} = 0,$$

or

$$bz^2 + a\left(x + \frac{\varepsilon}{2b}\right)^2 + \alpha + \frac{\varepsilon^2 a}{4b^2} = 0.$$

If $ab > 0$ then this equation describes an ellipse (see Fig. 8.2 (b)) when

$$\frac{\alpha}{b} + \frac{\varepsilon^2 a}{4b^3} < 0.$$

When $a = b = 1$ we get Example 3.

In the case $ab < 0$ we obtain a hyperbola. Let $\alpha - \varepsilon^2/4 > 0$. The right branch of this hyperbola in Fig. 8.3 (a) is a canard (an attractive/repelling trajectory), the left one is a false canard (a repelling/attractive trajectory). If $\alpha - \varepsilon^2/4 < 0$ then the upper (lower) branch of the hyperbola is a repelling (an attractive) slow integral manifolds, see Fig. 8.3 (b).

Example 6. The case $\alpha = \varepsilon^2/4$ is illustrated by Fig. 8.4.

Example 7. ($\dim y = 1$).

Consider the system

$$\dot{x} = 1, \ \ \dot{y} = 0, \ \ \varepsilon\dot{z} = 2xz + \alpha - y.$$

If α is a parameter then the different canards are determined by

$$\dot{x} = 1, \ \ y = y_0, \ \ z = 0,$$

that is, they pass through the unique gluing point $x = 0$, $y = y_0$, $z = 0$ on the breakdown curve $x = 0$ of the slow surface $2xz + y_0 - y = 0$ for $\alpha = y_0$.

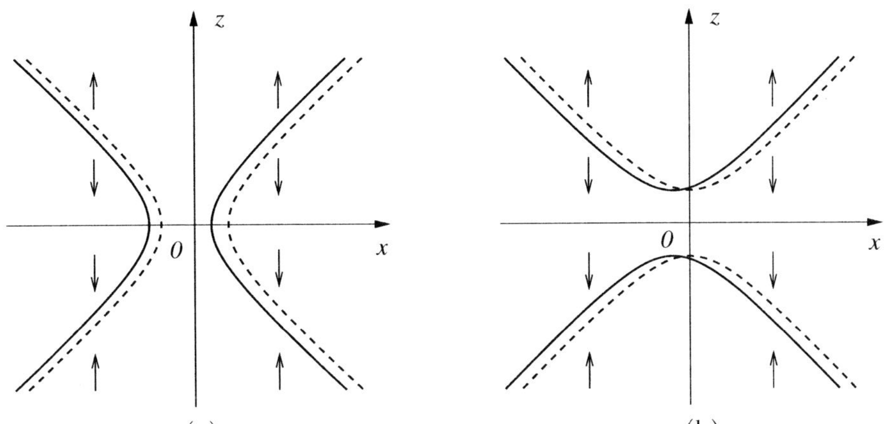

Figure 8.3. *The case $b > 0$*

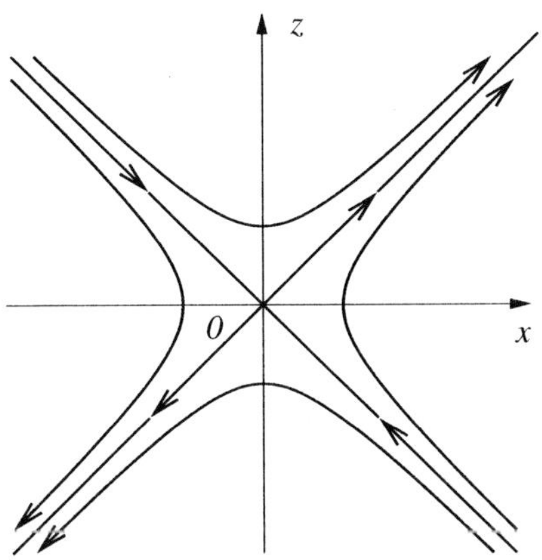

Figure 8.4.

If α is a function of the variable y then for $\alpha = y$ the integral manifold $z = 0$ is stable for $x < 0$ and unstable for $x > 0$.

Example 8.

In the system

$$\dot{x} = x^2 + z^2, \quad \varepsilon \dot{z} = xz,$$

the straight line $z = 0$ plays the role of a black swan on a plane. It should be noted that this line represents an invariant manifold, but it is not a canard trajectory because it is not a trajectory; it consists of three trajectories: $x < 0, z = 0$; $x = z = 0$, and $x > 0, z = 0$.

Example 9. The same straight line $z = 0$ is a canard trajectory of the system

$$\dot{x} = 1, \quad \varepsilon\dot{z} = z\sin x,$$

with a countable point set allowing exchange of stability.

8.1 Integral Manifolds and Canards

In this section the authors review the main results of the integral manifold method and canard theory, including theorems of existence and asymptotic expansions for canards. The concepts and main results of integral manifold theory are considered in the Chapter 7 of this book.

We consider the system of ordinary differential equations:

$$\dot{x} = f(x, y, z, \varepsilon), \tag{8.4}$$

$$\dot{y} = g(x, y, z, \varepsilon), \tag{8.5}$$

$$\varepsilon\dot{z} = p(x, y, z, \varepsilon). \tag{8.6}$$

Here $x \in R$, y and z are vectors in Euclidean spaces R^n and R^m, and ε is a small positive parameter. Vector-functions f, g and p are sufficiently smooth for all $x \in R,\ \ y \in R^n,\ \ z \in D \subset R^m,\ \ \varepsilon \in [0, \varepsilon_0]$.

It is assumed that the values of f, g and p are comparable to unity. System (8.4)–(8.6) is considered as a multi-scale system with the small parameter ε. The slow and fast subsystems are represented by (8.4)–(8.5) and (8.6), respectively. By substituting $\varepsilon = 0$ into (8.4)–(8.6) we obtain the so-called degenerate system:

$$\dot{x} = f(x, y, z, 0), \tag{8.7}$$

$$\dot{y} = g(x, y, z, 0), \tag{8.8}$$

$$0 = p(x, y, z, 0). \tag{8.9}$$

Equation (8.9) determines an $(n+1)$–dimensional surface S, called a slow surface. The intersection of S and the surface given by

$$det\left|\frac{\partial p}{\partial z}(x, y, z, 0)\right| = 0 \tag{8.10}$$

is an n–dimensional surface Γ. This surface Γ divides S into foliations on which

$$det\left|\frac{\partial p}{\partial z}(x, y, z, 0)\right| \neq 0.$$

By the implicit function theorem, a leaf of the slow surface is determined by a well-defined vector-function: $z = h^{(0)}(x, y)$. A slow surface can consist of several foliations determined by different functions $z = h_i^{(0)}(x, y)$, domains of which can intersect, depending on the structure of the slow surface.

It should be noted that the conditions of the implicit function theorem do not hold on the leaf boundary given by (8.10). Therefore the asymptotic methods discussed in Chapter 7 do not hold there.

The aim of this chapter is to investigate the behaviour of slow integral manifolds for systems when the condition (8.10) holds.

8.1.1 Canards of two-dimensional systems

Some special solutions of singularly perturbed ordinary differential equations are called *canard or duck–trajectories.* This term has been introduced by French mathematicians [4, 5, 9]. Let us consider the following two-dimensional autonomous system:

$$\dot{x} = f(x, z, \alpha), \tag{8.11}$$

$$\varepsilon \dot{z} = p(x, z, \alpha), \tag{8.12}$$

where x, z are scalar functions of time, ε is a scalar, f and p are sufficiently smooth scalar functions. The set of points

$$S_\alpha = \{(x, z) : p(x, z, \alpha) = 0\}$$

of the phase plane is called a *slow curve* of the system (8.11), (8.12).

We will need the following assumptions:

1) The curve S_α consists of regular points, i.e. at every point $(x, z) \in S_\alpha$

$$[p_x(x, z, \alpha)]^2 + [p_z(x, z, \alpha)]^2 > 0.$$

2) Singular points, i.e. points at which

$$p_z(x, z, \alpha) = 0,$$

are isolated on S_α.

3) At singular points the following inequality is satisfied:

$$p_{zz} \neq 0.$$

Definition 1.1. A singular point $\mathcal{A}$ of the slow curve S_α is called a *jump point* [33] if

$$sgn \quad [p_z(\mathcal{A}) p_x(\mathcal{A}) f(\mathcal{A})] = 1.$$

Definition 1.2. Parts of S_α which contains only regular points are called *regular.* A regular part of S_α, all points of which satisfy the inequality

$$p_z(x, z, \alpha) < 0 \qquad (p_z(x, z, \alpha) > 0)\,,$$

is called *stable (unstable).*

Stable and unstable parts of the slow curve are zeroth order approximations of corresponding stable and unstable integral manifolds. The integral manifolds lie in an ε–neighborhood of the slow curve, except for the jump points at which the theorems of Chapter 7 do not hold.

To illustrate a situation near jump points, consider the following system:

$$\dot{x} = z - \alpha,$$

$$\varepsilon \dot{z} = \nu(z) - x, \tag{8.13}$$

where $\nu(z) = -1/3z^3 + z$ in some bounded part of the phase plane. The jump points $(-1, -2/3)$ and $(1, 2/3)$ divide the slow curve $x = \nu(z)$ into the stable and the unstable parts, see Fig. 8.5 (a). System (8.13) has a singular point at $z = \alpha$, $x = \nu(\alpha)$. Elementary analysis shows that the singular point is unstable when $-1 < \alpha < 1$ and stable when $\alpha > 1$ or $\alpha < -1$. When $\alpha \in (-1, 1)$, heuristic reasoning leads to the expectation that there will be a limit cycle, see Fig. 8.5 (a). When $\alpha > 1$ or $\alpha < -1$ there will be no a limit cycle. The question is how does the limit cycle disappear when α passes through the value -1 or 1. In what follows we shall concentrate on the case when α passes -1, the other case is entirely similar. It has been shown [4] that there exists a value $\alpha = \alpha_c(\varepsilon)$ such that for α in a small neighborhood of α_c the limit cycle deforms into a curve, see Fig. 8.5 (b). A humorous hand added a few lines to the figure, producing a duck as given in Fig. 8.5 (c). As α diminishes (still in the neighborhood of α_c) the head of the duck gets smaller and at the next stage one has a duck without a head, as given in Fig. 8.5 (d). The duck continues to shrink as α tends to -1 (see Fig. 8.5 (e)) and disappears. For $\alpha < -1$ all solutions of the system tend to the stable steady state, see Fig. 8.5 (f).

Definition 1.3. Trajectories which at first pass along the stable integral manifold and then continue for a while along the unstable integral manifold are called *canards or duck-trajectories.*

The existence of canards for the system (8.11), (8.12) was proved originally by tools from non-standard analysis [4, 5, 9, 63]. A standard interpretation of the main results can be summarized briefly as follows [10, 15, 17, 32]:

The canards and corresponding values of the parameter α allow asymptotic expansions in powers of the small parameter ε. Near the slow curve the canards are exponentially close, and have the same asymptotic expansion in powers of ε. An analogous assertion is true for corresponding parameter values α. Namely, any two values of the parameter α for which canards exist have the same asymptotic expansions, and the difference between them is given by $\exp(-1/c\varepsilon)$ where c is some positive number.

8.1.2 Canards of three-dimensional systems

In this subsection we discuss the existence of canards for some special types of three-dimensional systems.

Let us consider the following autonomous system of three ordinary differential equations:

$$\dot{x} = f(x, y, z, \varepsilon), \tag{8.14}$$

$$\dot{y} = g(x, y, z, \alpha, \varepsilon), \tag{8.15}$$

$$\varepsilon\dot{z} = p(x, y, z, \alpha, \varepsilon), \tag{8.16}$$

where the dot denotes first derivative with respect to the time, f, g, p are scalar functions, α and ε are scalars.

The question is whether the system (8.14)–(8.16) has a canard. To answer this question, we will investigate the two-dimensional system obtained from (8.14)–(8.16) by eliminating the variable t. It is assumed that this two-dimensional system

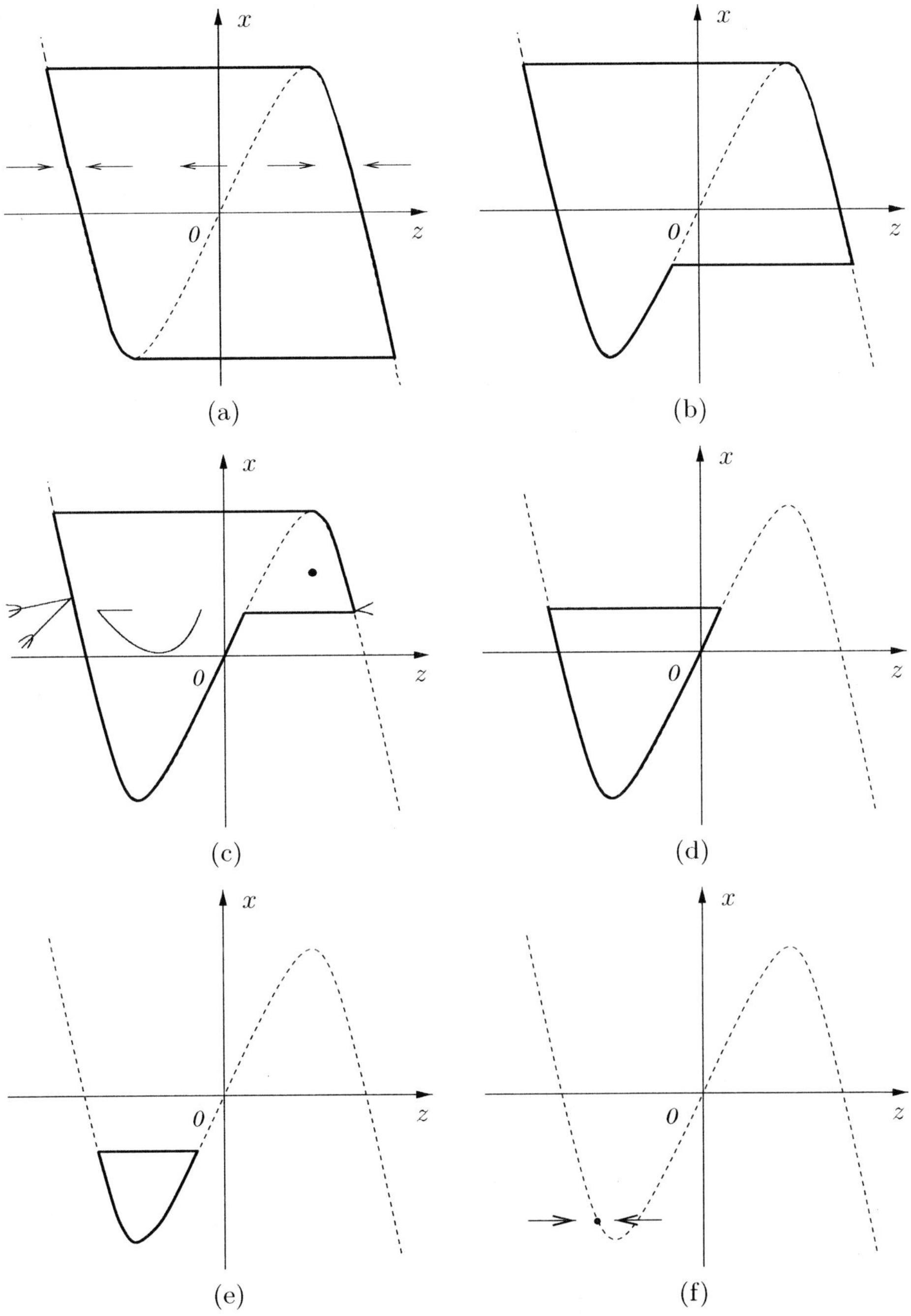

Figure 8.5.

can be represented as:

$$y\prime = Y(x, y, z, a, \varepsilon), \tag{8.17}$$

$$\varepsilon z\prime = 2xz + Z(x, y, z, a, \varepsilon), \tag{8.18}$$

where a is a scalar and the function $Z(x, y, z, a, \varepsilon)$ has the following form:

$$Z(x, y, z, a, \varepsilon) = Z_1(x, y, z) + \varepsilon(C + aC_0) + \varepsilon Z_2(x, y, z, a, \varepsilon), \tag{8.19}$$

and prime represents a derivative with respect to x. Here C, C_0 are some constants, functions $Y(x, y, z, a, \varepsilon)$, $Z_1(x, y, z)$ and $Z_2(x, y, z, a, \varepsilon)$ are defined, bounded and continuous in

$$\Omega = \{x \in R,\ \ y \in R,\ \ |a + CC_0^{-1}| \leq \nu,\ \ \varepsilon \in [0, \varepsilon_0]\},$$

and satisfy the following conditions in Ω:

$$|Y(x, y, z, a, \varepsilon) - Y(x, \bar{y}, \bar{z}, \bar{a}, \varepsilon)| \leq M\left(|y - \bar{y}| + |z - \bar{z}|\right) + \mu|a - \bar{a}|, \tag{8.20}$$

$$|Z_1(x, y, z)| \leq M|z|^2, \tag{8.21}$$

$$|Z_1(x, y, z) - Z_1(x, \bar{y}, \bar{z})| \leq M\left(|z| + |\bar{z}|\right)^2 |y - \bar{y}| + \frac{M}{2}\left(|z| + |\bar{z}|\right)|z - \bar{z}|, \tag{8.22}$$

$$|Z_2(x, y, z, a, \varepsilon)| \leq M\mu, \tag{8.23}$$

$$|Z_2(x, y, z, a, \varepsilon) - Z_2(x, \bar{y}, \bar{z}, \bar{a}, \varepsilon)| \leq M\left(|y - \bar{y}| + |z - \bar{z}|\right) + \mu|a - \bar{a}|, \tag{8.24}$$

where M, ν are positive constants and μ is a sufficiently small positive constant.

The slow surface of system (8.17), (8.18) is defined by the equation $z = 0$, due to the identity

$$\{2xz + Z(x, y, z, a, 0)\}_{z=0} \equiv 0.$$

It is well known (see for instance [34, 58]) that in an ε-neighborhood of stable and unstable foliations of the slow surface there are stable and unstable slow integral manifolds

$$z = h(x, y, a, \varepsilon).$$

The parameter a ensures the existence of the gluing point of these integral manifolds. By fixing the jump point $(0, y^*)$, we can single out the trajectory

$$y = \phi(x, a) \quad (\phi(0, a) = y^*)$$

on the integral manifold $h(x, y, a, \varepsilon)$, which passes along the stable leaf to the jump point and then continues for a while along the unstable leaf. For convenience we use the same term 'trajectory' for both systems (8.14)–(8.16) and (8.17), (8.18). The following theorem holds.

Theorem 1.1. *Let conditions (8.20)–(8.24) hold. Then there is ε_0, such that for every $\varepsilon \in (0, \varepsilon_0)$ there exist $a = a^*(\varepsilon)$ and a canard corresponding to this parameter value which passes through the point $(0, y^*)$.*

The reader is referred to [17, 55] for a proof of this theorem.

Remark. Usually, conditions (8.20)–(8.24) hold only for

$$|x| \le r_1, \quad |y| \le r_2.$$

In this case canards are local. To prove the existence of these we need to continue the right-hand sides of the system to $x \in R$, $y \in R$ with the corresponding properties preserved.

8.1.3 Asymptotic expansions for canards

In this subsection the asymptotic expansions for the canards of the system (8.17), (8.18) are obtained.

It is assumed that functions Y and Z in (8.17), (8.18) have sufficient continuous and bounded partial derivatives with respect to all variables. For simplicity we exclude the ε-dependence of the functions Y and Z_2. Then the canard and the parameter value a^* (corresponding to this trajectory) allow the asymptotic expansions in powers of the small parameter ε:

$$a^* = \sum_{i \ge 0} \varepsilon^i a_i,$$

$$y = \phi(x, a^*) = \sum_{i \ge 0} \varepsilon^i \phi_i(x), \tag{8.25}$$

$$z = \psi(x, a^*, \varepsilon) = h\left(x, \sum_{i \ge 0} \varepsilon^i \phi_i(x), \sum_{i \ge 0} \varepsilon^i a_i, \varepsilon\right) = \sum_{i \ge 0} \varepsilon^i \psi_i(x).$$

We can calculate these asymptotic expansions from (8.17), (8.18). To do so, we first obtain the asymptotic expansion for Y, Z_1, Z_2 (these exist due to the properties of Y, Z_1 and Z_2).

$$Y(x, y, z, a) = Y\left(x, \sum_{i \ge 0} \varepsilon^i \phi_i(x), \sum_{i \ge 0} \varepsilon^i \psi_i(x), \sum_{i \ge 0} \varepsilon^i a_i\right)$$

$$= Y(x, \phi_0, \psi_0, a_0) + \varepsilon\left[\phi_1 \frac{\partial}{\partial y} Y(x, \phi_0, \psi_0, a_0)\right.$$

$$\left. + \psi_1 \frac{\partial}{\partial z} Y(x, \phi_0, \psi_0, a_0) + a_1 \frac{\partial}{\partial a} Y(x, \phi_0, \psi_0, a_0)\right]$$

$$+ \frac{1}{2}\varepsilon^2\left[\phi_1^2 \frac{\partial^2}{\partial y^2} Y(x, \phi_0, \psi_0, a_0) + 2\phi_2 \frac{\partial}{\partial y} Y(x, \phi_0, \psi_0, a_0)\right.$$

$$+ \psi_1^2 \frac{\partial^2}{\partial z^2} Y(x, \phi_0, \psi_0, a_0) + 2\psi_2 \frac{\partial}{\partial z} Y(x, \phi_0, \psi_0, a_0)$$

$$\left. + a_1^2 \frac{\partial^2}{\partial a^2} Y(x, \phi_0, \psi_0, a_0) + 2a_2 \frac{\partial}{\partial a} Y(x, \phi_0, \psi_0, a_0)\right]$$

$$
\begin{aligned}
&+\frac{1}{6}\varepsilon^3\left[\phi_1^3\frac{\partial^3}{\partial y^3}Y(x,\phi_0,\psi_0,a_0)+6\phi_1\phi_2\frac{\partial^2}{\partial y^2}Y(x,\phi_0,\psi_0,a_0)\right.\\
&\quad+6\phi_3\frac{\partial}{\partial y}Y(x,\phi_0,\psi_0,a_0)+\psi_1^3\frac{\partial^3}{\partial z^3}Y(x,\phi_0,\psi_0,a_0)\\
&\quad+6\psi_1\psi_2\frac{\partial^2}{\partial z^2}Y(x,\phi_0,\psi_0,a_0)+6\psi_3\frac{\partial}{\partial z}Y(x,\phi_0,\psi_0,a_0)\\
&\quad+a_1^3\frac{\partial^3}{\partial a^3}Y(x,\phi_0,\psi_0,a_0)+6a_1a_2\frac{\partial^2}{\partial a^2}Y(x,\phi_0,\psi_0,a_0)\\
&\quad\left.+6a_3\frac{\partial}{\partial a}Y(x,\phi_0,\psi_0,a_0)\right]+\dots\\
&=Y(x,\phi_0,\psi_0,a_0)+\sum_{i\geq 1}\varepsilon^i\left[\phi_i\frac{\partial}{\partial y}+\psi_i\frac{\partial}{\partial z}+a_i\frac{\partial}{\partial a}\right]Y(x,\phi_0,\psi_0,a_0)\\
&\quad+\sum_{i\geq 2}\varepsilon^iY_i(x,\phi_0,\dots,\phi_{i-1},\psi_1,\dots,\psi_{i-1},a_0,\dots,a_{i-1}).
\end{aligned}
$$

$$
\begin{aligned}
&Z_2(x,y,z,a)\\
&=Z_2(x,\phi_0,\psi_0,a_0)+\sum_{i\geq 1}\varepsilon^i\left[\phi_i\frac{\partial}{\partial y}+\psi_i\frac{\partial}{\partial z}+a_i\frac{\partial}{\partial a}\right]Z_2(x,\phi_0,\psi_0,a_0)\\
&\quad+\sum_{i\geq 2}\varepsilon^iZ_i^{(2)}(x,\phi_0,\dots,\phi_{i-1},\psi_1,\dots,\psi_{i-1},a_0,\dots,a_{i-1}).
\end{aligned}
$$

$$
\begin{aligned}
Z_1(x,y,z)&=Z_1\left(x,\sum_{i\geq 0}\varepsilon^i\phi_i(x),\sum_{i\geq 0}\varepsilon^i\psi_i(x)\right)=Z_1(x,\phi_0,\psi_0)\\
&\quad+\varepsilon\left[\phi_1\frac{\partial}{\partial y}Z_1(x,\phi_0,\psi_0)+\psi_1\frac{\partial}{\partial z}Z_1(x,\phi_0,\psi_0)\right]\\
&\quad+\frac{1}{2}\varepsilon^2\left[\phi_1^2\frac{\partial^2}{\partial y^2}Z_1(x,\phi_0,\psi_0)+2\phi_2\frac{\partial}{\partial y}Z_1(x,\phi_0,\psi_0)\right.\\
&\quad\left.+\psi_1^2\frac{\partial^2}{\partial z^2}Z_1(x,\phi_0,\psi_0)+2\psi_2\frac{\partial}{\partial z}Z_1(x,\phi_0,\psi_0)\right]\\
&\quad+\frac{1}{6}\varepsilon^3\left[\phi_1^3\frac{\partial^3}{\partial y^3}Z_1(x,\phi_0,\psi_0)+6\phi_1\phi_2\frac{\partial^2}{\partial y^2}Z_1(x,\phi_0,\psi_0)\right.\\
&\quad+6\phi_3\frac{\partial}{\partial y}Z_1(x,\phi_0,\psi_0)+\psi_1^3\frac{\partial^3}{\partial z^3}Z_1(x,\phi_0,\psi_0)\\
&\quad\left.+6\psi_1\psi_2\frac{\partial^2}{\partial z^2}Z_1(x,\phi_0,\psi_0)+6\psi_3\frac{\partial}{\partial z}Z_1(x,\phi_0,\psi_0)\right]+\dots\\
&=Z_1(x,\phi_0,\psi_0)+\sum_{i\geq 1}\varepsilon^i\left[\phi_i\frac{\partial}{\partial y}+\psi_i\frac{\partial}{\partial z}\right]Z_1(x,\phi_0,\psi_0)
\end{aligned}
$$

$$+\sum_{i\geq 2}\varepsilon^i Z_i^{(1)}\left(x,\phi_0,\ldots,\phi_{i-1},\psi_1,\ldots,\psi_{i-1}\right).$$

Next, we substitute (8.25) and the expressions obtained for Y, Z_1, Z_2 into (8.17), (8.18):

$$\sum_{i\geq 0}\varepsilon^i\frac{d\phi_i}{dx}=Y\left(x,\phi_0,\psi_0,a_0\right)+\sum_{i\geq 1}\varepsilon^i\left[\phi_i\frac{\partial}{\partial y}+\psi_i\frac{\partial}{\partial z}+a_i\frac{\partial}{\partial a}\right]Y\left(x,\phi_0,\psi_0,a_0\right)$$

$$+\sum_{i\geq 2}\varepsilon^i Y_i\left(x,\phi_0,\ldots,\phi_{i-1},\psi_1,\ldots,\psi_{i-1},a_0,\ldots,a_{i-1}\right), \tag{8.26}$$

$$\varepsilon\sum_{i\geq 0}\varepsilon^i\frac{d\psi_i}{dx}=2x\sum_{i\geq 0}\varepsilon^i\psi_i+Z_1\left(x,\phi_0,\psi_0\right)+\sum_{i\geq 1}\varepsilon^i\left[\phi_i\frac{\partial}{\partial y}+\psi_i\frac{\partial}{\partial z}\right]Z_1\left(x,\phi_0,\psi_0\right)$$

$$+\sum_{i\geq 2}\varepsilon^i Z_i^{(1)}\left(x,\phi_0,\ldots,\phi_{i-1},\psi_1,\ldots,\psi_{i-1}\right)+\varepsilon\left(C+C_0\sum_{i\geq 0}\varepsilon^i a_i\right)$$

$$+\varepsilon Z_2\left(x,\phi_0,\psi_0,a_0\right)+\varepsilon\sum_{i\geq 1}\varepsilon^i\left[\phi_i\frac{\partial}{\partial y}+\psi_i\frac{\partial}{\partial z}+a_i\frac{\partial}{\partial a}\right]Z_2\left(x,\phi_0,\psi_0,a_0\right)$$

$$+\varepsilon\sum_{i\geq 2}\varepsilon^i Z_i^{(2)}\left(x,\phi_0,\ldots,\phi_{i-1},\psi_1,\ldots,\psi_{i-1},a_0,\ldots,a_{i-1}\right). \tag{8.27}$$

Setting $\varepsilon=0$ in (8.26), (8.27) we have:

$$\frac{d\phi_0}{dx}=Y\left(x,\phi_0,\psi_0,a_0\right), \tag{8.28}$$

and

$$2x\psi_0+Z_1\left(x,\phi_0,\psi_0\right)=0.$$

Applying (8.19), (8.21) and (8.22) yields

$$\psi_0(x)\equiv 0, \tag{8.29}$$

$$Z_1\left(x,\phi_0,0\right)=0,$$

$$\frac{\partial Z_1}{\partial y}\left(x,\phi_0,0\right)=\frac{\partial Z_1}{\partial z}\left(x,\phi_0,0\right)=0. \tag{8.30}$$

Applying (8.28) and additional initial data of (8.17), (8.18) we get

$$\phi_0=\phi_0(x,a_0).$$

Next, equating terms in ε in (8.26)–(8.27) and applying (8.28), (8.30) we obtain

$$\frac{d\phi_1}{dx}=\phi_1\frac{\partial Y}{\partial y}\left(x,\phi_0,0,a_0\right)+\psi_1\frac{\partial Y}{\partial z}\left(x,\phi_0,0,a_0\right)$$

$$+a_1 \frac{\partial Y}{\partial a}(x, \phi_0, 0, a_0), \tag{8.31}$$

$$0 = 2x\psi_1 + C + a_0 C_0 + Z_2(x, \phi_0, 0, a_0).$$

From the last equation it follows that

$$\psi_1(x) = -\left[C + a_0 C_0 + Z_2(x, \phi_0, 0, a_0)\right]/2x. \tag{8.32}$$

The denominator of the right-hand side of (8.32) is equal to zero at the gluing point of the integral manifold. By continuity of the function $\psi_1(x)$ we require the following condition:

$$a_0 = -C_0^{-1}\big[Z_2(x, \phi_0, 0, a_0) + C\big]. \tag{8.33}$$

We now note that the mapping $K(a)$, defined by the right-hand side of (8.33), has a unique fixed point. It follows that a_0 is well defined by equation (8.33).

Substituting the expressions obtained for a_0, ϕ_0 into (8.32) we can find the function

$$\psi_1 = \psi_1(x).$$

Next, from (8.31), we calculate the function

$$\phi_1 = \phi_1(x, a_1).$$

Further, we equate terms in ε^2 in (8.26), (8.27), taking into account (8.29) and (8.30), to give

$$\frac{d\phi_2}{dx} = \phi_2 \frac{\partial Y}{\partial y}(x, \phi_0, 0, a_0) + \psi_2 \frac{\partial Y}{\partial z}(x, \phi_0, 0, a_0)$$

$$+a_2 \frac{\partial Y}{\partial a}(x, \phi_0, 0, a_0) + Y_2(x, \phi_0, \phi_1, 0, \psi_1, a_0, a_1), \tag{8.34}$$

$$\frac{d\psi_1}{dx} = 2x\psi_2 + a_1\left(C_0 + \frac{\partial Z_2}{\partial a}(x, \phi_0, 0, a_0)\right) + \phi_1 \frac{\partial Z_2}{\partial y}(x, \phi_0, 0, a_0)$$

$$+\psi_1 \frac{\partial Z_2}{\partial z}(x, \phi_0, 0, a_0) + Z_2^{(1)}(x, \phi_0, \phi_1, 0, \psi_1).$$

From the last equation it follows that

$$\psi_2(x) = \left[\frac{d\psi_1}{dx} - C_0 a_1 - \phi_1 \frac{\partial Z_2}{\partial y}(x, \phi_0, 0, a_0) - \psi_1 \frac{\partial Z_2}{\partial z}(x, \phi_0, 0, a_0)\right.$$

$$\left. -a_1 \frac{\partial Z_2}{\partial a}(x, \phi_0, 0, a_0) - Z_2^{(1)}(x, \phi_0, \phi_1, 0, \psi_1)\right]/2x. \tag{8.35}$$

To avoid a discontinuity in the function $\psi_2(x)$ at the point $x = 0$ we require the following condition

$$\frac{d\psi_1(0)}{dx} - a_1\left(C_0 + \frac{\partial Z_2}{\partial a}(0, \phi_0(0), 0, a_0)\right) - \phi_1(0)\frac{\partial Z_2}{\partial y}(0, \phi_0(0), 0, a_0)$$

$$-\psi_1(0)\frac{\partial Z_2}{\partial z}(0,\phi_0(0),0,a_0) - Z_2^{(1)}(0,\phi_0(0),\phi_1(0),0,\psi_1(0)) = 0. \tag{8.36}$$

It can be shown that a_1 is well defined by (8.36) in a way similar to that for a_0 in the previous case.

Further, from (8.35) and (8.34) we get $\psi_2 = \psi_2(x)$ and $\phi_2 = \phi_2(x, a_2)$, respectively. At the kth step we have

$$\frac{d\phi_k}{dx} = \phi_k \frac{\partial Y}{\partial y}(x,\phi_0,0,a_0) + \psi_k \frac{\partial Y}{\partial z}(x,\phi_0,0,a_0)$$

$$+a_k \frac{\partial Y}{\partial a}(x,\phi_0,0,a_0) + Y_k(x,\phi_0,\ldots,\phi_{k-1},0,\ldots,\psi_{k-1},a_0,\ldots,a_{k-1}),$$

$$\psi_k(x) = \Bigg[\frac{d\psi_{k-1}}{dx} - a_{k-1}\left(C_0 + \frac{\partial Z_2}{\partial a}(x,\phi_0,0,a_0)\right) - \phi_{k-1}\frac{\partial Z_2}{\partial y}(x,\phi_0,0,a_0)$$

$$-\psi_{k-1}\frac{\partial Z_2}{\partial z}(x,\phi_0,0,a_0) - Z_k^{(1)}(x,\phi_0,\ldots,\phi_{k-1},0,\ldots,\psi_{k-1})$$

$$-Z_{k-1}^{(2)}(x,\phi_0,\ldots,\phi_{k-2},0,\ldots,\psi_{k-2},a_0,\ldots,a_{k-2})\Bigg]/2x,$$

$$\frac{d\psi_{k-1}(0)}{dx} - a_{k-1}\left(C_0 + \frac{\partial Z_2}{\partial a}(0,\phi_0(0),0,a_0)\right)$$

$$-\phi_{k-1}(0)\frac{\partial Z_2}{\partial y}(0,\phi_0(0),0,a_0) - \psi_{k-1}(0)\frac{\partial Z_2}{\partial z}(0,\phi_0(0),0,a_0)$$

$$-Z_k^{(1)}(0,\phi_0(0),\ldots,\phi_{k-1}(0),0,\ldots,\psi_{k-1}(0)) = 0.$$

The following theorem is useful in obtaining approximations to canards.

Theorem 1.2. *Let the assumptions of Theorem 1.1 hold. Moreover, let the functions Y, Z_1 and Z_2 have sufficient continuous and bounded partial derivatives with respect to all variables. Then the canard and the parameter value a (corresponding to this trajectory) can be represented as*

$$a^* = \sum_{i=0}^{k} \varepsilon^i a_i + a_{k+1}(\varepsilon),$$

$$y = \phi(x,a^*) = \sum_{i=0}^{k} \varepsilon^i \phi_i(x) + \phi_{k+1}(x,\varepsilon),$$

$$z = \psi(x,a^*,\varepsilon) = \sum_{i=0}^{k} \varepsilon^i \psi_i(x) + \psi_{k+1}(x,\varepsilon),$$

where the continuous functions $a_{k+1}(\varepsilon)$, $\phi_{k+1}(x,\varepsilon)$ and $\psi_{k+1}(x,\varepsilon)$ satisfy the inequalities

$$|a_{k+1}(\varepsilon)| \le \varepsilon^k R_1,\ \ |\phi_{k+1}(x,\varepsilon)| \le \varepsilon^{k+1} R_2,\ \ |\psi_{k+1}(x,\varepsilon)| \le \varepsilon^{k+1/2} R_3,$$

with $R_1 > 0$, $R_2 > 0$, $R_3 > 0$.

Note that Theorem 1.1 and Theorem 1.2 can be generalised to the cases $y \in R^n, z \in R$ and $y \in R^n, z \in R^m$.

8.2 The Stable/Unstable Slow Integral Manifolds

In this paper we use the standard approach to study slow integral surfaces of variable stability (or black swans). These surfaces are considered as natural generalizations of the notion of a canard.

We suggest the term "black swan" for two reasons. The first is that a swan is a bird of the family of ducks. The second is connected with the usual meaning of "black swan" in the sense of a rare phenomenon. It should be noted also that the French term "canard" is used in the sense of a false rumour[1] in English.

In order to glue the stable and unstable parts of a canard an additional parameter is used. To glue integral manifolds whose dimension is greater than one we need an additional function. The argument of this function is a vector variable parameterizing the breakdown surface.

8.2.1 Black swans

Let us reduce the system (8.1) to the form

$$\frac{dy}{dx} = Y(x, y, z, \varepsilon), \qquad y \in R^n, \qquad x \in R; \tag{8.37}$$

$$\varepsilon\frac{dz}{dx} = 2xz + a + Z(x, y, z, a, \varepsilon), \qquad |z| \leq r, \tag{8.38}$$

where r and a_0 are positive constants. It is supposed that the functions Y, Z are continuous and satisfy the following inequalities for $x \in R$, $y \in R^n$, $|z| \leq r$, $|a| \leq a_0$, $\varepsilon \in [0, \varepsilon_0]$:

$$\|Y(x, y, z, \varepsilon)\| \leq k, \quad |Z(x, y, z, a, \varepsilon)| \leq M\left(\varepsilon^2 + \varepsilon|z| + |z|^2\right), \tag{8.39}$$

$$\begin{aligned} &\|Y(x, y, z, \varepsilon) - Y(x, \bar{y}, \bar{z}, \varepsilon)\| \leq M(\|y - \bar{y}\| + |z - \bar{z}|), \\ &|Z(x, y, z, a, \varepsilon) - Z(x, \bar{y}, \bar{z}, \bar{a}, \varepsilon)| \leq M\left\{(\varepsilon + |\tilde{z}|)|z - \bar{z}|\right. \\ &\left.+(\varepsilon^2 + \varepsilon|\tilde{z}| + |\tilde{z}|^2)\|y - \bar{y}\| + \varepsilon|a - \bar{a}|\right\}, \quad |\tilde{z}| = \max\{|z|, |\bar{z}|\}, \end{aligned} \tag{8.40}$$

where $\|\cdot\|$ denotes the usual norm in R^n and $|\cdot|$ denotes the absolute value of a scalar, k and M are positive constants.

Let us consider a as a function: $a = a(y, \varepsilon)$.

Let F be the complete metric space of functions $a(y, \varepsilon)$ continuous with respect to y and satisfying

$$|a(y, \varepsilon)| \leq \varepsilon^2 K, \quad |a(y, \varepsilon) - a(\bar{y}, \varepsilon)| \leq \varepsilon^2 L\|y - \bar{y}\|,$$

for $\varepsilon \in (0, \varepsilon_0]$, where K and L are positive constants, with the metric defined by

$$\rho(a, \bar{a}) = \sup_{y \in R^n} |a(y, \varepsilon) - \bar{a}(y, \varepsilon)|.$$

[1] "An absurd story circulated as a hoax", see Shorter Oxford English Dictionary.

Let H be the complete metric space of functions $h(x, y, \varepsilon)$ mapping $R \times R^n$ to R continuous with respect to x, y and satisfying

$$|h(x, y, \varepsilon)| \leq \varepsilon^{\frac{3}{2}} q,$$

$$|h(x, y, \varepsilon) - h(x, \bar{y}, \varepsilon)| \leq \varepsilon^{\frac{3}{2}} \delta \|y - \bar{y}\|,$$

for $\varepsilon \in (0, \varepsilon_0]$, where q and δ are positive constants, with the metric

$$\rho(h, \bar{h}) = \sup_{x \in R, y \in R^n} |h(x, y, \varepsilon) - \bar{h}(x, y, \varepsilon)|.$$

On the space H we define an operator T by the formula

$$Th(x, y, \varepsilon) = \begin{cases} -\varepsilon^{-1} \int\limits_x^\infty e^{(x^2 - s^2)/\varepsilon} [Z(\cdot) + a(\varphi(s, x), \varepsilon)] ds & , \quad x \geq 0 \\ \varepsilon^{-1} \int\limits_{-\infty}^x e^{(x^2 - s^2)/\varepsilon} [Z(\cdot) + a(\varphi(s, x), \varepsilon)] ds & , \quad x < 0. \end{cases}$$

where $Z(\cdot) = Z(s, \varphi(s, x), h(s, \varphi(s, x), \varepsilon), a(\varphi(s, x), \varepsilon), \varepsilon)$, and $\varphi(s, x)$ is defined as follows. For any element $h \in H$ the initial value problem

$$\frac{d\varphi}{ds} = Y(s, \varphi, h(s, \varphi, \varepsilon), \varepsilon),$$

$$\varphi(x) = y,$$

is considered. This problem is obtained from (8.37) if we replace z by an element $h \in H$. The solution of this problem is denoted by $\Phi(s, x, y, \varepsilon | h) = \varphi(s, x)$. When the operator T possesses a fixed point $h(x, y, \varepsilon)$ in H then the surface $z = h(x, y, \varepsilon)$ is a slow integral manifold with changing stability (black swan).

It should be noted that we use a modification of the usual technique of integral manifold theory [34, 58]. The detailed proof can be found in [17, 45, 46]. The following theorem gives sufficient conditions for the system (8.37), (8.38) to have a black swan.

Theorem 2.1. *Let the conditions (8.39)–(8.40) be satisfied. Then there are numbers $\varepsilon_0 > 0$ and K, L, q, δ such that, for all $\varepsilon \in (0, \varepsilon_0)$, there exist functions $a(y, \varepsilon) \in F$ and $h(x, y, \varepsilon) \in H$ such that $z = h(x, y, \varepsilon)$ is a slow integral manifold (the black swan).*

Remark 1. Usually the conditions (8.39)–(8.40) are fulfilled for $|x| \leq r_1$, $\|y\| \leq r_2$ only. Integral manifolds in this case are local.

Let the functions Y and Z in (8.37)–(8.38) be sufficiently smooth, then asymptotic expansions can be derived for the functions h and $a(y, \varepsilon)$.

Theorem 2.2. *Let the assumptions of Theorem 2.1. hold. Moreover, the functions Y and Z have sufficient continuous and bounded partial derivatives with respect to all variables. Then the function h describing the black swan and the function $a(y, \varepsilon)$ (corresponding to this black swan) can be represented as*

$$a(y, \varepsilon) = \sum_{i=0}^N \varepsilon^i a_i(y) + a_{N+1}(y, \varepsilon),$$

$$h(x,y,\varepsilon) = \sum_{i=0}^{N} \varepsilon^i h_i(x,y) + h_{N+1}(x,y,\varepsilon),$$

where the continuous functions $a_{N+1}(y,\varepsilon)$ *and* $h_{N+1}(x,y,\varepsilon)$ *satisfy the inequalities*

$$|a_{N+1}(y,\varepsilon)| \le \varepsilon^{N+1}K_1, \quad K_1 > 0, \; |h_{N+1}(x,y,\varepsilon)| \le \varepsilon^{N+1/2}q_1, \quad q_1 > 0.$$

Remark 2. Systems of the type (8.1) can be reduced to the form (8.37)–(8.38) in a neighborhood of the first approximation ($N = 1$) of the slow integral manifold.

Note that Theorem 2.1 and Theorem 2.2 can be generalized to the case when z is a vector variable.

Let us consider now the system

$$\frac{dy}{dx} = Y(x,y,z_1,z_2,\varepsilon), \; y \in R^n, \; x \in R, \tag{8.41}$$

$$\varepsilon\frac{dz_1}{dx} = 2xz_1 + a(y,\varepsilon) + Z_1(x,y,z_1,z_2,\varepsilon), \; z_1 \in R, \tag{8.42}$$

$$\varepsilon\frac{dz_2}{dx} = A(x)z_2 + a(y,\varepsilon)B + Z_2(x,y,z_1,z_2,\varepsilon), \; z_2 \in R^m, \tag{8.43}$$

where $A(x)$ is a bounded matrix, satisfying a Lipschitz condition, with eigenvalues $\lambda_i(x)$:

$$Re\lambda_i(x) \le -2\beta < 0 \quad (i = 1, 2, \dots, m),$$

B is a constant vector, the continuous functions Y, Z_1, Z_2, a satisfy the inequalities

$$\|Y(x,y,z_1,z_2,\varepsilon)\| \le k, \tag{8.44}$$

$$|Z_1(x,y,z_1,z_2,\varepsilon)| \le M\left(\varepsilon^2 + \varepsilon\|z\| + \|z\|^2\right), \tag{8.45}$$

$$\|Z_2(x,y,z_1,z_2,\varepsilon)\| \le M\left(\varepsilon^2 + \varepsilon\|z\| + \|z\|^2\right), \tag{8.46}$$

$$\|Y(x,y,z_1,z_2,\varepsilon) - Y(x,\bar{y},\bar{z}_1,\bar{z}_2,\varepsilon)\| \le M(\|y-\bar{y}\| + \|z-\bar{z}\|), \tag{8.47}$$

$$\begin{aligned}|Z_1(x,y,z_1,z_2,\varepsilon) - Z_1(x,\bar{y},\bar{z}_1,\bar{z}_2,\varepsilon)| \le M\{&(\varepsilon + \|\tilde{z}\|)\|z-\bar{z}\| \\ &+(\varepsilon^2 + \varepsilon\|\tilde{z}\| + \|\tilde{z}\|^2)\|y-\bar{y}\|\},\end{aligned} \tag{8.48}$$

$$\begin{aligned}\|Z_2(x,y,z_1,z_2,\varepsilon) - Z_2(x,\bar{y},\bar{z}_1,\bar{z}_2,\varepsilon)\| \le M\{&(\varepsilon + \|\tilde{z}\|)\|z-\bar{z}\| \\ &+(\varepsilon^2 + \varepsilon\|\tilde{z}\| + \|\tilde{z}\|^2)\|y-\bar{y}\|\},\end{aligned} \tag{8.49}$$

where

$$z = \begin{pmatrix} z_1 \\ z_2 \end{pmatrix}, \quad \bar{z} = \begin{pmatrix} \bar{z}_1 \\ \bar{z}_2 \end{pmatrix}, \quad \|\tilde{z}\| = \max\{\|z\|, \|\bar{z}\|\}, k > 0, M > 0.$$

We consider $a(y,\varepsilon)$ to be a continuous function satisfying the inequalities

$$|a(y,\varepsilon)| \le \varepsilon^2 K, \; |a(y,\varepsilon) - a(\bar{y},\varepsilon)| \le \varepsilon^2 L\|y-\bar{y}\|, \tag{8.50}$$

for $\varepsilon \in (0, \varepsilon_0]$ with some positive constants K and L.

Theorem 2.3. *Let the conditions (8.44)–(8.49) be satisfied. Then there are numbers $\varepsilon_0 > 0$ and K, L, q, δ such that, for all $\varepsilon \in (0, \varepsilon_0)$, there exist functions $a(y, \varepsilon)$ satisfying (8.50) and a continuous function $h(x, y, \varepsilon)$ satisfying*

$$|h_1(x,y,\varepsilon)| \le \varepsilon^{3/2} q, \ |h_1(x,y,\varepsilon) - h_1(x,\bar{y},\varepsilon)| \le \varepsilon^{3/2}\delta\|y - \bar{y}\|, \tag{8.51}$$

$$\|h_2(x,y,\varepsilon)\| \le \varepsilon^2 q, \ \|h_2(x,y,\varepsilon) - h_2(x,\bar{y},\varepsilon)\| \le \varepsilon^2\delta\|y - \bar{y}\|, \tag{8.52}$$

$$h(x,y,\varepsilon) = \begin{pmatrix} h_1(x,y,\varepsilon) \\ h_2(x,y,\varepsilon) \end{pmatrix}, \quad h_1 \in R, \quad h_2 \in R^m,$$

such that $z = h(x, y, \varepsilon)$ is a black swan of the system (8.41)–(8.43).

The proof of Theorem 2.3 follows the lines of the proof of Theorem 2.1, but now H is the complete metric space of functions $h(x, y, \varepsilon)$ mapping $R \times R^n$ to R^{m+1}, continuous with respect to x, y and satisfying (8.51), (8.52) for $\varepsilon \in (0, \varepsilon_0]$. The operator T is defined by the formula $T = \begin{pmatrix} T_1 \\ T_2 \end{pmatrix}$:

$$T_1 h(x,y) = \begin{cases} -\varepsilon^{-1} \int\limits_x^\infty e^{(x^2-s^2)/\varepsilon}[Z_1(\cdot) + a(\varphi(s,x),\varepsilon)]ds &, \quad x \ge 0, \\ \varepsilon^{-1} \int\limits_{-\infty}^x e^{(x^2-s^2)/\varepsilon}[Z_1(\cdot) + a(\varphi(s,x),\varepsilon)]ds &, \quad x < 0, \end{cases}$$

$$T_2 h(x,y) = \varepsilon^{-1} \int\limits_{-\infty}^x W(x,s,\varepsilon)[Z_2(\cdot) + Ba(\varphi(s,x),\varepsilon)]ds,$$

where

$$Z_{1,2}(\cdot) = Z_{1,2}(s, \varphi(s,x), h_1(s,\varphi(s,x),\varepsilon), h_2(s,\varphi(s,x),\varepsilon), \varepsilon)$$

and $W(x, s, \varepsilon)$ $(W(s, s, \varepsilon) = I)$ is the fundamental matrix of the equation

$$\varepsilon \frac{dz_2}{dx} = A(x) z_2.$$

Corollary. The proof of Theorem 2.3 can be extended to the case when the system (8.41)–(8.43) has the form

$$\frac{dy}{dx} = Y(x,y,z_1,z_2,\varepsilon), \ y \in R^n, \ x \in R,$$

$$\varepsilon\frac{dz_1}{dx} = 2xb(x,y)z_1 + a(y,\varepsilon) + Z_1(x,y,z_1,z_2,\varepsilon), \ z_1 \in R,$$

$$\varepsilon\frac{dz_2}{dx} = A(x,y)z_2 + a(y,\varepsilon)B(x,y,\varepsilon) + Z_2(x,y,z_1,z_2,\varepsilon), \ z_2 \in R^m.$$

Here the matrix $A(x, y)$ and functions Y, Z_1, Z_2, a satisfy the conditions of Theorem 2.3. Moreover, the matrix $A(x, y)$, the scalar function $b(x, y)$ $(0 < c_1 \le b(x, y) \le c_2)$ and the vector function $B(x, y, \varepsilon)$ are continuous and bounded, and satisfy a Lipschitz condition.

Continuing in the same manner as in Subsection 8.1.3 we can prove the asymptotic representations for black swans and the corresponding functions $a(y, \varepsilon)$.

8.2.2 Black swans and canards

We now discuss a connection between slow integral manifolds and canards. At first, we consider system (8.1) in the case $\dim y = 0$. This system can be reduced to the form (8.38), where Z does not depend on y, and a is a parameter. In this case the slow integral manifold is one–dimensional. If the variables t and x increase simultaneously and x passes through zero then this integral manifold contains a canard.

If a gluing function $a(y, \varepsilon)$ exists, then every trajectory on the slow integral manifold is a canard if it crosses the surface $x = 0$ from the stable part ($x < 0$) to unstable one ($x > 0$). Thus, in Example 7 for $\alpha = y$, every trajectory on the slow integral manifold $z = 0$ is a canard. An analogous situation takes place for the model of combustion in an inert porous medium. It follows from physical reasons, however, that the gluing function has to be constant: $a(y, \varepsilon) = a(y_0, \varepsilon)$. In this case, the stable and unstable parts of the integral manifold can be glued at one point $y = y_0$ only. The canard passes only through this point. A natural generalization of this situation is possible. Let the gluing function $a = a(y, \varepsilon)$ be given. On the n–dimensional breakdown surface let some n_1–dimensional surface $y = \chi(u), u \in R^{n_1}$ of lower dimension ($n_1 < n$) be given. If the gluing function $a(y, \varepsilon)$ is restricted to $y = \chi(u)$, then the gluing of the stable and unstable parts of slow integral surfaces can be realized only at points of the surface $y = \chi(u)$. This permits us to construct slow integral manifolds with changing stability of various forms and dimensions.

Example 10. Consider the following system

$$\dot{x} = 1,$$

$$\dot{y} = 0, \;\; y \in R^n,$$

$$\varepsilon \dot{z} = xz + a(y, \varepsilon) + p(y) + xq(y) + x^2 r(y).$$

Here p, q, r are scalar continuous functions of the vector variable y. By setting $a(y, \varepsilon) = -p(y) - \varepsilon r(y)$, we obtain $h = -q(y) - xr(y)$. Let $y = \chi(u), u \in R^{n_1}$ be any surface, then the system

$$\dot{x} = 1,$$

$$\dot{y} = 0, \;\; y \in R^n,$$

$$\varepsilon \dot{z} = xz - p(\chi(u)) - \varepsilon r(\chi(u)) + p(y) + xq(y) + x^2 r(y),$$

possesses the higher–dimensional cylindrical slow integral surface $z = -q(\chi(u)) - xr(\chi(u))$, and every element of this cylindrical surface is a canard.

In conclusion a higher–dimension generalization of the Example 3 will be given.

Example 11. Consider the differential system

$$\dot{x} = z,$$

$$\dot{y}_i = z, \;\; i = 1, \ldots, n,$$

$$\varepsilon \dot{z} = x^2 + \sum_{i=1}^{n} y_i^2 + z^2 - a^2.$$

It is a straightforward exercise to see that the higher–dimensional sphere

$$(x+\varepsilon/2)^2+\sum_{i=1}^{n}(y_i+\varepsilon/2)^2+z^2=a^2-\frac{n+1}{4}\varepsilon^2$$

is a slow integral manifold, one part ($z<0$) is stable and other ($z>0$) is unstable. This black swan lives for all $a^2>\frac{n+1}{4}\varepsilon^2$.

8.3 Chemical Models

In this section we shall consider the relationship between canards and black swans and critical phenomena in different chemical systems. We shall show that canards play the role of *the separating solutions.* This means that canards simulate the critical regimes separating the basic types of chemical regimes.

The application of black swans consisting entirely of canards to the modelling of critical phenomena permits us to take into account small perturbations in the chemical systems. Moreover we can use black swans for the modelling of critical phenomena in chemical problems without fixed initial conditions.

Before we consider the combustion models, we first give some relatively simple examples of other physical systems.

8.3.1 Lang-Kobayashi equations

External cavity semiconductor lasers present many interesting features for both technological applications and fundamental non-linear science. Their dynamics has been the subject of numerous studies for the last twenty years. Motivations for these studies vary from the need for stable tunable laser sources, for laser cooling or multiplexing, to the general understanding of their complex stability and chaotic behavior. The typical experiment is usually described by a set of delay differential equations introduced by Lang and Kobayashi [25]:

$$\begin{aligned}\dot{E}&=\kappa\,(1+i\alpha)\,(N-1)\,E+\gamma e^{-i\varphi_0}E\,(t-\tau)\,,\\ \dot{N}&=-\gamma_{\|}\left(N-J+|E|^2N\right).\end{aligned}\tag{8.53}$$

Here E is the complex amplitude of the electric field, N is the carrier density, J is pumping current, κ is the field decay rate, $1/\gamma_{\|}$ is the spontaneous time scale, α is the linewidth enhancement factor, γ represents the feedback level, φ_0 is the phase of the feedback if the laser emits at the solitary laser frequency and τ is the external cavity round trip time. Numerical simulations of these equations have successfully reproduced many experimental observations such as mode hopping between external cavity modes [36], and the period doubling route to chaos [26]. However there are few analytical results since delay equations are nonlocal.

However, this model was recently reduced to a 3D dynamical system describing the temporal evolution of the laser power $P=|E|^2$, carrier density N and phase difference $\eta(t)=\varphi\,(t)-\varphi\,(t-\tau)$. This was achieved by assuming $P\,(t-\tau)=P\,(t)$ together with the approximation $\dot{\varphi}=\eta/\tau+\dot{\eta}/2$. This expression remains valid when the phase fluctuates on a time scale much shorter than the re-injection time τ. Under these approximations, the Lang-Kobayashi equations (8.53) reduces to [20]:

$$\dot{P} = 2\left(\kappa\left(N-1\right)+\gamma\cos\left(\eta+\varphi_0\right)\right)P,$$
$$\dot{N} = -\gamma_{\|}\left(N-J+PN\right),$$
$$\dot{\eta} = -\frac{2}{\tau}\eta + 2\kappa\alpha\left(N-1\right) - 2\gamma\sin\left(\eta+\varphi_0\right).$$

This model was successfully used to describe low frequency fluctuations commonly observed in semiconductor lasers with optical feedback [20]. Note that complicated behaviour in this model is considered in Chapter 10.

Suppose that the following relations hold

$$\gamma = o(1), 1/\tau = o(1), \gamma_{\|} = o(1); \quad \kappa = O(1), \alpha = O(1).$$

In this case, the system under consideration possesses the black swan $P \equiv 0$. The exchange of stability is carried out on the curve

$$\kappa(N-1) + \gamma\cos(\eta+\varphi_0) = 0.$$

8.3.2 Exchange of stability in a high–gain control problem

Consider the control system

$$\dot{x} = f(x) + B(x)u, \;\; x(0) = x_0,$$

where $x \in R^n$, $u \in R^r$, $t \geq 0$, from Chapter 7 (Subsection 2.4) of this book.

The change of stability of $N(x,t)$ on some subsurface of $S(x) = 0$ will create serious complications, and the phenomenon of black swans is possible in this case.

Example 12. Consider the control system

$$\dot{x} = y,$$

$$\dot{y} = -x - \nu y + u,$$

with $S : x^2 + y^2 - 1 = 0$. In this case $N = 2Ky$, where K is a scalar. Using the modified control law

$$u = \varepsilon^{-1}K(S - \varepsilon\nu K^{-1}y)$$

leads to the system

$$\dot{x} = y,$$

$$\varepsilon\dot{y} = -x - \varepsilon^{-1}Kz,$$

$$\varepsilon\dot{z} = -2K\varepsilon^{-1}yz,$$

with $z = S$. It is clear that the system

$$\dot{x} = y,$$

$$\varepsilon \dot{y} = -x - K z_1,$$

$$\varepsilon \dot{z} = -2K y z_1,$$

with $z_1 = \varepsilon z, K < 0$ possesses the black swan described by $z_1 = 0$. It should be noted that the controlled system

$$\dot{x} = y,$$

$$\varepsilon \dot{y} = -x - \varepsilon^{-1} K(x^2 + y^2 - 1)$$

possesses the canard $x^2 + y^2 = 1$.

8.3.3 The simple laser

The nonlinear first-order equation

$$\dot{y} = k y^p + \lambda(t) y + \delta, \quad 0 < \varepsilon < 1, \quad \delta \leq 0,$$

with $k = \pm 1$, $p = 2, 3$ and the control parameter

$$\lambda(t) = \lambda_0 + \varepsilon t, \quad \lambda_0 < 0$$

is typical model of simple lasers, and lasers with saturable absorbers [29]. Both ε and δ are small quantities, and $\lambda_0 = O(1)$. Note that this equation may be written in the form

$$\dot{\lambda} = \varepsilon,$$

$$\dot{y} = k y^p + \lambda y + \delta.$$

For $\delta = 0$ this system has the canard $y = 0$. Physically the canard simulates the critical regime: for $p = 2$ it corresponds to a chemical reaction separating the domain of self–accelerating reactions and the domain of slow reactions; for $p = 3$, $k = -1$ it separates two types of slow regime; in the case $p = 3$, $k = 1$ the canard describes the unique slow regime.

8.3.4 The classical combustion models

Thermal explosion occurs when chemical reactions produce heat too rapidly for a stable balance between heat production and loss. The exothermic oxidation reaction is usually modelled as a single step reaction obeying an Arrhenius temperature dependence. The first model for the self-ignition was constructed by Semenov in 1928 (see, for example [41]). The basic idea of the model was a competition between heat production in the reactant vessel (due to an exothermic reaction) and heat losses on the vessel's surface. Heat losses were assumed proportional to the temperature excess over the ambient temperature (Newtonian cooling). The main assumption was that there is no reactant conversion during the fast highly exothermic reaction. This assumption implies the absence of the energy conservation law in the model. This gave the possibility of constructing an extremely simple and attractive mathematical model. Spatial uniformity of the temperature was also assumed so that the governing equation was one first-order ordinary differential equation for the temperature changes:

$$c\rho V\frac{dT}{dt} = QV\left(-\frac{dC}{dt}\right) - \chi S(T - T_0),$$

$$-\frac{dC}{dt} = \Psi(C)Aexp\left(-\frac{E}{RT}\right),$$

where Ψ expresses the dependence of reaction rate on reactant concentration. Here Q is an exothermicity per mole reactant; C and C_0 are a reactant concentration and its initial value; A is constant which is known as a pre-exponential rate factor; c is specific heat capacity; ρ is reactant density; χ is the heat-transfer coefficient; E is the Arrhenius activation energy; R is the universal gas constant; V is the reactant vessel volume; S is the surface area of the reactant vessel; t is a time variable; T is absolute temperature; T_0 is ambient temperature. The initial temperature is assumed to be equal to the ambient temperature T_0.

Dimensionless variables τ, η, Θ are introduced by

$$\tau = tC_0^{n-1}Aexp\left(-\frac{E}{RT_0}\right), \quad \eta = 1 - C/C_0, \quad \Theta = \frac{E}{RT_0}(T - T_0),$$

(n is the order of the chemical reaction) and we obtain the classical model of thermal explosion with reactant consumption [19, 62]:

$$\varepsilon\frac{d\Theta}{d\tau} = \Psi(\eta)exp(\Theta/(1 + \beta\Theta)) - \alpha\Theta, \tag{8.54}$$

$$\frac{d\eta}{d\tau} = \Psi(\eta)exp(\Theta/(1 + \beta\Theta)), \tag{8.55}$$

$$\eta(0) = \eta_0/(1 + \eta_0) = \bar{\eta}_0, \quad \Theta(0) = 0.$$

Here η_0 is the criterion for autocatalyticity, where the small dimensionless parameters

$$\beta = \frac{RT_0}{E} \text{ and } \varepsilon = \frac{c\rho}{QC_0}\frac{E}{RT_0^2}$$

characterize the physical properties of gas mixture, and

$$\alpha = \frac{\chi S}{VQC_0^n A}\frac{RT_0^2}{E}exp\left(\frac{E}{RT_0}\right)$$

is the dimensionless heat loss parameter.

The following cases are examined:

$$\Psi(\eta) = \begin{cases} 1 - \eta, & \text{first-order reaction } (\eta_0 = 0) \\ \eta(1 - \eta), & \text{autocatalytic reaction.} \end{cases}$$

It should be noted that the system (8.54), (8.55) is singularly perturbed. According to the standard approach to such systems the limiting case $\varepsilon \to 0$ is examined, and discontinuous solutions of the reduced system are analyzed. This makes it possible to determine some critical values of initial conditions, which provide a jump transition from the slow regime to the explosive ones. The study of transitional regimes requires the application of higher approximations in the asymptotic analysis of the systems of the type given in equation (8.54), (8.55). The integral manifold technique [54, 58] is applied below to the qualitative analysis of critical and transitional regimes for both types of chemical reaction.

First-order reaction

We begin the analysis of the system (8.54), (8.55) with the first–order reaction case, when $\Psi(\eta) = 1 - \eta$ and the dimensionless concentration $\bar{\eta} = 1 - \eta$ replaces η. The system (8.54), (8.55) in this case is

$$\varepsilon\frac{d\Theta}{d\tau} = \bar{\eta}exp(\Theta/\left(1 + \beta\Theta\right)) - \alpha\Theta, \tag{8.56}$$

$$\frac{d\bar{\eta}}{d\tau} = -\bar{\eta}exp(\Theta/\left(1 + \beta\Theta\right)). \tag{8.57}$$

The initial conditions are

$$\bar{\eta}(0) = 1, \quad \Theta(0) = 0. \tag{8.58}$$

The parameter α characterizes the initial state of the chemical system. Depending on its value the chemical reaction either changes to a slow regime with decay of the reaction, or into a regime of self–acceleration which leads to an explosion. For some value of α (we call it critical) the reaction is maintained and gives rise to a rather sharp transition from slow motions to explosive ones. The transition region from slow regimes to explosive ones exists due to the continuous dependence of the system (8.56), (8.57) on the parameter α. To find the critical value of the parameter α, it is possible to use special asymptotic formulae [33]. That approach was used in [2, 3, 16, 27].

The equation

$$\bar{\eta}exp(\Theta/\left(1 + \beta\Theta\right)) - \alpha\Theta = 0$$

gives the slow curve S_α of the system (8.56), (8.57). The curve S_α has two jump points given by the equation

$$\frac{\partial}{\partial\Theta}\left(\bar{\eta}exp(\Theta/\left(1 + \beta\Theta\right)) - \alpha\Theta\right) = 0.$$

The jump points divide the slow curve into three parts $S^s_{1,\alpha}$, $S^u_{2,\alpha}$, $S^s_{3,\alpha}$ (see Fig. 8.6) which are zeroth order approximations for the corresponding slow integral manifolds $S^s_{1,\alpha,\varepsilon}$, $S^u_{2,\alpha,\varepsilon}$ and $S^s_{3,\alpha,\varepsilon}$. Manifolds $S^s_{1,\alpha,\varepsilon}$ and $S^s_{3,\alpha,\varepsilon}$ are stable and $S^u_{2,\alpha,\varepsilon}$ is unstable. It is clear that each value of α has a corresponding slow curve but these curves merge in the domain of critical values. Each manifold $S^s_{1,\alpha,\varepsilon}$, $S^u_{2,\alpha,\varepsilon}$ and $S^s_{3,\alpha,\varepsilon}$ is at the same time part of some trajectory of the system (8.56), (8.57).

With some values of α, trajectories of equations (8.56)–(8.58) move along the manifold $S^u_{2,\alpha,\varepsilon}$, sooner or later either falling into an explosive regime, or rapidly

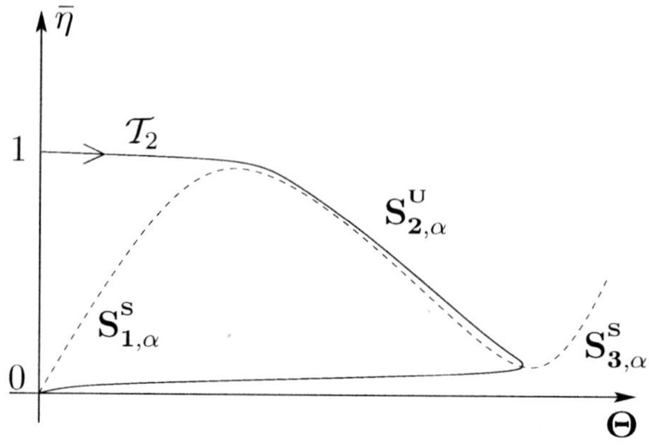

Figure 8.6. *The slow curve (the dashed line) and the trajectory $\mathcal{T}_2$ (the solid line)*

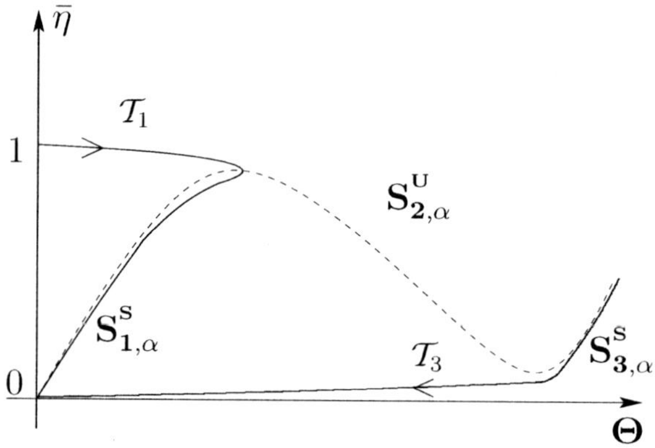

Figure 8.7. *The slow curve (the dashed line) and the trajectories $\mathcal{T}_1$ and $\mathcal{T}_3$ (the solid line)*

passing into a slow regime (see Fig. 8.6). The value of α_2, at which the trajectory $\mathcal{T}_2$ of (8.56)–(8.58) contains manifold $S^u_{2,\alpha,\varepsilon}$, is supposed to be critical. This regime is not slow, since $\Theta > 1$, and is not explosive, as the temperature increases at the tempo of the slow variable. The value α_1, at which the trajectory $\mathcal{T}_1$ contains the manifold $S^s_{1,\alpha,\varepsilon}$ (see Fig. 8.7), is called the slow critical value. The trajectory $\mathcal{T}_3$ contains the manifold $S^s_{3,\alpha,\varepsilon}$ and does not determine any critical regime, since it does not intersect the axis $\bar{\eta}$. We point out that any trajectory of the system starting at the point $\bar{\eta} = 1$, $\Theta = 0$ runs to the left from $\mathcal{T}_3$.

Thus the value of α_1 gives the critical trajectory. It separates the transition region from slow regimes which are characterized by a slowdown of the reaction with small degrees of conversion and heating up is limited from above by $\Theta < 1$.

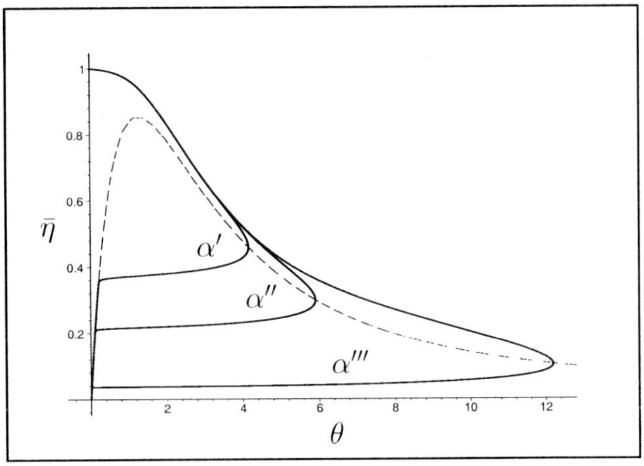

Figure 8.8. *The slow curve and the trajectories of (8.56)–(8.58) for* $\varepsilon = 0.01$, $\beta = 0.1$, $\alpha' = 2.08039$, $\alpha'' = 2.0803865$, $\alpha''' = 2.080386$

The region of slow transitional trajectories corresponds to the interval (α_2, α_1). They are characterized by a comparatively rapid (but not explosive) flow of the reaction till the essential degree of conversion takes place and then a jump slow-down and a transition to the slow flow of the reaction, Fig. 8.8.

The critical value α_2 was obtained by means of the asymptotic expansion technique given in [33]:

$$\alpha_2 = e(1-\beta)\left[1 - \Omega_0 \sqrt[3]{2}\left(1 + \frac{7}{3}\beta\right)\varepsilon^{2/3} + \frac{4}{9}(1+6\beta)\varepsilon \ln\frac{1}{\varepsilon}\right] + O(\varepsilon + \beta^2),$$

where $\Omega_0 = 2.338107$.

Autocatalytic reaction

The system showing autocatalytic features of the reaction is [19]

$$\varepsilon \frac{d\Theta}{d\tau} = \eta(1-\eta)exp(\Theta/(1+\beta\Theta)) - \alpha\Theta, \tag{8.59}$$

$$\frac{d\eta}{d\tau} = \eta(1-\eta)exp(\Theta/(1+\beta\Theta)). \tag{8.60}$$

To simplify the demonstration of the main qualitative effects we use a wide-spread assumption, $\beta = 0$, in thermal explosion theory. In this case the slow curve S_α of the system (8.59), (8.60) is described by the equation

$$\eta(1-\eta)e^\theta - \alpha\theta = 0.$$

The curve S_α has a different form depending on whether $\alpha > e/4$ or $\alpha < e/4$ (see Fig. 8.9). In the region $\Theta < 1$ some part of the curve S_α will be stable and in the

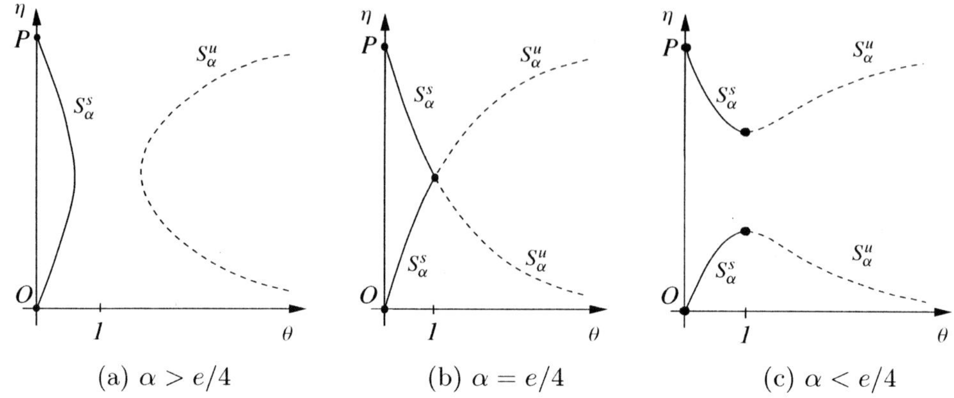

Figure 8.9. *The slow curve of the system (8.59), (8.60)*

region $\Theta > 1$ it will be unstable. We shall denote a stable part S_α as S^s_α and an unstable part as S^u_α. There exist integral manifolds $S^s_{\alpha,\varepsilon}$ and $S^u_{\alpha,\varepsilon}$ at a distance of $O(\varepsilon)$ from the curve S_α, corresponding to S^s_α and S^u_α.

As in the first–order reaction we shall give a qualitative description of the behavior of the system (8.59), (8.60) with the changing parameter α. When $\alpha > e/4$ the trajectories of the system in the phase plane move along the stable branch S^s_α and the value of Θ does not exceed 1. These trajectories correspond to the slow regimes.

With $\alpha < e/4$ the stable part S^s_α of the curve S_α consists of two separated branches and the system's trajectories, having reached the jump point at the tempo of the slow variable along S^s_α, jump into the explosive regime.

Due to the continuous dependence of the right–hand side of (8.59), (8.60) on the parameter α we can consider that there are some intermediate trajectories in the region between those shown above in the neighborhood of $\alpha = e/4$, and a critical one also. But with $\alpha = e/4$ the slow curve S^s_α has a self–intersection point $(1, 1/2)$, and it makes it impossible to apply the approach used in 8.3.4 to this case.

The canard, passing along the stable part of slow curve S^s_α and then along the unstable part S^u_α at some value of α (see Fig. 8.10), is taken as a mathematical object to model the critical trajectory in the autocatalytic case. The critical value of the parameter α^* corresponding to this trajectory is found in the form

$$\alpha^* = \alpha_0 + \varepsilon\alpha_1 + \varepsilon^2\alpha_2 + \ldots, \quad \alpha_0 = e/4.$$

The coefficients of this asymptotic series are found from recurrent formulas obtained by the methods of [15].

The critical value of α is given by

$$\alpha^* = e/4(1 - 2\sqrt{2}\varepsilon - 49/9\varepsilon^2) + O(\varepsilon^3).$$

Note that there is one more trajectory passing along $S^u_{\alpha,\varepsilon}$ and $S^s_{\alpha,\varepsilon}$. The part $S^s_{\alpha,\varepsilon}$ of this trajectory plays the same role as a manifold $S^s_{1,\alpha,\varepsilon}$ in Subsection 8.3.4

and separates the slow trajectories from transition ones. The value

$$\alpha^{**} = e/4(1 + 2\sqrt{2}\varepsilon - 49/9\varepsilon^2) + O(\varepsilon^3)$$

corresponds to this trajectory.

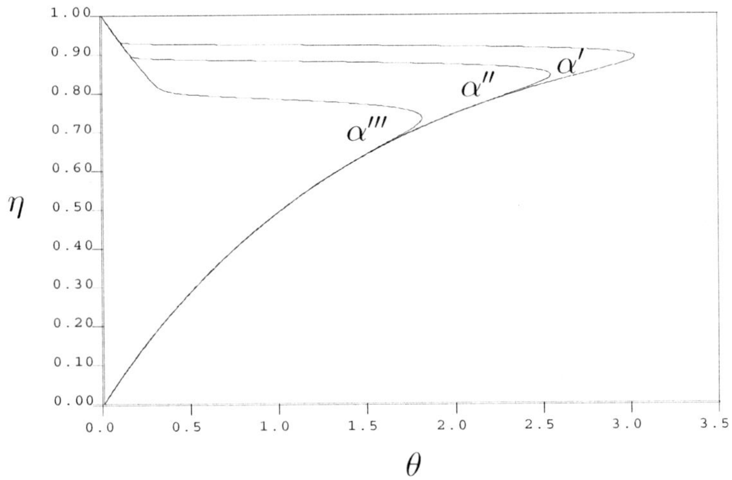

Figure 8.10. *Canard trajectories of system for* $\varepsilon = 0.05$, $\alpha' = 0.659941603$, $\alpha'' = 0.659941646$, $\alpha''' = 0.659952218$

The transition trajectories between $S^s_{\alpha,\varepsilon}$ and $S^u_{\alpha,\varepsilon}$ correspond to the interval (α^*, α^{**}). The canard corresponding to α^* is given by the formulas, see [15, 16]

$$\eta = H(\theta, \varepsilon) \equiv H_0(\theta) + \varepsilon H_1(\theta) + \dots .$$

$$H_0(\theta) = \frac{1}{2} \pm \sqrt{\frac{1}{4} - \alpha_0 \theta e^{-\theta}},$$

$$H_1(\theta) = \frac{\theta(\alpha_1 H_0' + \alpha_0)}{H_0'(1 - 2H_0)e^{\theta}},$$

$$H_2(\theta) = \frac{\theta\left(\alpha_1 H_1' + \alpha_2 H_0'\right) + H_0' H_1^2 e^{\theta} + H_1\left(1 - H_1'\right)\left(1 - 2H_0\right) e^{\theta}}{H_0'(1 - 2H_0)e^{\theta}}.$$

8.3.5 Canard travelling waves

In this subsection we shall consider the problem of thermal explosion in the case of an autocatalytic combustion reaction taking into account heat conductivity and diffusion of the reacting substances ([13, 16, 62]). We suppose for simplicity that $\beta = 0$, i.e. we shall investigate the system

$$\varepsilon \frac{\partial \theta}{\partial t} = \eta(1-\eta)e^{\theta} - \alpha\theta + \delta \frac{\partial^2 \theta}{\partial \xi^2} \,,$$

$$\varepsilon \frac{\partial \eta}{\partial t} = \varepsilon\eta(1-\eta)e^{\theta} + \mu \frac{\partial^2 \eta}{\partial \xi^2} \,. \tag{8.61}$$

The goal of this subsection is to study travelling wave solutions of (8.61) connecting the steady states $O(\eta = 0, \theta = 0)$ and $P(\eta = 1, \theta = 0)$. Analyzing the corresponding boundary value problem we will show that it is possible to choose the parameters in such a way that the projection of the associated heteroclinic trajectory onto the θ, η–plane is located in a small neighborhood of the canard trajectory characterizing the occurrence of a critical regime. We call the corresponding travelling wave solution a canard travelling wave solution.

Note that combustion waves have been extensively studied over the last three decades (see [6, 8, 12, 13, 31, 39, 50, 49, 51, 59, 61] and references therein). Most research has been focused on the adiabatic case for first order combustion reactions (for n-th order reactions see [6, 8]). The non-adiabatic case for a first order reaction has been studied in [60]. In the present subsection we investigate the non-adiabatic case ($\alpha > 0$) in case of an autocatalytic reaction.

We are interested in travelling wave solutions of (8.61) with constant speed c and which connect the steady states O and P. That means we are looking for solutions to (8.61) of the type

$$\theta(t,\xi) = \tilde{\theta}(\xi + ct) \equiv \theta(x), \quad \eta(t,\xi) = \tilde{\eta}(\xi + ct) \equiv \eta(x) \tag{8.62}$$

satisfying

$$\lim_{x \to -\infty} \eta(x) = \lim_{x \to -\infty} \theta(x) = 0,$$

$$\lim_{x \to +\infty} \eta(x) = 1, \quad \lim_{x \to +\infty} \theta(x) = 0,$$

and where $x = \xi + ct$ is the phase of the wave. Such solutions correspond to a one-dimensional flame propagating to the left with speed c.
Substituting (8.62) into (8.61) we get

$$\begin{aligned} \varepsilon c \frac{d\theta}{dx} &= \eta(1-\eta)e^{\theta} - \alpha\theta + \delta \frac{d^2\theta}{dx^2} \,, \\ \varepsilon c \frac{d\eta}{dx} &= \varepsilon\eta(1-\eta)e^{\theta} + \mu \frac{d^2\eta}{dx^2} . \end{aligned} \tag{8.63}$$

At first we consider the case of a travelling wave solution with speed $c = \nu/\varepsilon$ where ν does not depend on ε. Since ε is a small parameter we are looking for travelling waves with a high speed. In that case, (8.63) takes the form

$$\nu \frac{d\theta}{dx} = \eta(1-\eta)e^{\theta} - \alpha\theta + \delta \frac{d^2\theta}{dx^2} ,$$
$$\nu \frac{d\eta}{dx} = \varepsilon\eta(1-\eta)e^{\theta} + \mu\frac{d^2\eta}{dx^2}.$$

This system is equivalent to the system

$$\begin{aligned} \frac{d\eta}{dx} &= \varepsilon p , \\ \frac{d\theta}{dx} &= q , \\ \delta \frac{dq}{dx} &= \nu q - \eta(1-\eta)e^{\theta} + \alpha\theta , \\ \mu \frac{dp}{dx} &= \nu p - \eta(1-\eta)e^{\theta} . \end{aligned} \tag{8.64}$$

Introducing the new independent variable s by $s = \varepsilon x$ ($\varepsilon \neq 0$) we obtain

$$\begin{aligned} \frac{d\eta}{ds} &= p , \\ \varepsilon \frac{d\theta}{ds} &= q , \\ \varepsilon\delta \frac{dq}{ds} &= \nu q - \eta(1-\eta)e^{\theta} + \alpha\theta , \\ \varepsilon\mu \frac{dp}{ds} &= \nu p - \eta(1-\eta)e^{\theta} . \end{aligned} \tag{8.65}$$

Since ε is assumed to be small, (8.65) is a singularly perturbed system with the slow variable η and the fast variables θ, p, q. We are interested in a solution of (8.65) satisfying the boundary conditions

$$\lim_{s\to-\infty} \eta(s) = \lim_{s\to-\infty} p(s) = \lim_{s\to-\infty} \theta(s) = 0, \ \lim_{s\to-\infty} q(s) = 0,$$
$$\lim_{s\to+\infty} \eta(s) = 1, \ \lim_{s\to+\infty} p(s) = \lim_{s\to+\infty} \theta(s) = 0, \ \lim_{s\to+\infty} q(s) = 0,$$

that is, we are looking for a heteroclinic trajectory of the singularly perturbed system (8.65) connecting the equilibria O and P.

The degenerate equations of (8.65) read

$$\begin{aligned} 0 &= q , \\ 0 &= \nu q - \eta(1-\eta)e^{\theta} + \alpha\theta , \\ 0 &= \nu p - \eta(1-\eta)e^{\theta} , \end{aligned}$$

or, in more convenient form,

$$\begin{aligned} 0 &= q , \\ 0 &= -\eta(1-\eta)e^{\theta} + \alpha\theta , \\ \nu p &= \alpha\theta . \end{aligned} \tag{8.66}$$

The system (8.66) defines the slow surface $\tilde{S}_\alpha$ of the system (8.65) in R^4. It is easy to see that $\tilde{S}_\alpha$ is a differentiable curve located in the plane $q = 0, p = \alpha\theta/c$ and that its projection onto the θ, η-plane coincides with S_α introduced in the previous subsection (see 8.3.4). $\tilde{S}_\alpha$ represents the set of equilibria of the so called *layer subsystem*

$$\begin{aligned} \frac{d\theta}{d\tau} &= q \, , \\ \delta \frac{dq}{d\tau} &= \nu q - \eta(1-\eta)e^\theta + \alpha\theta \, , \\ \mu \frac{dp}{d\tau} &= \nu p - \eta(1-\eta)e^\theta \, . \end{aligned} \tag{8.67}$$

It can be checked that for $\nu < 0$ the slow surface $\tilde{S}_\alpha$ consists of stable (unstable) equilibria of (8.67) in the region $\theta > 1$ $(\theta < 1)$.

We now find the critical value of the parameter $\alpha = \tilde{\alpha}^*(\varepsilon)$

$$\tilde{\alpha}^* = \tilde{\alpha}_0 + \varepsilon\tilde{\alpha}_1 + \varepsilon^2\tilde{\alpha}_2 + \ldots$$

corresponding to the canard trajectory of (8.64) (see Fig. 8.11), which can be approximated by the asymptotic series

$$\begin{aligned} \eta &= \tilde{H}(\theta,\varepsilon) = \tilde{H}_0(\theta) + \varepsilon\tilde{H}_1(\theta) + \varepsilon^2\tilde{H}_2(\theta) + \ldots \, , \\ q &= \varepsilon\tilde{Q}(\theta,\varepsilon) = \varepsilon\tilde{Q}_1(\theta) + \varepsilon^2\tilde{Q}_2(\theta) + \ldots \, , \\ p &= \tilde{P}(\theta,\varepsilon) = \tilde{P}_0(\theta) + \varepsilon\tilde{P}_1(\theta) + \varepsilon^2 P_2(\theta) + \ldots \end{aligned}$$

under fixed values of ν, δ, μ. For this purpose we shall use the usual technique [17].

To calculate the coefficients of the asymptotic expansions we substitute these series into (8.65) and equate the coefficients of the same power of ε. We get

$$\tilde{\alpha}_0 = \alpha_0, \;\; \tilde{\alpha}_1 = \alpha_1, \;\; \tilde{\alpha}_2 = \alpha_2 + \frac{e^2}{2c^2}\left(\frac{5}{3}\delta - \mu\right),$$

$$\tilde{H}_0 = H_0, \; \tilde{H}_1 = H_1,$$

where α_i and H_i are defined in 8.3.4.

We return to system (8.63). In what follows we assume

$$\delta = \kappa\mu, \tag{8.68}$$

where κ is some positive constant.

Introducing the new variable z by $x = -cz$ we get from (8.63), taking into account (8.68),

$$\begin{aligned} -\varepsilon \frac{d\theta}{dz} &= \eta(1-\eta)e^\theta - \alpha\theta + \frac{\kappa\mu}{c^2}\frac{d^2\theta}{dz^2} \, , \\ -\varepsilon \frac{d\eta}{dz} &= \varepsilon\eta(1-\eta)e^\theta + \frac{\mu}{c^2}\frac{d^2\eta}{dz^2} . \end{aligned} \tag{8.69}$$

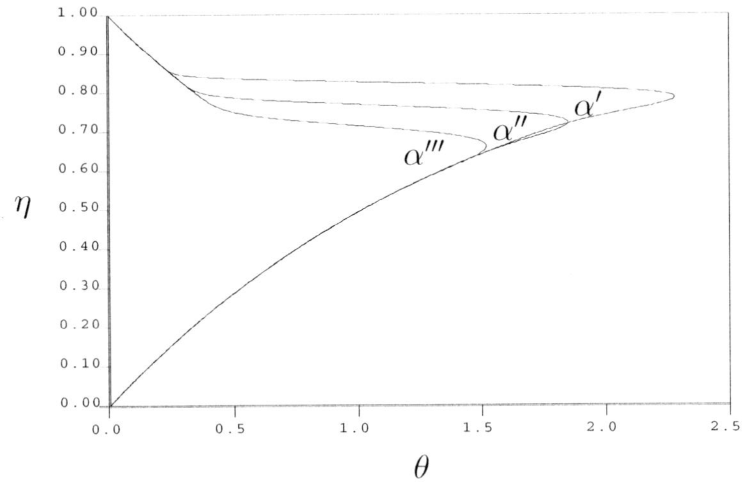

Figure 8.11. *Projection of canard trajectories of the system (8.64) for* $\varepsilon = 0.05$, $\alpha' = 0.66022803$, $\alpha'' = 0.66024835$, $\alpha''' = 0.66025281$

From now on we set

$$\varepsilon := \mu/c^2$$

and assume ε (dimensionless) to be small, that is, we assume that the quotient of the diffusivity and the square of the velocity is small. System (8.69) is equivalent to the singularly perturbed system

$$\begin{aligned}
\frac{d\eta}{dz} &= -p, \\
\frac{d\theta}{dz} &= -q \ , \\
\kappa\varepsilon \frac{dq}{dz} &= -\varepsilon + q\eta(1-\eta)e^{\theta} - \alpha\theta \ , \\
\varepsilon \frac{dp}{dz} &= -\varepsilon p + \varepsilon\eta(1-\eta)e^{\theta}.
\end{aligned} \tag{8.70}$$

Assuming $\alpha > 0$, the system (8.70) has the equilibria $O_1 := (p = q = \eta = \theta = 0)$ and $P_1 := (p = q = \theta = 0, \eta = 1)$ which do not depend on any parameter.

The corresponding degenerate equations are

$$\begin{aligned}
0 &= \varepsilon q - \eta(1-\eta)e^{\theta} + \alpha\theta \ , \\
0 &= p - \eta(1-\eta)e^{\theta} \ ,
\end{aligned}$$

and their solution set S_α can be represented in the form

$$\begin{aligned}
q &= \varepsilon^{-1}(\eta(1-\eta)e^{\theta} - \alpha\theta) \ , \\
p &= \eta(1-\eta)e^{\theta} \ .
\end{aligned}$$

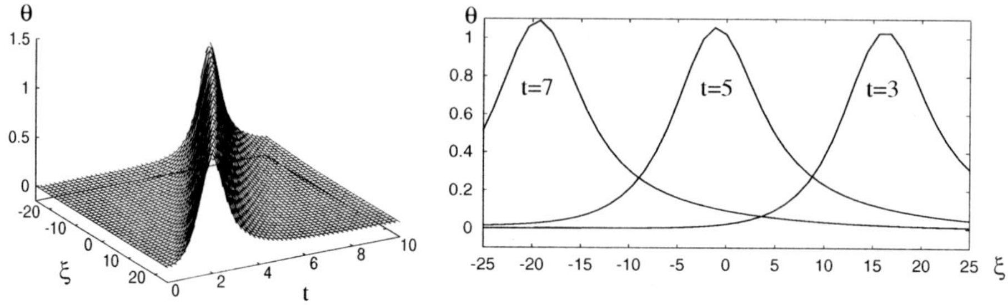

Figure 8.12. *θ–profiles of the canard travelling wave solution of system (8.61) for $\varepsilon = 0.05, \alpha = 0.58443, \delta = \mu = 1$*

It is easy to check that S_α contains no jump point. According to a fundamental result of the geometric theory of singularly perturbed differential equations [11, 58], there exists, for sufficiently small ε, a smooth invariant manifold $S_{\alpha,\varepsilon}$ of (8.70) which is close to S_α, it contains the equilibria O_1 and P_1 and can be represented in the form

$$q = \varphi(\eta, \theta, \varepsilon) = \varepsilon^{-1}\Big[\eta(1-\eta)e^\theta - \alpha\theta + \varepsilon\varphi_1(\eta,\theta) + O(\varepsilon^2)\Big],$$
$$p = \psi(\eta, \theta, \varepsilon) = \eta(1-\eta)e^\theta + \varepsilon\psi_1(\eta,\theta) + O(\varepsilon^2)\ .$$

On $S_{\alpha,\varepsilon}$, the system (8.70) can be written as

$$\frac{d\eta}{dz} = \eta(1-\eta)e^\theta + \varepsilon\psi_1(\eta,\theta) + O(\varepsilon^2),$$
$$\varepsilon\frac{d\theta}{dz} = \eta(1-\eta)e^\theta - \alpha\theta + \varepsilon\varphi_1(\eta,\theta) + O(\varepsilon^2).$$

With $\alpha = \tilde{\alpha}^*$ the system (8.61) has a canard travelling wave solution connecting the equilibria O and P [40]. The canard value $\tilde{\alpha}^*(\varepsilon)$ separates two types of waves corresponding to the slow combustion regime and to the thermal explosion (self-ignition) one, respectively. As in 8.3.4, the case $\alpha > \tilde{\alpha}^*(\varepsilon)$ corresponds to slow combustion profiles, while the case $\alpha < \tilde{\alpha}^*(\varepsilon)$ characterizes self-ignition profiles.

The value $\tilde{\alpha}^*$ permits us to write the asymptotic expansion for the speed of the travelling wave solution $c = \tilde{c}$ for a fixed value of the parameter α which differs from $\alpha^*(\varepsilon)$ by $\Delta = O(\varepsilon^2)$, i. e. $\Delta = \alpha - \alpha^*$. For

$$\Delta(\frac{5}{3}\delta - \mu) > 0$$

we have

$$\tilde{c}^2 = v^2/\varepsilon^2 = \frac{e^2}{2\Delta}(\frac{5}{3}\delta - \mu) + O(1).$$

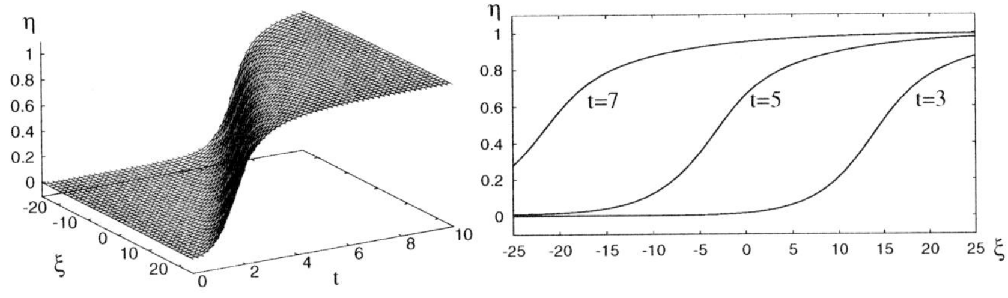

Figure 8.13. *η–profiles of the canard travelling wave solution of system (8.61) for $\varepsilon = 0.05, \alpha = 0.58443, \delta = \mu = 1$*

The speed of the travelling wave solution satisfying the inequality $c^2 < \tilde{c}^2$ ($c^2 > \tilde{c}^2$) corresponds to the slow combustion (self-ignition) travelling wave solutions.

Figures 8.12, 8.13 show numerical investigations of the travelling wave solution of the system (8.61) in the case of the critical regime.

8.3.6 Gas combustion in a dust–laden medium

We now consider models of combustion of a rarefied gas mixture in an inert porous or in a dusty medium. We assume that the temperature distribution and phase–to–phase heat exchange are uniform. The chemical conversion kinetics are represented by a one–stage, irreversible reaction. The dimensionless model in this case has the form [3, 14]

$$\varepsilon\dot{\Theta} = \Psi(\eta)exp(\Theta/\,(1+\beta\Theta)) - \alpha(\Theta - \Theta_c) - \delta\Theta,$$

$$\gamma_c\dot{\Theta}_c = \alpha(\Theta - \Theta_c),$$

$$\dot{\eta} = \Psi(\eta)exp(\Theta/\,(1+\beta\Theta)),$$

$$\eta(0) = \eta_0/\,(1+\eta_0) = \bar{\eta}_0, \;\; \Theta(0) = \Theta_c(0) = 0.$$

Here, Θ and Θ_c are the dimensionless temperatures of the reactant phase and of the inert one; η is the depth of conversion; η_0 is the criterion of autocatalyticity; the small parameters β and ε characterize the physical properties of a gas mixture. The terms $-\delta\Theta$ and $-\alpha(\Theta - \Theta_c)$ reflect the external heat dissipation and phase-to-phase heat exchange. The parameter γ_c characterizes the physical features of the inert phase. Depending on the relation between values of the parameters, the chemical reaction either changes to a slow regime with decay of reaction, or into a regime of self–acceleration which leads to an explosion. So, if we change the value of one parameter with fixed values of the other parameters we can change the type of chemical reaction. Thus, it is possible to consider this problem as a special control problem. For example, if we take a heat loss from the gas phase as a control action, we consider δ as a control variable. If the control variable is γ_c it means a regulation of the dust level in the reactant vessel.

The following cases are considered:

$$\Psi(\eta) = \begin{cases} 1-\eta, & \text{first-order reaction } (\eta_0 = 0) \\ \eta(1-\eta), & \text{autocatalytic reaction.} \end{cases}$$

Autocatalytic reaction

Let us consider the combustion model for the case of autocatalytic reaction ($\Psi(\eta) = \eta(1-\eta)$).

In the absence of external heat dissipation ($\delta = 0$) the system of differential equations possesses a first integral

$$\varepsilon\Theta + \gamma_c\Theta_c - \eta = \bar{\eta}_0,$$

and therefore we obtain $\dim y = 0$ in (8.1). The dependence of the slow curve S_α

$$\eta(1-\eta)\exp\left(\Theta/\left(1+\beta\Theta\right)\right) - \alpha(\Theta - (\eta - \bar{\eta}_0)/\gamma_c) = 0$$

on the relation between parameter values gives different forms, as in the case of the classical model (see Subsection 8.3.4).

In the first case each set S_α^s and S_α^u of S_α consists of a single connected curve. Hence the system has an stable integral manifold $S_{\alpha,\varepsilon}^s$ and a unstable integral manifold $S_{\alpha,\varepsilon}^u$ near S_α^s and S_α^u, respectively.

Since the initial point $(0, \bar{\eta}_0)$ belongs to the basin of attraction of the set $S_{\alpha,\varepsilon}^s$, after a short time the trajectory follows the stable slow integral manifold $S_{\alpha,\varepsilon}^s$ and tends to the equilibrium P as t tends to ∞. This behavior corresponds to the slow combustion regime.

In the other case, each set S_α^s and S_α^u consists of two different components and the system has an stable integral manifold $S_{\alpha,\varepsilon}^s$ (unstable integral manifold $S_{\alpha,\varepsilon}^u$) near each component of S_α^s (S_α^u). For ε sufficiently small and after a short time, the solution will follow the component of $S_{\alpha,\varepsilon}^s$ to the breakdown point. After this time, $\Theta(t)$ will increase rapidly. This behavior characterizes the explosive regime.

The transition region from the slow regime to the explosive one exists due to the continuous dependence of our system on the parameters α and γ_c ($\gamma_c > 0$). In this special case the slow curve has an intersection point. Here the system has an stable integral manifold $S_{\alpha,\varepsilon}^s$ (unstable integral manifold $S_{\alpha,\varepsilon}^u$) near each component of the slow curve S_α^s (S_α^u).

We can observe the existence of canard solutions which describe the following regime: the temperature increases as high as is possible but without explosion, and that may be the aim of a technological process. We note that this regime is critical, and it corresponds to a chemical reaction separating the domain of self–accelerating reactions and the domain of slow reactions.

If we take α as a control parameter we can find the canard solution and corresponding value of α by following asymptotic expansions

$$\alpha^* = \alpha_0 + \varepsilon\alpha_1 + \varepsilon^2\alpha_2 + ...,$$

$$\eta(\Theta,\varepsilon) = H_0(\Theta) + \varepsilon H_1(\Theta) + \varepsilon^2 H_2(\Theta) + ...$$

In this case, the asymptotic expansion of the canard value of parameter α is [55, 57] (we take the zero–approximation term with order $O(\beta)$ and the first–approximation term with order $O(\varepsilon)$)

$$\alpha^* = \frac{1-\beta\Theta_{00}^2}{2+\sqrt{4+\gamma_c^{-2}}}e^{\Theta_{00}}\left[1-\varepsilon\left(\frac{1}{2}\gamma_c^{-2}+\frac{1}{2}\gamma_c^{-1}\left(2+\sqrt{4+\gamma_c^{-2}}\right)+\right.\right.$$

$$\left.\left.+\sqrt[4]{4+\gamma_c^{-2}}\sqrt{2+\sqrt{4+\gamma_c^{-2}}}\right)\right], \quad \Theta_{00} = \frac{1}{2}\left(\gamma_c^{-1}+\sqrt{4+\gamma_c^{-2}}\right).$$

In the case $\delta \neq 0$ we can observe the analogous situation in R^3.

Let us consider γ_c as a control parameter and recall that it means a regulation of a dust level in the reactant vessel. In this case we construct a special type of feedback control.

We find the critical function $\gamma_c(\Theta,\varepsilon)$ and the black swan $\Theta_c = \Theta_c(\eta,\Theta,\varepsilon)$ in the form of asymptotic expansions:

$$\gamma_c = \Gamma_0(\Theta) + \varepsilon\Gamma_1(\Theta) + O(\varepsilon^2),$$

$$\Theta_c = P_0(\eta,\Theta) + \varepsilon P_1(\eta,\Theta) + O(\varepsilon^2).$$

It should be noted that we use the black swan for the following reasons. We construct the canard modelling the critical regime with fixed initial point (or gluing one). However during a chemical process perturbations are possible. Due to the perturbations the trajectory of the system deviates from the canard and as a result a qualitative change of system behaviour is possible.

Using the method of integral manifolds we obtain (to simplify the calculations we ignore the small parameter β)

$$P_0(\eta,\Theta) = \alpha^{-1}[(\alpha+\delta)\Theta - \eta(1-\eta)e^{\Theta}],$$

$$P_1(\eta,\Theta) = \left[\alpha(\Theta-P_0) - \eta(1-\eta)e^{\Theta})\frac{\partial P_0}{\partial\eta}\Gamma_0\right]/\alpha\Gamma_0\frac{\partial P_0}{\partial\Theta},$$

$$\Gamma_0(\Theta) = \frac{\alpha(\alpha+\delta-\delta\Theta)}{(\alpha+\delta)\sqrt{e^{2\Theta}-4e^{\Theta}(\alpha+\delta)}},$$

$$\Gamma_1(\Theta) = -\frac{\alpha P_1 + \Gamma_0\left[\eta(1-\eta)e^{\Theta}\dfrac{\partial P_1}{\partial\eta} + \alpha P_1\dfrac{\partial P_1}{\partial\Theta}\right]}{\eta(1-\eta)e^{\Theta}\dfrac{\partial P_0}{\partial\eta}}.$$

First-order reaction

The case of the first-order reaction ($\Psi(\eta) = (1-\eta)$) is studied now. For simplicity we introduce the dimensionless concentration $\bar{\eta} = 1 - \eta$.

In the absence of external heat dissipation ($\delta = 0$) the system

$$\varepsilon\dot{\Theta} = \bar{\eta}\exp\left(\Theta/\left(1+\beta\Theta\right)\right) - \alpha(\Theta - \Theta_c),$$

$$\gamma_c\dot{\Theta}_c = \alpha(\Theta - \Theta_c),$$

$$\dot{\bar{\eta}} = -\bar{\eta}\exp\left(\Theta/\left(1+\beta\Theta\right)\right),$$

with initial conditions

$$\bar{\eta}(0) = 1, \quad \Theta(0) = \Theta_c(0) = 0,$$

possesses a first integral

$$\varepsilon\Theta + \gamma_c\Theta_c + \bar{\eta} = 1,$$

and we obtain $\dim y = 0$ in (8.1).

The slow curve S_α is defined by the equation

$$\bar{\eta}\exp\left(\Theta/\left(1+\beta\Theta\right)\right) - \alpha\big(\Theta - \gamma_c^{-1}(1-\bar{\eta})\big) = 0.$$

With different relations between values of the parameters α, δ and γ_c we can observe the following chemical regime types:

— the slow combustion regime;

— the classic thermal explosion;

— thermal explosion with delay [3, 14].

The last regime consists of three stages: fast initial, slow (delay) and explosive, see Fig. 8.14. This regime is characterized by a rather long induction period and a significant time for reactant conversion before a thermal explosion.

It should be noted that there are two types of slow regimes: the slow regime with essential initial heating (EIH, see Fig. 8.15) and the slow regime with nonessential initial heating (NIH, see Fig. 8.16).

Thus, in this case there are two critical regimes which separate fast explosive, explosive with delay and non-explosive regimes, see Fig. 8.17. The first critical regime takes place when, after a short time, the trajectory reaches the jump point and then follows the unstable slow integral manifold. This trajectory and the corresponding value of the control parameter can be found by Mishchenko–Rozov asymptotics [33], and if we take α as a control parameter we get [17]

$$\alpha^* = (1-\beta)e - \varepsilon^{2/3}\Omega_0\sqrt[3]{2(1-\gamma_c^{-1})^2 e}\left[1+\beta\left(1+\frac{4\gamma_c^{-1}}{3(1-\gamma_c^{-1})}\right)\right] +$$

$$+\frac{4}{9}\varepsilon\ln\frac{1}{\varepsilon}e\left(1-\gamma_c^{-1}\right) + O\left(\beta + \varepsilon\ln\frac{1}{\varepsilon}\right).$$

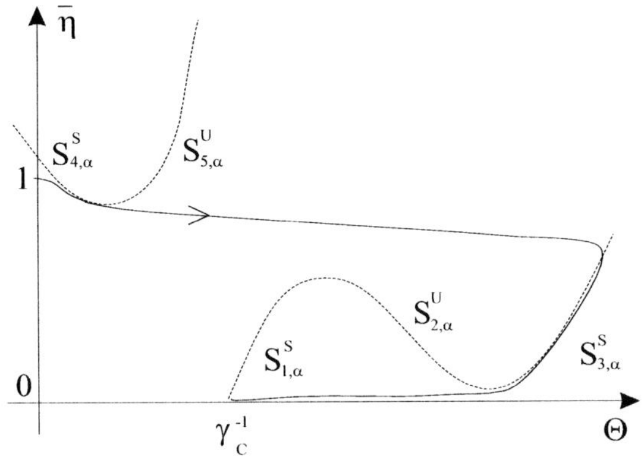

Figure 8.14. *The slow curve (the dashed line) and the trajectory (the solid line) in the case of thermal explosion with delay*

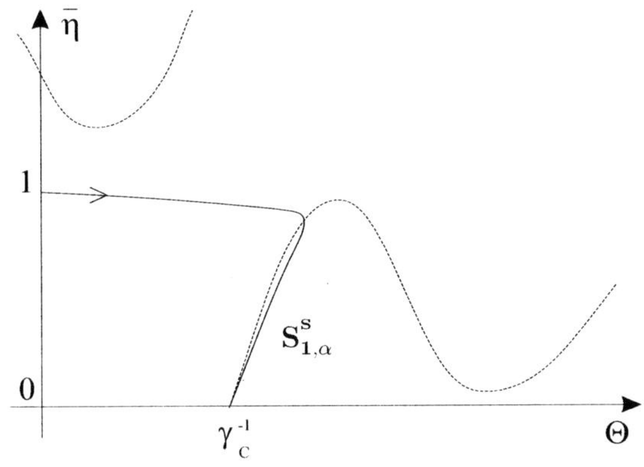

Figure 8.15. *The slow curve (the dashed line) and the trajectory (the solid line) in the case of slow combustion regime with essential initial heating*

The second critical regime is modelled by a canard, see Fig. 8.18. The asymptotic expansion of the canard value α^{**} is [14, 17]

$$\alpha^{**} = \exp\left(\frac{\gamma_c^{-1}}{1+\beta\gamma_c^{-1}}\right)\left[\gamma_c - \varepsilon\big(2+\gamma_c^{-1}+\beta(4-2\gamma_c^{-1})\big)\right] + o(\varepsilon+\beta).$$

If we investigate the system in a more general case ($\delta \neq 0$) we can construct a black swan which consists of canards simulating the second type of critical regimes.

Let us take $\gamma_c(\Theta, \varepsilon)$ as control function. Then it and the black swan $\Theta_c = \Theta_c(\bar{\eta}, \Theta, \varepsilon)$ have asymptotic expansions of the form:

$$\gamma_c = \Gamma_0(\Theta) + \varepsilon\Gamma_1(\Theta) + O(\varepsilon^2),$$

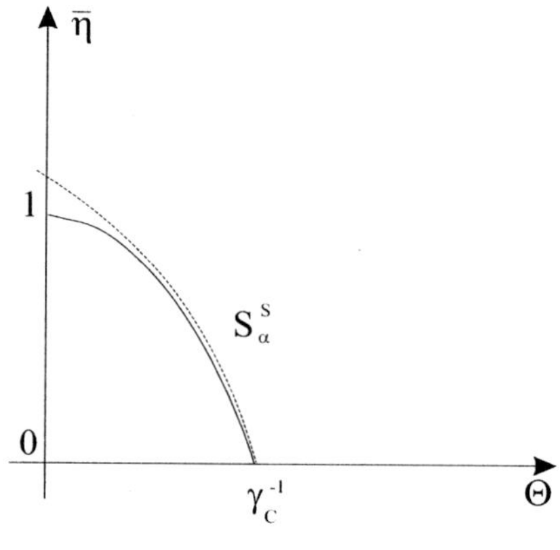

Figure 8.16. *The slow curve (the dashed line) and the trajectory (the solid line) in the case of slow combustion regime with nonessential initial heating*

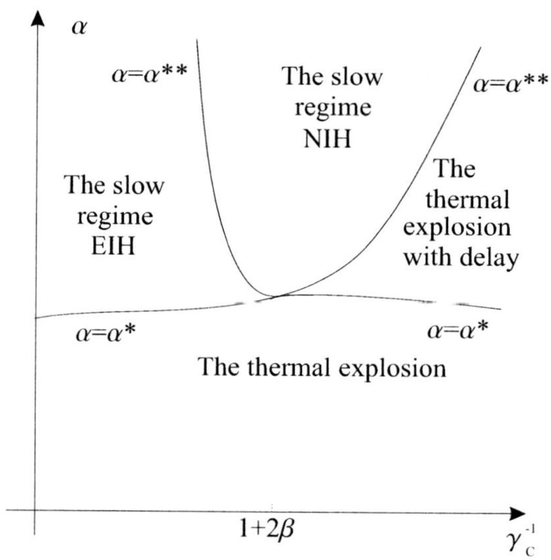

Figure 8.17. *The domains of parameters and the associated types of combustion regimes*

$$\Theta_c = P_0(\bar{\eta}, \Theta) + \varepsilon P_1(\bar{\eta}, \Theta) + O(\varepsilon^2),$$

where

$$P_0(\bar{\eta}, \Theta) = (\delta\Theta - \bar{\eta}e^{\Theta})/\alpha + \Theta,$$

$$P_1(\bar{\eta}, \Theta) = -\delta\Theta\bar{\eta}e^{\Theta}/(\alpha + \delta - \delta\Theta),$$

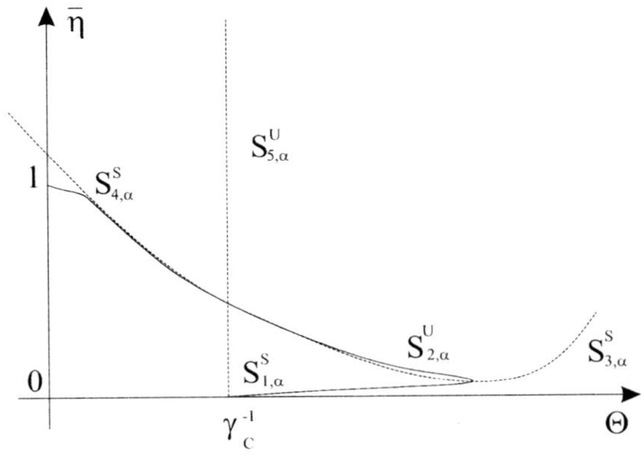

Figure 8.18. *The slow curve (the dashed line) and the canard (the solid line) in the case of second critical regime*

$$\Gamma_0(\Theta) = \alpha\frac{\alpha+\delta-\delta\Theta}{(\alpha+\delta)e^{\Theta}},$$

$$\Gamma_1(\Theta) = -\frac{\alpha^2\delta\Theta\big[(\alpha+\delta-\delta\Theta)(\alpha\delta\Theta-\alpha-\delta)+\alpha\delta(\alpha+\delta)\big]}{(\alpha+\delta)^2(\alpha+\delta-\delta\Theta)^2}.$$

In the case when δ is a control function (it means that we control the combustion process by regulating the external heat dissipation) we get the following asymptotic expansion for and δ [46]

$$\delta = \delta(\Theta,\varepsilon) = \Theta^{-1}\Big[\alpha\frac{\left(\Theta - \ln\alpha\gamma_c^{-1} - 1\right)e^{\Theta} + \alpha\gamma_c^{-1}}{\alpha\gamma_c^{-1} - e^{\Theta}} +$$

$$+\varepsilon(\alpha\gamma_c^{-1} - e^{\Theta}) + O(\varepsilon^2)\Big]$$

corresponding to the black swan $\Theta_c = \Theta_c(\bar{\eta},\Theta,\varepsilon)$ of the system.

For the fixed point $\Theta = \Theta^*$ of the breakdown curve we can find the value δ^* from the last expression which corresponds to the canard of the system. This trajectory passes through the point Θ^* of the breakdown curve and simulates the critical regime. It should be noted that the choice of the gluing point Θ^* is equivalent to the choice the starting point $\Theta(0)$ of the trajectory. For example, with $\Theta(0) = 0$, $\gamma_c = 1/6$, $\varepsilon = 0.01$, $\alpha = 2.34$ the critical regime corresponds to $\delta^* = 1.10797$.

8.4 Acknowledgements

The authors are grateful to Grigory Barenblatt, Michael Mortell and Robert O'Malley for helpful discussions. This work was supported in part by the Boole Centre for Research in Informatics, University College Cork, by the Program "Development of Higher School Scientific Potential" 3.3 no. 75364 and by the Presidium of the Russian Academy of Science Program no. 19.

Bibliography

[1] V. I. Arnold, V. S. Afraimovich, Yu. S. Il'yashenko, and L. P. Shil'nikov, *Theory of Bifurcations*, in Dynamical Systems, 5, Encyclopedia of Mathematical Sciences, V. Arnold, ed., Springer Verlag, New York, 1994.

[2] V. I. Babushok and V. M. Goldshtein, *Structure of the thermal explosion limit*, Combust. Flame, 72 (1988), pp. 221–224 .

[3] V. I. Babushok, V. M. Goldshtein, and V. A. Sobolev, *Critical condition for the thermal explosion with reactant consumption*, Combust. Sci. and Tech., 70 (1990), pp. 81–89.

[4] E. Benoit, J. L. Callot, F. Diener, and M. Diener, *Chasse au canard*, Collect. Math., 31–32(1–3) (1981–1982), pp. 37–119.

[5] E. Benoit, *Systèmes lents-rapides dans R^3 et leurs canards*, Société Mathématique de France. Astérisque, 1983, pp. 109–110, 159–191.

[6] H. Berestycki, B. Nikolaenko, and B. Scheurer, *Travelling wave solutions to combustion models and their singular limits*, SIAM J. Math. Anal., 16 (1985), pp. 1207–1242.

[7] M. Brøns and K. Bar Eli, *Asymptotic analysis of canards in the EOE equations and the role of the inflection line*, Proc. R. Soc. Lond. A, 445 (1994), pp. 305–322.

[8] J. D. Buckmaster and G. S. S. Ludford, *Theory of Laminar Flames*, Cambridge Univ. Press, Cambridge, 1982.

[9] M. Diener, *Nessie et Les Canards*, Publication IRMA, Strasbourg, 1979.

[10] W. Eckhaus, *Relaxation oscillations including a standard chase on French ducks*, Lecture Notes Math., 925 (1983), pp. 449–494.

[11] N. Fenichel, *Geometric singular perturbation theory for ordinary differential equations*, J. Diff. Eq., 31 (1979), pp. 53–98.

[12] P. C. Fife and B. Nikolaenko, *The singular perturbation approach to flame theory with chain and competing reactions*, Ordinary and Partial Differential Equations, W. N. Everitt and B. D. Sleeman, eds., Lect. Notes in Math., 962 (1980), pp. 232–250.

[13] D. A. FRANK-KAMENETSKII, *Diffusion and Heat Transfer in Chemical Kinetics*, Plenum Press, New York, 1969.

[14] V. GOL'DSHTEIN, A. ZINOVIEV, V. SOBOLEV, AND E. SHCHEPAKINA, *Criterion for thermal explosion with reactant consumption in a dusty gas*, Proc. R. Soc. Lond. A, 452 (1996), pp. 2103–2119.

[15] G. N. GORELOV AND V. A. SOBOLEV, *Duck-trajectories in a thermal explosion problem*, Appl. Math. Lett., 5(6) (1992), pp. 3–6.

[16] ——, *Mathematical modelling of critical phenomena in thermal explosion theory*, Combust. Flame, 87(1991), pp. 203–210.

[17] G. GORELOV, V. SOBOLEV AND E. SHCHEPAKINA, *The Singularly Perturbed Models of Combustion.* Russian Academy of Natural Sciences, SamVien, Samara, 1999 (in Russian).

[18] J. GRASMAN AND J. J. WENTZEL, *Co-existence of a limit cycle and an equilibrium in Kaldor's business cycle model and its consequences*, J. of Economic Behavior and Organization, 24 (1994), pp. 369–377.

[19] B. F. GRAY, *Critical behaviour in chemical reacting systems: 2. An exactly soluble model*, Combust. Flame, 21 (1973), pp. 317–325.

[20] G. HUYET, P. A. PORTA, S. P. HEGARTY, J. G. MCINERNEY, AND F. HOLLAND, *A low-dimensional dynamical system to describe low-frequency fluctuations in a semiconductor laser with optical feedback*, Optics Communications, 180 (2000), pp. 339–344,.

[21] A. KELLY, *The Stable, Centre-Stable, Centre, Centre-Unstable and Unstable Manifolds*, J. of Differential Equations, 3 (1967), pp. 546–570.

[22] L. I. KONONENKO AND V. A. SOBOLEV, *Asymptotic expansion of slow integral manifolds, Sib. Math. J.*, 35 (1994), pp. 1119–1132.

[23] M. A. KRASNOSEL'SKII AND P. P. ZABREIKO, *Geometrical Methods of Nonlinear Analysis*, Springer-Verlag, Berlin, 1984.

[24] M. KRUPA AND P. SMOLYAN, *Extending slow manifolds near transcritical and pitchfork singularities*, Nonlinearity, 14 (2001), pp. 1473–1491.

[25] R. LANG AND K. KOBAYASHI, *External optical feedback effects on semiconductor injection laser properties*, IEEE J. Quantum Electron., QE-16 (1980), pp. 347–355.

[26] H. LI, J. YE, AND J. G. MCINERNEY, *Detailed analysis of coherence collapse in semiconductor lasers*, IEEE J. Quantum Electron., QE-29 (1993), pp. 2421–2432.

[27] A. LINAN AND D. K. KASSOY, *The influence of reacting consumption on the critical conditions for homogeneous thermal explosion*, A. J. Mech. Appl. Math. Electron., 31(1) (1978), pp. 99-111.

[28] A. M. LYAPUNOV, *The General Problem of the Stability of Motion*, Tayor & Francis, London, Washington, DC, 1992.

[29] P. MANDEL AND T. ERNEUX, *The slow passage through a steady bifurcation: delay and memory effects*, J. of Statistical Physics, 48(5–6) (1987), pp. 1059–1070.

[30] A. C. MCINTOSH , B. F. GRAY, AND G. C. WAKE, *Analysis of the bifurcational behaviour of a simple model of vapour ignition in porous material*, Proc. R. Soc. Lond. A, 453 (1997), pp. 281–301.

[31] A. G. MERZHANOV AND B. I. KHAIKIN, *Theory of Combustion Waves in a Homogeneous Medium*, Russian Academy of Sciences, Chernogolovka, 1992 (in Russian).

[32] E. F. MISHCHENKO, YU. S. KOLESOV, A. YU. KOLESOV, AND N. KH. ROZOV, *Asymptotic Methods in Singularly Perturbed Systems*, Plenum Press, New York, 1995.

[33] E. F. MISHCHENKO AND N. KH. ROZOV, *Differential Equations with Small Parameters and Relaxation Oscillations*, Plenum Press, New York, 1980.

[34] YU. A. MITROPOL'SKII AND O. B. LYKOVA, *Integral Manifolds in Nonlinear Mechanics*, Nauka, Moscow, 1975 (in Russian).

[35] J. MOEHLIS, *Canards in a surface oxidation reaction*, J. Nonlinear Sci, 12 (4) (2002), pp. 319-345.

[36] J. MOERK AND B. TROMBORG, *Stability analysis and the route to chaos for laser diodes with optical feedback*, IEEE Phot. Tech. Lett., 2 (1990), pp. 21–23.

[37] A. I. NEISHTADT, *On delayed stability loss under dynamical bifurcation, I, II*, Differential Equations, 23 (1987), pp. 2060–2067; 24 (1988), pp. 226–233.

[38] R. E. O'MALLEY, J. G. L. LAFORGUE, AND M. WARD, *Metastable travelling wave solutions of singularly perturbed reaction–diffusion equations*, Euro. J. Appl. Math., 9 (1998) pp. 399-416.

[39] K. SCHNEIDER, *A note on the existence of periodic travelling wave solutions with large periods in generalized reaction-diffusion systems*, Z. Angew. Math. Phys., 34 (1983), pp. 236-240.

[40] K. SCHNEIDER, E. SHCHEPAKINA, AND V. SOBOLEV, *A new type of travelling wave*, Mathematical Methods in the Applied Sciences, 26 (2003), pp. 1349–1361.

[41] N. N. SEMENOV, *Zur theorie des verbrennungsprozesses*, Z. Physik. Chem., 48 (1928), pp. 571–581.

[42] E. A. SHCHEPAKINA, *Attracting-repelling integral surfaces in combustion problems*, Mat. Model., 14(3) (2002), pp. 30–42 (in Russian).

[43] ——, *Integral manifolds, duck trajectories, and thermal blast*, Vestn. Samar. Gos. Univ. Mat. Mekh. Fiz. Khim. Biol., Special Issue (1995), pp. 49–58 (in Russian, MR 1789170).

[44] ——, *Black swans and canards in self-ignition problem*, Nonlinear Anal. Real World Appl., 4(1) (2003), pp. 45–50.

[45] E. SHCHEPAKINA AND V. SOBOLEV, *Attracting/repelling invariant manifolds*, Stab. Control Theory Appl., 3(3) (2000), pp. 263–274.

[46] ——, *Integral manifolds, canards and black swans*, Nonlinear Analysis. Ser. A: Theory Methods, 44(7) (2001), pp. 897–908.

[47] ——, *Standard chase on black swans and canards*, Weierstraß–Institut für Angewandte Analysis und Stochastik. Preprint 426, Berlin, 1998.

[48] A. R. SHOUMAN, *Solution to the dusty gas explosion problem with reactant consumption. Part I: the adiabatic case*, Combust. Flame, 119(1-2) (1999), pp. 189-194.

[49] G. I. SIVASHINSKY AND C. GUTFINGER, *Applications of asymptotic methods to laminar flame theory*, in Current Topics in Heat and Mass Transfer, C. Gutfinger, ed., Hemispheric Publishing Corporation, Washington, D.C., 1975, pp. 555–607.

[50] G. I. SIVASHINSKY, *On a steady corrugated flame front*, Astronautica Acta, 18 (1974), pp. 253–260.

[51] J. SMOLLER, *Shock Waves and Reaction-Diffusion Equations*, Springer Verlag, New York, 1983.

[52] P. SMOLYAN AND M. WECHSELBERGER, *Canards in R^3*, J. Diff. Equations, 177(2) (2001), pp. 419–453.

[53] V. A. SOBOLEV, *Geometrical theory of singularly perturbed control systems*, Proc. 11th Congress of IFAC, Tallinn, 6 (1990), pp. 163–168.

[54] ——, *Integral manifolds and decomposition of singularly perturbed systems*, System and Control Lett., 5 (1984), pp. 169–179.

[55] V. A. SOBOLEV AND E. A. SHCHEPAKINA, *Duck trajectories in a problem of combustion theory*, Differential Equations, 32 (1996), pp. 1177–1186.

[56] ——, *Integral surfaces of variable stability and duck-trajectories*, Transactions of RANS, series MMMIC, 1(3) (1997), pp. 151–175 (in Russian, Zbl 0927.34031).

[57] ——, *Self-ignition of laden medium*, J. Combustion, Explosion and Shock Waves, 29(3) (1993), pp. 378–381.

[58] V. V. Strygin and V. A. Sobolev, *Separation of Motions by the Integral Manifolds Method*, Nauka, Moskow, 1988 (in Russian, MR 89k:34071).

[59] A. I. Volpert, V. A. Volpert, and V. A. Volpert, *Traveling Wave Solutions of Parabolic Equations*, AMS Translations of Math. Monographs 140, 1994.

[60] R. O. Weber, G. N. Mercer, B. F. Gray, and S. D. Watt, *Combustion waves: non-adiabatic*, in Modeling in Combustion Science. J. Buckmuster, T. Takeno, eds., Springer Lect. Notes in Physics, 449, Springer-Verlag, Berlin, 1995.

[61] F. Williams, *Combustion Theory*, Addison-Wesley, Reading MA, 1983.

[62] Ya. B. Zeldovich, G. I. Barenblatt, V. B. Librovich, and G. M. Makhviladze , *The Mathematical Theory of Combustion and Explosions*, Consultants Bureau, New York, 1985.

[63] A. K. Zvonkin and M. A. Shubin, *Non-standard analysis and singular perturbations of ordinary differential equations*, Russian Math. Surveys, 39(2) (1984), pp. 69–131.

Chapter 9

Multi-Scale Analysis of Pressure Driven Flames

V. Bykov, I. Goldfarb, and V. Gol'dshtein

The present paper considers and examines the evolution of asymptotic tools applied to the problem of pressure driven flames in inert porous media to deal with the growth in the level of the complexity of the mathematical models of the phenomena.

It was shown recently that the hydrodynamic conditions of gas flow through a porous medium filled with flammable gas may play the crucial role in the determination of the flame speed. A number of models of combustion waves driven by a local elevation of pressure in inert porous media filled with a combustible gaseous mixture were suggested and analyzed. To study a typical problem a number of steps are made. The original complex system of PDEs is reduced to a system of ODEs by introducing the automodel time-like coordinate. The introduction of new auxiliary variables permits the derivation of a system of ODEs, which correspond to a singularly perturbed system of ODEs. To determine these new variables, one divides the flame front into two distinct sub-zones (preheat and reaction) and derives approximate laws of conservation (energy and momentum) within each of the sub-zones. The new variables are defined as deviations from the laws of conservation. The dynamics of the singularly perturbed system is then analyzed using a novel mathematical technique, namely, a geometric version of the method of integral (invariant) manifolds (MIM). This approach permits us to explore analytically the fine internal structure of the reaction zone, to derive explicit formulae for the parameters of the burnt products, and to obtain an explicit formula for the flame speed. To check the accuracy of the analytical results obtained from the analysis of the model, a series of direct numerical simulations are performed, and these are in a good agreement with theoretical predictions.

The paper also presents the most recent results in the development of the asymptotic tool. A solution of a model example is given to exhibit the essence of the suggested improvement.

9.1 Introduction

We begin with a short history of, and a motivation for, the present work. The study of pressure driven flames in porous media was initiated some years ago by some intriguing experimental results obtained by Prof. Vyacheslav Babkin and his group [1]. They showed that a dynamical picture of possible regimes of propagating flames in porous media is complex, and that it contains both low- and high-speed combustion waves. Their results and numerous discussions with Prof. Babkin drew our attention to the fact that the effect of pressure is possibly underestimated and that the problem of flame propagation through porous media should be revisited.

Traditionally the study of premixed gas flames in an open space ignores pressure perturbations. This is because the flame propagation velocity is significantly less than the speed of sound. This assumption is well founded in many scientific and engineering problems. It means that pressure disturbances leave the reaction zone and do not influence the intensity of the thermal processes. In this research we deal with gas phase combustion in inert porous media. The distinguishing feature of such a medium is the well-known fact that under specific conditions the speed of pressure perturbations may be significantly lower than the sound velocity in open space. Under these conditions a local elevation of the pressure may lead to the formation of a self-sustaining combustion wave controlled by pressure diffusion. In works published as a result of numerous considerations [3, 4, 5, 14], a new physical model of the phenomenon was suggested and analyzed. In one of the key publications on this theme [14], a new mechanism for flame spreading through porous media was advanced and a corresponding physical model was developed and analyzed for the isentropic approximation. The main idea of the proposed mechanism is that under certain conditions the propagation of a combustion wave is governed mainly by the diffusion of pressure in porous media. Using the δ-function approximation of the reaction zone (high activation energy assumption) the authors demonstrated that the flame velocity may be significantly higher than that in open space (caused by the conventional thermal conductivity). Later the existence of the travelling wave solution was proved in [3] under the additional simplifying assumption of weak heat release. The detailed analysis of the internal structure of the flame was presented in [3] and the transition from low velocity regimes (governed by the conventional thermal conductivity) to high velocity regimes (controlled by pressure diffusion) was analyzed.

Later there were a number of attempts to adapt the method of inner and outer asymptotic expansions to the problem (under the high activation energy assumption). Our colleagues and we intended to investigate the fine structure of the flame front and to derive analytical formulae for the wave velocity. To our surprise all these attempts failed and our joint attempts to obtain the results along the lines of the conventional machinery adopted in the field were not successful. While analyzing the possible reasons for this, we concluded that our troubles were conceptual. The study of a new type of combustion problem may lead to new mathematical models, and may demand new mathematical approaches as well.

We (a group of applied mathematicians at Ben-Gurion University of the Negev) continued to look for an appropriate mathematical approach to the problem. As a result of our research we suggested a new formal basis for asymptotic methods in the mathematical theory of combustion, namely, the theory of integral (invariant) manifolds. This technique is different from the conventional machinery used in the field. We have found a way to apply this technique to the problem of pressure driven flames in porous media. The mathematical formulation involves a multi-scale problem, but the corresponding system of ODEs is not written as a singular perturbation system (SPS).

To reach the desired results we exploit the data of the preliminary analysis of the problem under investigation along the lines of well-known Zel'dovich method [23]. The classical Zel'dovich approach allows us to single out the preheat and reaction sub-zones within the flame front. The fine structure of these sub-zones and the relations between the main parameters of the reaction (temperature, pressure and concentration) within them are subject to analytical investigation. In addition, the Zel'dovich approach allows us to detect approximate laws of conservation which are valid within the sub-zones. To reformulate the mathematical problem and rewrite the set of ODEs in the form of the SPS of equations, we introduce new auxiliary variables, which are defined as deviations from the appropriate conservation laws (partial integrals). Due to the method of choice, it is reasonable to expect that their rates of change will be slow (compared to the rate of change of any other variable). Further detailed analysis confirms our expectations and gives us the possibility of applying a modified MIM (geometrical version) to the reformulated problem. The approach we develop allows us to investigate the fine structure of the flame front and to get analytical formulae for the main flame characteristics. Direct numerical simulations of the original set of equations are also performed and these results are in a good agreement with theoretical predictions.

In the present work we demonstrate how the asymptotic tools evolved as a function of the complexity of the models describing the phenomenon of flame propagation through a porous medium. Additionally, we present new results on combustion waves in porous media driven by the local pressure elevation and suggest a prototype of a novel asymptotic tool (our final results), which is currently in the very early stages of its development. This latter elaboration allows us to solve the problem under investigation without the subdivision of the flame front into two sub-zones.

The structure of the paper is as follows. Section 2 contains the formulation of the generic problem of the pressure-driven flame propagation in porous media. Section 3 exhibits briefly the chosen asymptotic tool - the method of integral manifolds (MIM). Section 4 demonstrates how the simplest problem (non-inertial, linear friction force - an approximation of the original system of ODEs) is solved along the lines of Zel'dovich's classical approach. The same problem is solved in Section 5 using a modified version of the MIM, developed especially for this problem. Section 6 includes a solution of a more complicated problem (non-inertial, the friction force being proportional to the square of the local gas velocity). The full original system described in Section 2 is analyzed in Section 7, where we demonstrate the current version of the asymptotic tool and its successful application. The last Section, Section 8, summarizes our results and formulates directions for further research. An Appendix contains our most recent work on improving the asymptotic method we have illustrated here.

9.2 Problem Statement - General Description

We describe the main assumptions of the generic model. We restrict ourselves to a one-dimensional approach as it is sufficient to give us conceptual qualitative information about the dynamics of the process. The porous medium is considered as a set of evenly spread parallel capillaries of the same inner radius (the so called capillary model of solid foam), filled with a premixed combustible gas mixture (the solid matrix is inert). The natural assumption is that within all the capillaries the same processes take place and the flame propagates under the same conditions and at the same velocity in all the capillaries (the cell model). The capillaries may be considered as thermally insulated (the detailed justification of this assumption may be found in a recent paper of the authors [9] devoted to the problem of thermal runaway in solid foam, where the capillary model of solid foam was considered). The conventional one-temperature approach and cell model is applied. The relationship between the inner capillary radius and the gas mixture viscosity may lead to the appearance of so called creeping flow of the reactant mixture. This regime corresponds to low Reynolds number flow.

To elucidate the impact of the pressure effect on the wave characteristics (the front structure, propagation velocity, etc) and to make the problem tractable analytically, some additional assumptions are made. In particular, the conventional mechanism of the combustion wave propagation (thermal diffusion) is excluded from our consideration. With the above assumptions, the generic system of governing equations includes six equations: the energy equation (9.1), the concentration equation (9.2), the momentum equation (9.3), the continuity equation (9.4), the equation of the state for the ideal gas (9.5), and the Arrhenius reaction rate of chemical reaction (one-step, bimolecular reaction of the first order) (9.6).

$$\frac{d}{dx}\left(\rho\left(u-D\right)\left(c_vT+\frac{1}{2}u^2\right)+pu\right)=QW, \tag{9.1}$$

$$\frac{d}{dx}\left(\rho\left(u-D\right)C_f\right)=-W, \tag{9.2}$$

$$\frac{d}{dx}\left(\rho\left(u-D\right)+p\right)=F, \tag{9.3}$$

$$\frac{d}{dx}\left(\rho\left(u-D\right)\right)=0, \tag{9.4}$$

$$P=\frac{\rho}{\mu}RT=\left(c_p-c_v\right)\rho T, \tag{9.5}$$

$$W = AC_f \rho \exp\left(-\frac{E}{RT}\right). \tag{9.6}$$

The expression for the friction force F can be taken as proportional to the first power of the gas velocity (Darcy's law, subscript D) or the multiplication $u\,|u|$ (Forcheimer's law, subscript F)

$$F_F = -K_F \rho u\,|u|\,; \qquad F_D = -K_D \rho u\,. \tag{9.7}$$

The following notation was used: T - temperature (K); P - pressure (Pa); E - activation energy ($J/kmol$); D - velocity of the flame front in the laboratory system of coordinates (m/s); C_f - concentration of the deficient reactant; c - specific heat capacity ($J/kg/K$); u - gas velocity in the laboratory frame of reference (m/s); Q - combustion energy (J/kg); W - reaction rate ($kg/(sm^3)$); ρ - density (kg/m^3); K - permeability of the medium (m^2); μ - kinematic viscosity (m^2/s); A - pre-exponential (frequency) factor ($1/s$); R - universal gas constant. Subscripts mean: f - combustible component of the gas mixture; p - under constant pressure; v - under constant volume; 0 - undisturbed state; b - burnt (behind the combustion wave front), F - related to the case of quadratic dependence of the friction force on gas velocity. The system (9.1)-(9.7) is subject to boundary conditions (fresh mixture far ahead of the flame front)

$$\begin{aligned} &T(x \to +\infty) = T_0;\ C_f(x \to +\infty) = C_{f0}, \\ &P(x \to +\infty) = P_0;\ \rho(x \to +\infty) = \rho_0. \end{aligned} \tag{9.8}$$

The system (9.1)-(9.7) together with the boundary conditions (9.8) are a mathematical description of the problem.

To simplify our further analysis we introduce the following dimensionless variables

$$\xi = -\frac{x}{D} A \exp\left(\frac{1}{2\beta}\right); \quad \beta = \frac{RT_0}{E};$$

$$\eta = \frac{C_f}{C_{f0}}; \quad \theta = \frac{1}{\beta}\frac{T - T_0}{T_0}; \quad \Pi = \frac{1}{\beta}\frac{P - P_0}{P_0},$$

where θ, Π, η are dimensionless temperature, pressure and concentration, respectively, ξ is a dimensionless coordinate, and β is a reduced initial temperature.

Further dimensionless parameters which will be used are

$$\begin{gathered} \varepsilon_1 = \frac{C_{f0}Q}{C_p T_0 \beta} \exp\left(-\tfrac{1}{2\beta}\right); \quad \varepsilon_2 = \exp\left(-\tfrac{1}{2\beta}\right); \\ \varepsilon_1 \ll 1;\ \ \varepsilon_2 \ll 1,\ \text{since}\,\beta \ll 1. \\ \sigma = 1 - \tfrac{1}{\gamma};\ \ \gamma = \tfrac{c_p}{c_v}. \end{gathered} \tag{9.9}$$

The initial conditions are

$$\theta(\xi \to -\infty) = \theta_0 = 0;\ \eta(\xi \to -\infty) = \eta_0 = 1;\ \Pi(\xi \to -\infty) = \Pi_0 = 0. \tag{9.10}$$

9.3 Method of Integral Manifolds (MIM) – Asymptotic Method of Analysis

The theory of integral manifolds was developed for nonlinear mechanics in a number of works [2, 7, 17, 18, 22]. This approach was adopted for problems of chemical kinetics and combustion in [15]. It has been developed for problems of gaseous combustion [15], mechanics and non-linear control theory [20, 21, 22], self-ignition in multiphase media [10, 11, 12], and flame propagation in porous media [9].

Suppose that the initial value problem can be rewritten as a singularly perturbed system with the small parameter ε in general form as:

$$\varepsilon \frac{d\vec{X}}{dt} = F(\vec{X}, \vec{Y}, \varepsilon), \tag{9.11}$$

$$\frac{d\vec{Y}}{dt} = G(\vec{X}, \vec{Y}, \varepsilon), \tag{9.12}$$

$$\vec{X}(t_0) = \vec{X_0}; \ \vec{Y}(t_0) = \vec{Y_0}.$$

Here $\vec{X} \in R^m$, $\vec{Y} \in R^n$, $t \in (-\infty, +\infty)$, $0 < \varepsilon << 1$. The functions $F : R^m \times R^n \to R^m$, $G : R^m \times R^n \to R^n$are supposed to be sufficiently smooth for all $\vec{X} \in R^m$, $\vec{Y} \in R^n$, ε is a small parameter, and the values $F(\vec{X_0}, \vec{Y_0}, \varepsilon)$, $G(\vec{X_0}, \vec{Y_0}, \varepsilon)$ are assumed to be comparable to unity for small values of the parameter ε.

A smooth manifold (surface) in the phase space $M \in R^m \times R^n$ is called a global invariant manifold of the system (9.11)-(9.12), if any phase trajectory $(\vec{X}(t, \varepsilon), \vec{Y}(t, \varepsilon))$ such that $(\vec{X}(t_1, \varepsilon), \vec{Y}(t_1, \varepsilon)) \in M$ belongs to M for any $t > t_1$.

It is possible to define a local integral manifold in a similar way. The proposed approach deals mostly with a special class of invariant manifolds, which leads to considerable simplifications in the analysis of the system. More specifically, the system (9.11)-(9.12) is considered as a multi-scale system with the small parameter ε, with 'slow' and 'fast' subsystems. By setting $\varepsilon = 0$ into (9.11)-(9.12) we obtain so-called degenerate system:

$$0 = F(\vec{X}, \vec{Y}, 0), \tag{9.13}$$

$$\frac{d\vec{Y}}{dt} = G(\vec{X}, \vec{Y}, 0).$$

Equation (9.13) determines the so-called slow (quasi-stationary) surface. It is assumed that equation (9.13) has an isolated smooth solution $\vec{X} = h_0(\vec{Y})$, so that the surface has the dimension of the slow variable Y. It is also assumed that all the eigenvalues $\lambda_I(\vec{Y})$ of the matrix $F_{\vec{X}}(h_0(\vec{Y}), \vec{Y}, 0)$ have a non-zero real part.

Points on the surface determined by the Eq. (9.13) can be sub-divided into two types: standard points and turning points. A point (X,Y) is a standard point of the slow surface if in some neighborhood of this point the surface can be represented as the graph of a function $\vec{X} = h_0(\vec{Y})$ such that $\vec{X_0} = h_0(\vec{Y_0})$. Points where this condition is not satisfied are turning points of the slow surface. The slow surface has the dimension of the slow variable $\overrightarrow{Y}$.

The theory of integral (invariant) manifolds states that the system (9.11)-(9.12) has a unique integral (invariant) manifold $M := \{(\overrightarrow{X}, \overrightarrow{Y}) : \overrightarrow{X} = h(\overrightarrow{Y}, \varepsilon)\}$, that can be represented as

$$h(\overrightarrow{Y}, \varepsilon) = h_0(\overrightarrow{Y}) + \sum_{I \geq 1} \varepsilon^I h_I(\overrightarrow{Y}). \tag{9.14}$$

In the generic case, all eigenvalues λ_I of the matrix $F_{\overrightarrow{X}}(h_0(\overrightarrow{Y}), \overrightarrow{Y}, 0)$ are not zero. If, moreover, the eigenvalues λ_I satisfy the condition $Re(\lambda_I(\overrightarrow{Y})) \leq \alpha < 0$ for all $\overrightarrow{Y} \in R^n$, then the invariant manifold is stable, in the sense that it attracts the trajectories. Otherwise the invariant manifold is unstable, i.e. it repels some trajectories.

Note that the slow surface $\overrightarrow{X} = h_0(\overrightarrow{Y})$ is an $O(\varepsilon)$ approximation of the slow invariant manifold, except at the points at which the assumption on the eigenvalues does not hold (so-called turning points). We remark that the asymptotic series in (9.14) is not to be confused with the time-dependent asymptotic series. The invariant manifold (9.14) is called the invariant manifold of slow motions (or slow invariant manifold). The system dynamics on this manifold are described by

$$\frac{d\overrightarrow{Y}}{dt} = G(h(\overrightarrow{Y}, \varepsilon), \overrightarrow{Y}, \varepsilon). \tag{9.15}$$

If $\overrightarrow{Y}(t, \varepsilon)$ is a solution of (9.15) then the pair $(\overrightarrow{X}(t, \varepsilon), \overrightarrow{Y}(t, \varepsilon))$ with $\overrightarrow{X}(t, \varepsilon) = h(\overrightarrow{Y}(t, \varepsilon), \varepsilon)$ is a solution of the original system (9.11)-(9.12), since it determines a trajectory on the invariant manifold.

Thus, the analysis can be considerably simplified by reducing the dimension of the system to the dimension of the slow variables. In the $O(\varepsilon)$ approximation to the slow invariant manifold, the analysis of the original system can be reduced to the analysis on the slow surface $\overrightarrow{X} = h_0(\overrightarrow{Y})$. On the slow surface the changes of the slow and fast variables are comparable (i.e. fast and slow processes are balanced). Beyond the slow surface the slow variables are fixed (quasi-stationary) (to an $O(\varepsilon)$ approximation). The system behavior can be described by the typical system trajectories. Each trajectory can be decomposed into 'fast' parts (which are beyond the slow manifold) and 'slow' parts (which are on the slow manifold). The 'fast' and 'slow' parts of a trajectory can follow each other. The main types of system trajectories can be predicted by the use of the slow invariant manifold. The slow surface might consist of several branches: stable and unstable, in the sense that they attract or repel trajectories. In a generic case the "turning line" separates the stable branch from the unstable one. In the $O(\varepsilon)$ approximation of the slow invariant manifold, the turning line indicates (for typical situations) the end of the 'slow' part and the beginning of the 'fast' part of trajectories. In explosion problems, physically that means a transition from the slow process to the fast explosion. The slow motion on the slow manifold describes delay phenomena before the final ignition event. The time during which the trajectories stay on the slow surface gives a delay time before the final explosion, which is extremely important from a practical point of view.

We note that for systems with a complicated hierarchy (i.e. when there are several characteristic rates of change), the invariant manifold approach can be applied iteratively.

9.4 Linear Friction, No Inertia - Zel'dovich's Approach

One of the main goals of the present paper is to demonstrate the evolution of the asymptotic tools in line with the growth of the complexity level of the problems under investigation. The present section is devoted to the beginning of this process of matching mathematical tools to new problems.

At some stage of our research on the phenomenon (pressure driven flames in inert porous media) we were compelled to recognize the fact that existing methods applied to the problem failed to provide us with the desired information concerning the flame front (its structure, velocity, stability, etc).

While looking for a new line of attack, we applied Zel'dovich's [23] well-known approach to the analysis of the problem under investigation. We successfully applied this general approach to our specific problem. Moreover (and this is much more important) - the way in which we applied Zel'dovich's method prompted a direction for further development of the asymptotic tools we had used. Therefore, we will present a solution of the simplest version of the full problem (9.1) - (9.8) along the lines of Zel'dovich's approach as it was done initially in 1997, and give the conclusions of this work [9].

To focus on the essence of the approach, we start from the simplest version of the original system of PDEs (9.1)-(9.8) describing the phenomenon under consideration. We assume that the impact of the inertia effects is negligible and the friction force is proportional to the first power of the gas velocity (9.7). Non-dimensionalization and suitable integration allow us to re-write the original system in the form of the three ODES

$$\frac{d\theta}{d\xi} = \Lambda_D^2 \frac{\Pi - \theta}{1+\beta\theta} + \varepsilon_1 \eta \frac{1+\beta\Pi}{1+\beta\theta} \exp\left(\frac{\theta}{1+\beta\theta}\right), \tag{9.16}$$

$$\sigma \frac{d\Pi}{d\xi} = \Lambda_D^2 \frac{\Pi - \theta}{1+\beta\theta},$$

$$\frac{d\eta}{d\xi} = -\varepsilon_2 \eta \frac{1+\beta\Pi}{1+\beta\theta} \exp\left(\frac{\theta}{1+\beta\theta}\right), \tag{9.17}$$

where the dimensionless parameters ε_1, ε_2, β, σ are given by Eq. (9.9), and the initial conditions are determined by the Eq. (9.10). The expression for the flame velocity reads

$$\Lambda_D^2 = K_D \frac{D^2}{C_p T_0 A} \exp\left(\frac{1}{2\beta}\right).$$

The system (9.16)-(9.17) is adiabatic, and therefore the energy integral exists and can easily be derived. It reads

$$\eta - 1 + \frac{\varepsilon_2}{\varepsilon_1}(\theta - \sigma\Pi) = 0. \tag{9.18}$$

Using the high activation energy assumption, we single out two distinguishing sub-zones within the flame front: the preheat sub-zone and reaction sub-zone. The preheat sub-zone is characterized by a low reaction rate and the terms of the exothermic chemical reaction are negligible with respect to others. Hence, we can neglect all the reaction terms in the preheat sub-zone. Vice versa, within the reaction sub-zone the situation changes sharply. This sub-zone is typified by the extremely high reaction rate and the reaction comes to dictate "the rules of the game" in the system. All the non-reaction terms (except Arrhenius ones) are negligible here.

Therefore the original system of differential equations (9.16)-(9.17) may be significantly simplified in each of the specified sub-zones. For convenience in the further analysis, we introduce the point ξ_Q representing the boundary between the preheat and the reaction sub-zones. It is helpful here to use the following notation

$$\theta(\xi = \xi_Q) = \theta_Q;\ \eta(\xi = \xi_Q) = \eta_Q;\ \Pi(\xi = \xi_Q) = \Pi_Q.$$

9.4.1 Preheat sub-zone

Taking into account the above reasoning within the preheat sub-zone ($\xi < \xi_Q$) we can reduce the original system of governing equations (9.16)-(9.17) to the following form

$$\frac{d\theta}{d\xi} = \Lambda_D^2 \frac{\Pi - \theta}{1 + \beta\theta}, \tag{9.19}$$

$$\sigma\frac{d\Pi}{d\xi} = \Lambda_D^2 \frac{\Pi - \theta}{1 + \beta\theta}, \tag{9.20}$$

$$\frac{d\eta}{d\xi} = 0. \tag{9.21}$$

A first glance at the equations (9.19)-(9.21) allows us to conclude that the system is effectively decoupled: the equations (9.19)-(9.20) do not contain any reminder about the combustible gas component concentration, and the relation (9.21) does not include any trace of the pressure and temperature. Thus, the system is naturally separated and the last of these equations (9.21) shows that the concentration in the preheat sub-zone does not change. This variable conserves the initial value ($\eta = 1$) which it has within the fresh unburned mixture ($-\infty$). Further analysis shows that the growth of the dimensionless pressure should forestall the increase of the temperature because the coefficient in the right hand side of the equation (9.20) is greater than that in the equation (9.21).

Having excluded the dimensionless automodel variable η from the equations (9.19)-(9.20), we can integrate and derive the dependence of the temperature θ on the pressure Π in the preheat sub-zone

$$\theta - \sigma\Pi = 0. \tag{9.22}$$

Focusing on the point ξ_Q, which separates the preheat sub-zone from the reaction sub-zone, we relate the values of the temperature θ_Q and the pressure Π_Q at the inter-zone boundary

$$\theta_Q = \sigma\Pi_Q.$$

Note here, that a sub-zone of preliminary heating of the mixture at the head of the flame should be more precisely called a "pre-pressure" zone, because thermal diffusion is excluded from our consideration and heat is transferred due to pressure diffusion. Therefore, the term "preheat" is used conditionally to some extent.

9.4.2 Reaction sub-zone

The reaction sub-zone is characterized by the opposite hierarchy between the terms on the right hand side of the equation (9.16). The heat release terms are dominant in this sub-zone and the others are negligible. This permits us to rewrite the system (9.16)-(9.17) in the form

$$\frac{d\theta}{d\xi} = \varepsilon_1\eta\frac{1+\beta\Pi}{1+\beta\theta}\exp\left(\frac{\theta}{1+\beta\theta}\right), \tag{9.23}$$

$$\frac{d\Pi}{d\xi} = 0, \tag{9.24}$$

$$\frac{d\eta}{d\xi} = -\varepsilon_2\eta\frac{1+\beta\Pi}{1+\beta\theta}\exp\left(\frac{\theta}{1+\beta\theta}\right). \tag{9.25}$$

We note firstly, that the equation (9.24) implies no pressure changes within the reaction zone. We can conclude that the equality $\Pi(\xi) = \Pi_b$ is valid for $\xi > \xi_Q$,i.e., the pressure value $\Pi(\xi)$ within the reaction sub-zone equals the pressure Π_b behind the flame front (pressure of the reaction products). This means, in particular, that in the framework of this model the gas mixture pressure reaches its final value Π_b within the preheat sub-zone and does not changes within the reaction sub-zone.

The pair of equations (9.23), (9.25) in the reaction sub-zone allows us to establish an explicit functional relation between the temperature and the reactant concentration in this sub-zone (the pair have a first integral). To derive the integral, we multiply Equation (9.23) by $\frac{\varepsilon_2}{\varepsilon_1}$, sum the result with the expression (9.25), and integrate both right and left hand sides and use the boundary conditions. In accordance with our assumptions, the reactant is completely converted within the reaction sub-zone and its content behind the flame front is equal to zero. This means that when the temperature θ reaches its final value θ_b the concentration η has vanished. Hence, the first integral reads

$$\eta + \frac{\varepsilon_2}{\varepsilon_1}\theta = Const = 1 + \frac{\varepsilon_2}{\varepsilon_1}\sigma\Pi. \qquad (9.26)$$

The existence of the first integral of the pair (9.23), (9.25) can be physically interpreted in the following way: within the reaction sub-zone oxidation comes to be the single dominant mechanism. In particular, there are no significant energy losses in the reaction sub-zone (compared with the heat release) and the process is almost adiabatic (heat sinks are negligible). In turn, in such an approximation our physical system has an approximate conservation law - the energy integral (9.26).

This gives us the opportunity to connect the concentration and the temperature values at the inter-zone boundary point ξ_Q, and to derive values of the variables at the point ξ_Q

$$\Pi_Q = \Pi_b; \ \theta = \sigma\Pi_Q; \ \eta_Q = 1.$$

Comparing the energy integral (9.18) for the full system of equations (9.16)–(9.17) with the other relations ((9.22) and (9.26)), each is valid in the appropriate sub-zone of the flame. Thus, one can conclude that each of these relations represents an approximate conservation law within its domain of validity.

9.4.3 Flame velocity

The above analysis provides us with the internal structure of the flame front only. It does not answer one of the main questions of combustion theory - what is the velocity of the flame front. A formula for the propagation velocity can be obtained by introducing an additional assumption, which has been widely used in the theory of laminar flames [4]. The addition, which helps to determine the unknown flame speed, is the continuity of the heat flux at the point ξ_Q separating the preheat and the reaction zones. We recall here that the essence of the approximation (the subdivision of the actual distributions of the pressure, temperature and concentration into two sub-zones) employed in the present study is as follows. The gas characteristics in the preheat zone are determined by the local pressure elevation due to peculiarities of the hydrodynamic conditions (the exothermal oxidation reaction is unimportant here). Vice versa, in the reaction zone the combustion terms govern the behavior of the solution. To study the fine structure of the flame front we assumed that all the variables of the problem (temperature θ, concentration η and pressure Π, and, correspondingly, the local gas velocity u and the density ρ) are continuous through both the preheat and reaction zones.

The assumption that the heat flux at the point ξ_Q (inter-zone boundary) is continuous means that in addition to the continuity of the variables we demand the smoothness of the temperature $\theta(\xi)$. According to the conventional definition, the heat flux is proportional to the derivative of the temperature with respect to the corresponding coordinate (the automodel coordinate in our case). Hence, in the dimensionless variables used here the right hand sides of the equations (9.19) and (9.23) represent the heat fluxes in the preheat and reaction zones, respectively. Equating these two terms requires the equality of the two heat fluxes at the inter-

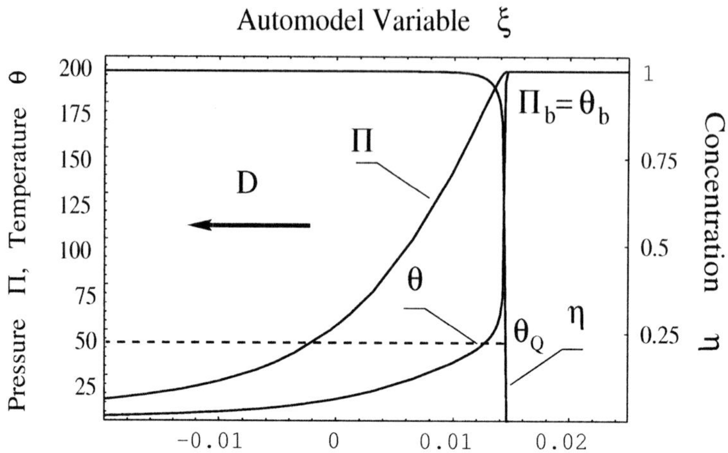

Figure 9.1. *Typical structure of the flame front. The system parameters are:* $\beta = 0.0295$; $\gamma = 1.3$; $\varepsilon_1 = 6.43 \times 10^{-6}$; $\varepsilon_2 = 4.35 \times 10^{-8}$. *The dimensionless flame velocity is* $\Lambda_D = 1.42$, *and the dimensional velocity is* $D = 72.1 m/s$

zone boundary, and consequently, the smoothness of the temperature profile $\theta(\xi)$. The result is

$$\Lambda_D^2 \frac{\Pi - \theta}{1 + \beta\theta} |_{\xi=\xi_Q} = \varepsilon_1 \eta(\theta, \Pi) \frac{1 + \beta\Pi}{1 + \beta\theta} \exp\left(\frac{\theta}{1 + \beta\theta}\right) |_{\xi=\xi_Q} . \tag{9.27}$$

Bearing in mind that at the point ξ_Q, separating the two zones, the temperature is θ_Q, the pressure is Π_Q and the concentration is $\eta_Q = 1$, (9.27) allows us to derive the expression for the flame propagation velocity Λ_D as a function of the temperature of the reaction products θ_b

$$\Lambda_D^2 = \varepsilon_1 \frac{1 + \beta\Pi_Q}{\Pi_Q - \theta_Q} \exp\left(\frac{\theta_Q}{1 + \beta\theta_Q}\right). \tag{9.28}$$

The formula (9.28) completes our study of the flame propagation in this framework.

9.4.4 Comparison with numerics

Typical results of the direct numerical simulations of the original system of equations (9.16)-(9.17) (time histories) are depicted in Fig. 9.1. The graphs (showing the dependence of the dimensionless temperature θ, pressure Π and concentration η on the coordinate ξ) permit us to easily distinguish two different parts of the flame front. The first one, where θ changes up to approximately the value 50, corresponds to the preheat zone. The gas pressure reaches its final value here, whereas the concentration almost does not change. The second part of this graph, where θ changes from 50 up to 200, matches the region of the fast motion (reaction zone).

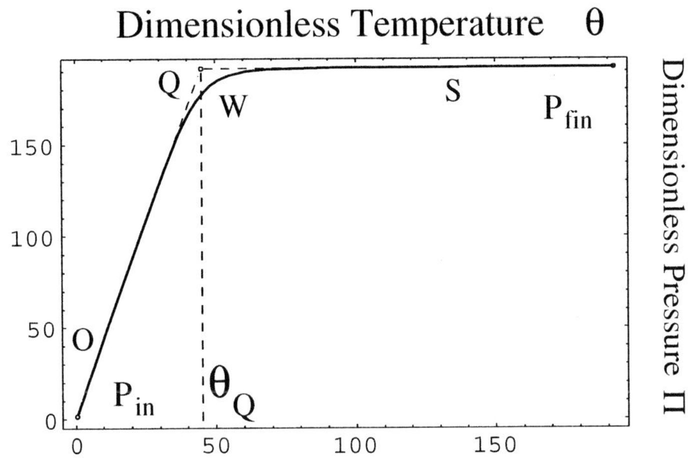

Figure 9.2. *Projection of the system's real trajectory on the planes* (θ, Π). *The system parameters are as in Fig. 9.1*

The graph of $\Pi(\xi)$ allows us to see that the pressure Π almost reaches its final value within the preheat zone. The function $\eta(\xi)$ shows that the concentration η of the combustible gaseous component hardly changes in the region of slow motion and almost full conversion occurs within the reaction zone. The fine structure of the flame is depicted in the Fig. 9.2. Fig. 9.2 presents analytical (dashed lines OQ and QS) and numerical (smooth solid curve OWS) solutions for the dimensionless pressure Π as a function of the dimensionless temperature θ. Fig. 9.3 compares analytical (dashed lines OQ and QS) and numerical solutions (smooth solid curve OWS) of the dimensionless concentration η as a function of the dimensionless temperature θ. One can see that the approximations obtained with Zel'dovich's classical approach describe with great accuracy both functions $\eta(\theta)$ and $\Pi(\theta)$ at any value of the temperature. The single exception is the narrow interval on the temperature scale near the point Q (separating the two sub-zones).

9.4.5 Conclusions

It is worthwhile to emphasize here, that there are two main conclusions. Firstly, we conclude that the pressure-driven flame is subdivided into two distinct sub-zones. Behind this subdivision of the flame front into two sub-zones there are rather deep physical principles (conservations laws). Within each of these sub-zones, the original system of governing equations can be simplified and reduced to a form which allows analytical solutions. The second conclusion is the existence of partial integrals within these sub-zones (each sub-zone is characterized by its own partial integral). Due to the approximate character of the subdivision, we can interpret the existence of the two partial integrals as the existence of approximate integrals of conservation (conservations laws). In other words, each sub-zone is characterized by a specific combination of variables which remains almost constant, while the phase trajectory

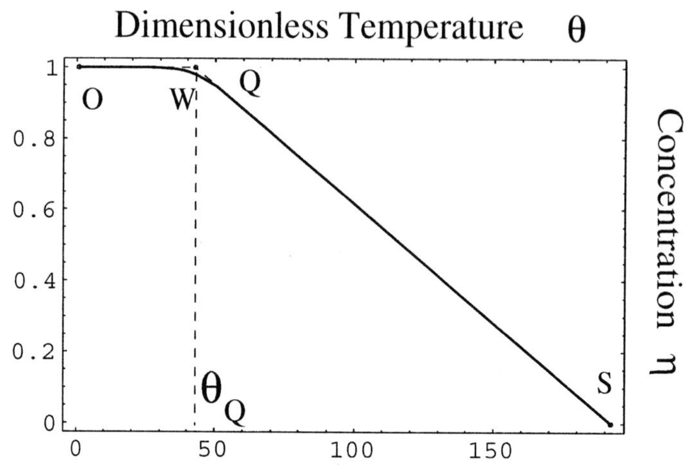

Figure 9.3. *Projection of the system's real trajectory on the planes (θ, η). The system parameters are as in Fig. 9.1*

of the system (9.16)-(9.17) belongs to the corresponding sub-zone.

From our point of view, these outcomes were of critical importance. They showed us the way forward and we used them to determine new auxiliary variables.

If we consider in detail either one of the sub-zones, the original system (9.16)-(9.17) or (9.29)-(9.30) of governing ODEs can be approximated within the selected sub-zone by a simpler one, which takes into account the peculiarities of the chosen sub-zone. The approximate system of ODEs provides us with the so-called partial integrals of the system ($\theta - \sigma\Pi = 0$ in the preheat sub-zone, $\eta + \frac{\varepsilon_2}{\varepsilon_1}\theta = const$ in the reaction sub-zone), which are evidently distinct from the energy integral of the original full system (9.1)-(9.8). The integrals, which are exact within the appropriate sub-zone of the flame, represent the approximate integrals of the original full system. The two approximate integrals are valid in distinct sub-zones of the flame. An expression "almost constant" means that the variable or a combination of the variables ($\theta - \sigma\Pi = 0$ or $\sigma\Pi = const$) changes slowly with respect to the other parameters involved in the problem. This slow change permits us to conclude that an expression representing the partial integral can be interpreted as a new variable, whose rate of change is known in advance - it is slower than that of any other variable in the problem. In other words - we found a way to determine a slow variable for each of the sub-zones.

9.5 Linear Friction, No Inertia - MIM Approach

The observations made during the process of looking for solutions of the original full system of ODEs along the lines of Zel'dovich's classical approach, revealed how we can rewrite the original system in the form of a singularly perturbed system of ODEs (recall, that the ODEs are not in the conventional SPS form). To implement this reformulation, we introduce new variables whose definitions are based on the outcome of the previous section. In what follows we present this approach, which was our first attempt [6] to modify the basic version of MIM [15] so that we could analyze the given system of ODEs. We return to the non-inertial, linear friction approximation (9.16)-(9.17) of the original full system (9.1)-(9.8) considered in the previous section. The dimensionless parameters of the system (9.16)-(9.17) are defined by the relations (9.9), the initial conditions are given by (9.10). The original system of three equations may be reduced to a system of the two ODEs due to the existence of the energy integral (9.18):

$$\frac{d\theta}{d\xi} = \Lambda_D^2 \frac{\Pi - \theta}{1 + \beta\theta} + \varepsilon_1 \eta(\theta, \Pi) \frac{1 + \beta\Pi}{1 + \beta\theta} \exp\left(\frac{\theta}{1 + \beta\theta}\right), \tag{9.29}$$

$$\sigma \frac{d\Pi}{d\xi} = \Lambda_D^2 \frac{\Pi - \theta}{1 + \beta\theta}. \tag{9.30}$$

We will exploit the existence of the approximate conservation laws within the sub-zones to determine new auxiliary variables and will proceed with our analysis in these variables.

9.5.1 Singularly perturbed system - SPS

Despite a significant difference in the characteristic times of the processes involved, the system (9.16)-(9.17) and the reduced system (9.29)-(9.30) do not represent a singularly perturbed system of ODEs in the conventional form. To make the reduced system of ODEs tractable to the chosen asymptotic approach (MIM), it is helpful to subdivide consideration of the problem into the two stages corresponding to the two distinct regions in the phase plain (θ, Π). As we will see, the suggested subdivision follows naturally the structure of the flame front as can be obtained from Zel'dovich's approach and represents an advance of the accepted approach, see [9].

In accordance with Zel'dovich's approach, we can single out two qualitatively different sub- zones within the flame front - preheat and reaction ones. Within the preheat zone the energy of the system $(\theta - \sigma\Pi\ )$ is almost constant and conserves its initial zeroth order value (due to the negligible impact of the exothermic chemical reaction at low temperatures in this sub-zone), whereas the system's momentum $(\sigma\Pi)$ changes significantly. Therefore, an approximate integral $\theta - \sigma\Pi = 0$ exists in the preheat sub-zone and a variable u, defined as $u = \theta - \sigma\Pi$, can be interpreted as a deviation from the approximate law of conservation. As a result, we expect that the rate of change of this variable will be less than that of any other variable. Thus we introduce two new auxiliary variables (recall, the third one, say concentration, can be excluded due to the existence of the energy integral) in the following way (the meaning of the second variable will be discussed later)

$$u = \theta - \sigma\Pi; \; v = \theta - 2\sigma\Pi. \tag{9.31}$$

The variable v has a simple physical interpretation. The RHS of the definition of v in (9.31) represents the difference between the two heat fluxes governing the dynamics of the temperature θ (see the RHS of the Eq. (9.16)). It will be shown

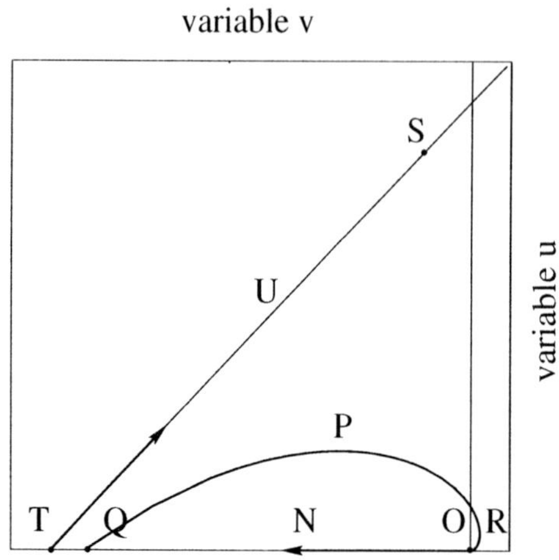

Figure 9.4. *The slow curve RPQ and two different stages of a trajectory for different values of flame velocity Λ_D. ON - stage of the initial fast motion, Q - point of intersection between the asymptotically fast motion ON and the slow curve RPQ, TU - the second stage of the trajectory (fast). Shape of the curve RPQ and location of the point Q depend on the unknown flame velocity Λ_D*

that the point where these two fluxes are equal plays a highly important role in the determination of the system dynamics.

The variables (9.31) allow us to rewrite system (9.29)-(9.30) in the conventional form of a singularly perturbed system of ordinary differential equations and to apply the appropriate methods of asymptotic analysis. Thus, the system of governing equations (9.29)-(9.30) now reads

$$\frac{1}{\Lambda_D^2}\frac{dv}{d\xi} = -H_{FD}\left(\theta(u,v),\Pi(u,v)\right) + \frac{1}{\Lambda_D^2}H_{RD}\left(\theta(u,v),\Pi(u,v)\right), \tag{9.32}$$

$$\frac{du}{d\xi} = H_{RD}\left(\theta(u,v),\Pi(u,v)\right), \tag{9.33}$$

where

$$H_{FD}\left(\theta,\Pi\right) = \frac{(\Pi-\theta)}{1+\beta\theta},$$

$$H_{RD}\left(\theta,\Pi\right) = \varepsilon_1\eta\left(\theta,\Pi\right)\frac{1+\beta\Pi}{1+\beta\theta}exp\left(\frac{\theta}{1+\beta\theta}\right).$$

9.5.2 Application of MIM

The system (9.32)-(9.33) represents a singularly perturbed system of ODEs, where the reciprocal of the square of the flame speed serves as a dimensionless small

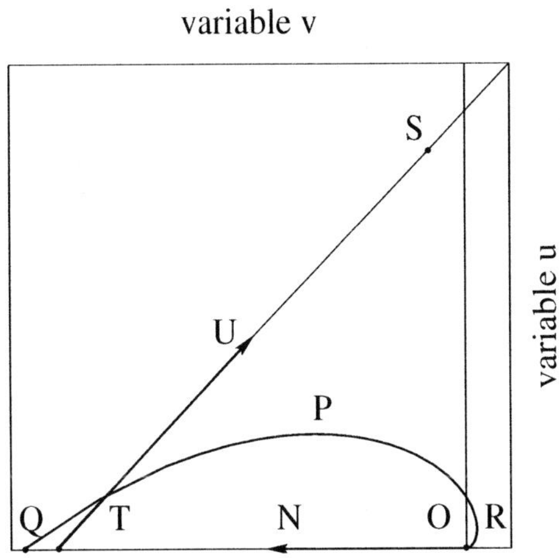

Figure 9.5. *The slow curve RPQ and two different stages of a trajectory for different values of flame velocity Λ_D. ON - stage of the initial fast motion, Q - point of intersection between the asymptotically fast motion ON and the slow curve RPQ, TU - the second stage of the trajectory (fast). Shape of the curve RPQ and location of the point Q depend on the unknown flame velocity Λ_D*

parameter. This makes an application of MIM legitimate. The singular character of the system implies that, at least initially (in the region of small temperatures θ, where the inequality $H_{FD}(v,u) >> H_{RD}(v,u)$ is valid), Eq. (9.32) describes the fast process, whereas the second equation (9.33) describes the slow one. Hence, we have the fast variable v and the slow variable u. Making use of MIM, one can subdivide an arbitrary trajectory of the system (9.32)-(9.33) in the phase space into fast and slow parts. The fast stage is characterized by a constant value of the slow variable u, where the slow part is located within a $\frac{1}{\Lambda_D^2}$ - neighborhood of the integral manifold. The exact location of the integral manifold of the system (9.32)-(9.33) is unknown and its resolution represents a separate complicated problem which is beyond the scope of the present work. Nevertheless, the general theory of integral manifolds [7], [15], [22] allows us to determine the zeroth order approximation (with respect to the small parameter $\frac{1}{\Lambda_D^2}$) to the exact integral manifold. Such an approximation is called the slow curve (see Section 3 for additional details).

The slow curve of Eqs. (9.32)-(9.33) is found by equating the RHS of Eq. (9.32) to zero, i.e.

$$\Omega(u,v) = H_{FD}(u,v) - \frac{1}{\Lambda_D^2} H_{RD}(u,v) = 0. \tag{9.34}$$

Due to the definitions of the variables v and u, the initial point O (in the (v,u) plane) coincides with the origin (Figs. 9.4). Recall that the exact location of the slow curve RPQ (Figs. 9.4, 9.5, 9.6) depends on the unknown parameter Λ_D. The trajectory starting at the origin moves along the v-axes (the slow variable u

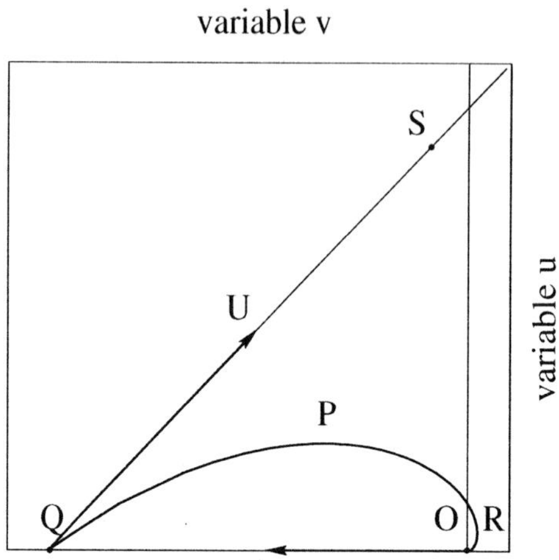

Figure 9.6. *The slow curve RPQ and two different stages of a trajectory for different values of flame velocity Λ_D. ON - stage of the initial fast motion, Q - point of intersection between the asymptotically fast motion ON and the slow curve RPQ, TU - the second stage of the trajectory (fast). Shape of the curve RPQ and location of the point Q depend on the unknown flame velocity Λ_D*

conserves its initial value $u = 0$). The branch RP of the slow curve is repulsive, therefore the trajectory moves in the negative direction (part ON, Fig. 9.4). We can show that this initial stage of the system dynamics is characterized by the constant value of the concentration η, which conserves its initial value $\eta_0 = 1$. Therefore, this part of the trajectory can be interpreted as the preheat zone of the flame, where only the temperature and pressure change, and the concentration is constant.

Asymptotically, the fast motion ON continues until the intersection with the attractive branch PQ of the slow curve. The location of the intersection point Q depends on the position of the slow curve RPQ which, in turn, depends on the value of the dimensionless flame speed Λ_D, and may vary (Figs. 9.4, 9.5 represent slow curves with various values of Λ_D). In accordance with the general approach (MIM) just after the intersection with the slow curve RPQ, the trajectory begins to move in close proximity to the slow curve and the slow part of the trajectory begins. This does not happen in this model! The explanation is intriguing: the system under consideration is unique (with respect to the numerous models analyzed earlier [9], [17]) and does not contain the conventional slow motion. The uniqueness of the slow curve RPQ resides in the following fact. When the trajectory (fast part ON) approaches the vicinity of the branch PQ, the value of the variable v becomes comparatively large (it can be about 20-40 depending on the parameters). These values of the variable v drastically affect the relation between various terms in the Eq. (9.34): they become closer and the distinction in the difference of rates of the two variables (u and v) is lost. As a result, the trajectory deviates from the v - axis and does not adhere to the branch PQ of the slow curve after intersection. In other words, the slow curve RPQ loses its important attractive properties and

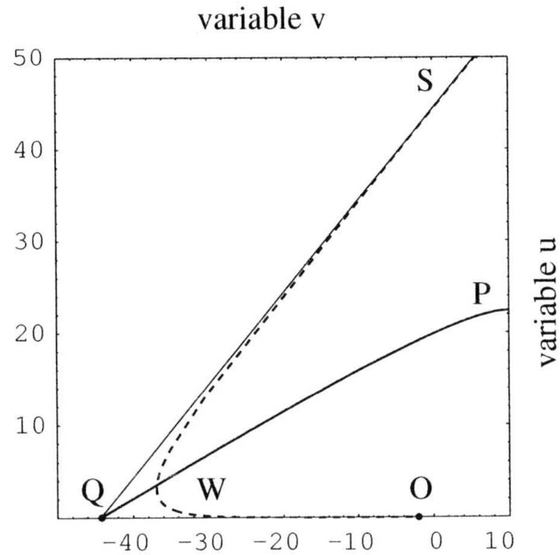

Figure 9.7. *Asymptotic trajectory OQS and real one OWS (data obtained as a result of a direct numerical simulation of the initial system of the governing equations). The system parameters read:* $\beta = 0.0295$; $\gamma = 1.3$; $\varepsilon_1 = 6.4\times10^{-6}$; $\varepsilon_2 = 4.35\times10^{-8}$; $q = 5.66$. *The dimensionless flame velocity (used for dashed trajectory, numerical data) is* $\Lambda_F = 1.42$ *the dimensional one is* $D = 72.1m/s$

ceases to be a real slow curve when the trajectory approaches it. Hence, just after intersection with the slow curve the trajectory begins the second stage of the motion, approaching the straight line TU (the equation of the line TU can be easily derived and is $u - v = \sigma\theta_b$). From the physical standpoint, the part TU represents the reaction zone of the flame and the point S describes the burnt mixture with the parameters θ_b, Π_b and $\eta_b = 0$ (far behind the flame front) and the coordinates in the (u,v) plane $v_b = (1-2\sigma)\theta_b$ and $u_b = (1-\sigma)\theta_b$.

To find the desired trajectory in this situation we should match the two different parts. The natural way to perform the matching is to suggest that the slow curve RPQ, the fast motion ON and the line TU have a single joint point Q (coincides with T, Fig. 9.6). Matching in this way permits us to determine the coordinates of the point Q, which are

$$\theta_Q = \sigma\Pi_Q;\ \Pi_Q = \Pi_b = \frac{\varepsilon_1}{\varepsilon_2(1-\sigma)};\ \eta_Q = 1, \tag{9.35}$$

and to uncover the internal structure of the flame. The point Q serves as a boundary between the preheat and reaction zones of the flame front. Fig. 9.7 presents a comparison between the asymptotic trajectory OQS, computed from MIM and the curve OWS, which is the result of the direct numerical simulation of the system. One can see that the approximation is reasonable and the main difference is observed in the neighborhood of the point Q.

9.5.3 Flame velocity

To derive an analytical expression for the wave velocity Λ_D, we assume additionally that both the temperature and its first derivative are continuous functions. This assumption means that we demand continuity of the heat flux across the flame front. The approximate trajectory (solution) consists of two stages (OQ and QS). The single point where the continuity might be problematic is the point Q. Equating the two terms on the RHS of the Eq. (9.29) at the point Q (with coordinates (9.35)) allows us to provide the smoothness of the temperature profile within the flame. Once this is done, we derive an analytical expression for the flame velocity Λ_D

$$\Lambda_D^2 = \varepsilon_1 \frac{1+\beta\Pi_Q}{\Pi_Q - \theta_Q} \exp\left(\frac{\theta_Q}{1+\beta\theta_Q}\right), \tag{9.36}$$

which coincides with the formula obtained earlier (9.28). This lent credence to the accepted modification of MIM and encouraged the authors to extend the region of its applications.

One can see that the structure of the expression (9.36) resembles the well known formula for the flame speed for the gaseous combustion suggested in [23], where the flame velocity is proportional to the square root of the Arrhenius exponent. The power of the exponent is determined by the temperature at the point Q serving as a boundary between two sub-zones of the flame front.

9.6 Non-linear Friction, No Inertia

We return now to the original full system (9.1)-(9.8) and consider the approximation characterized by an absence of the inertia effect but including the quadratic dependence of the friction force (9.7) on the velocity of the gaseous mixture. Thus, the present problem differs from the previous one only by the inclusion of the friction force. Then, we may rewrite the original full system (9.1)-(9.8) in the form:

$$\frac{d\theta}{d\xi} = \Lambda_F^3 \frac{(\Pi-\theta)\,|\Pi-\theta|}{(1+\beta\theta)(1+\beta\Pi)} + \varepsilon_1\eta\frac{1+\beta\Pi}{1+\beta\theta}\exp\left(\frac{\theta}{1+\beta\theta}\right), \tag{9.37}$$

$$\sigma\frac{d\Pi}{d\xi} = \Lambda_F^3 \frac{(\Pi-\theta)\,|\Pi-\theta|}{(1+\beta\theta)(1+\beta\Pi)},$$

$$\frac{d\eta}{d\xi} = -\varepsilon_2\eta\frac{1+\beta\Pi}{1+\beta\theta}\exp\left(\frac{\theta}{1+\beta\theta}\right). \tag{9.38}$$

In this simplified version of the original problem we use the following definition of the flame speed Λ_F (subscript F emphasizes the fact that the friction force is proportional to the square of the gas velocity)

$$\Lambda_F^3 = K_F \frac{\beta D^3}{C_p T_0 A} \exp\left(\frac{1}{2\beta}\right). \tag{9.39}$$

The system (9.37)-(9.38) is adiabatic, and therefore the energy integral exists and is derived easily to be

$$\eta - 1 + \frac{\varepsilon_2}{\varepsilon_1}(\theta - \sigma\Pi) = 0.$$

The adiabatic point (state of the burnt gas behind the flame front) is easily calculated to be

$$\eta_b = 0; \; \theta_b = \Pi_b = \frac{\varepsilon_1}{\varepsilon_2(1-\sigma)}.$$

The observations made above and our experience helped us to find an elegant way to determine the new auxiliary variables, which allows us to rewrite the original system in the classical form of a singularly perturbed system of ODEs and to give a physical explanation. We present this approach below, see also [6].

9.6.1 Singularly perturbed system - SPS

The system (9.37)-(9.38) can be reduced to two ODEs by eliminating the dimensionless concentration η. To make the reduced system of ODEs tractable to the chosen asymptotic approach (MIM), it is helpful to subdivide the consideration of the problem into the two stages corresponding to the two distinct regions in the phase plane (θ, Π). To perform this subdivision, it is useful to introduce two new pairs of auxiliary variables, which are based on Zel'dovich's approach (Section 4):

$$u = \theta - \sigma\Pi; \; v = \theta - 2\sigma\Pi. \tag{9.40}$$

The physical rationale lying behind such a choice of variables u and v was explained in detail in Section 5. The variables (9.40) allow us to rewrite system (9.37)-(9.38) in the conventional form of a singularly perturbed system of ordinary differential equations and to apply the appropriate asymptotic method of analysis.

The system of equations within the preheat sub-zone is

$$\frac{1}{\Lambda_F^3}\frac{dv}{d\xi} = -H_{FF}\left(\theta(u,v), \Pi(u,v)\right) + \frac{1}{\Lambda_F^3}H_{RF}\left(\theta(u,v), \Pi(u,v)\right), \tag{9.41}$$

$$\frac{du}{d\xi} = H_{RF}\left(\theta(u,v), \Pi(u,v)\right), \tag{9.42}$$

where the following notation is used

$$H_{FF}(u,v) = \frac{\left(\Pi(u,v) - \theta(u,v)\right)\left|\Pi(u,v) - \theta(u,v)\right|}{(1+\beta\theta(u,v))(1+\beta\Pi(u,v))}, \tag{9.43}$$

$$H_{RF}(u,v) = \varepsilon_1\eta\left(\theta(u,v), \Pi(u,v)\right)\frac{1+\beta\Pi(u,v)}{1+\beta\theta(u,v)}exp\left(\frac{\theta(u,v)}{1+\beta\theta(u,v)}\right). \tag{9.44}$$

9.6.2 Application of MIM

The system (9.41)-(9.42) represents a singularly perturbed system of ODEs, where the reciprocal of the cube of the flame speed serves as the dimensionless small parameter.

The slow curve of the Eqs. (9.41)-(9.42) is found by equating the RHS of Eq. (9.41) to zero, and reads

$$\Omega_{vu}(u,v) = -H_{FF}(u,v) + \frac{1}{\Lambda_F^3} H_{RF}(u,v) = 0. \tag{9.45}$$

Now, since the present system differs from the previous one by only the friction term,the reader might refer back to the analysis which was performed in the case of linear friction. Accordingly, the same phase picture can be obtained (see Fig. 9.6), and the approximate trajectory (asymptotic solution) consists of the two parts (OQ and QS Fig. 9.6). Matching made in this way permits us to determine the same coordinates of the point Q see (9.35) and to reveal the internal structure of the flame. The point Q acts as a matching point, distinguishing the preheat and reaction sub-zones of the flame front.

9.6.3 Flame velocity

To derive an analytical expression for the wave velocity Λ_F, we follow the procedure used for the derivation of formula (9.36). The point Q belongs to the slow curve (9.45); therefore, substituting the coordinates of the point Q (9.35) into (9.45), one readily gets an analytical expression for the flame velocity Λ_F:

$$\Lambda_F^3 = \varepsilon_1 \frac{(1+\beta\Pi_Q)^2}{(\Pi_Q - \theta_Q)^2} \exp\left(\frac{\theta_Q}{1+\beta\theta_Q}\right). \tag{9.46}$$

As before the physical rationale for (9.46) is rather simple. The condition that the point Q belongs to the slow curve (9.45) means that we demand continuity of the heat flux within the flame. Equating the two terms on the RHS of the Eq. (9.37) at the point Q (with coordinates (9.35)) allows us to provide the smoothness of the temperature profile within the flame.

The structure of the expression (9.46) resembles the well known formula for the flame speed for the gaseous combustion suggested in [23] where the flame velocity is proportional to the square root of the Arrhenius exponent. Unlike Zel'dovich's formula, the expression (9.46) gives an unusual cube root dependence of the flame velocity on the Arrhenius exponent, where the exponent is determined by the temperature at the point Q serving as a boundary between two sub-zones of the flame.

It is worthwhile to compare our prediction (9.46) with the results obtained in [16] on the basis of a very complicated modification of the conventional multi-scale approach. In the notation of [16] the flame velocity D determined by (9.46) can be re-written in the form

$$D^3 = \frac{1}{K_F} \frac{c_p T_0 A}{1} \frac{\gamma}{q} (1+q)^2 \exp\left(-\frac{E}{RT_+}\right), \tag{9.47}$$

$$T_+ = T_b - \frac{1}{\gamma}(T_b - T_0);\ T_b = T_0 + \frac{QC_{f0}}{c_v};\ q = \frac{T_b - T_0}{T_0}.$$

One can show that the dimensional equivalent of the formula (9.46) can be transformed into a form which coincides with the corresponding relation derived in [16]. This is further justification of our method.

9.6.4 Theory vs numerics

We can represent the trajectory analyzed in Section 5 in another way. Fig. 9.8 gives the flame velocity Λ_F as a function of the parameter γ, and compares the theoretical prediction (9.46) (estimate of the flame velocity Λ_F) and numerical data. The speed of the flame grows with increasing γ. The accuracy of the analytic prediction can be estimated by the formula

$$\frac{\Lambda_F^{theor} - \Lambda_F^{num}}{\Lambda_F^{num}} 100, \tag{9.48}$$

where Λ_F^{theor} is the theoretical prediction (9.46) for the flame velocity and Λ_F^{num} represents the result of the numerical simulations. We can see that the accuracy of the theoretical predictions depends significantly on the chosen parametric region. For our specific set of problem parameters, the relative error of the analytical formulae is about 4-5% and grows slowly when γ increases.

Fig. 9.9 graphs the flame velocity Λ_F versus the parameter β (reduced initial temperature of the system) and compares the theoretical prediction (9.46) with numerical data. The speed of the flame increases sharply in the region of small values of β and tends to some constant value when β increases. The accuracy (9.48) of the explicit expression (9.46) in our parameter region is about 3-7%. Note that the absolute error of the prediction (the difference between the asymptotic estimate and numerical data) looks almost constant and shows little dependence on β.

Fig. 9.10 graphs the flame velocity Λ_F versus the dimensionless parameter q (see Eq. (9.47)) and compares the theoretical prediction (9.46) and numerical data. The speed of the flame increases in the region. Both the absolute and relative values of disagreement grow with q. The accuracy of the explicit expression (9.46) in the parameter region used for the numerical simulations is within 2-6%.

9.7 Inertia

The most complex problem, which has been solved to date using the modified version of MIM, is the problem of inertia effects on the propagation of the pressure-driven flames in porous media. We note that by taking into account the inertia of the gaseous fluid we are no longer restricted to the sub-sonic region of the flame's velocity. The original system also describes supersonic flames (of both detonative and deflagrative natures).

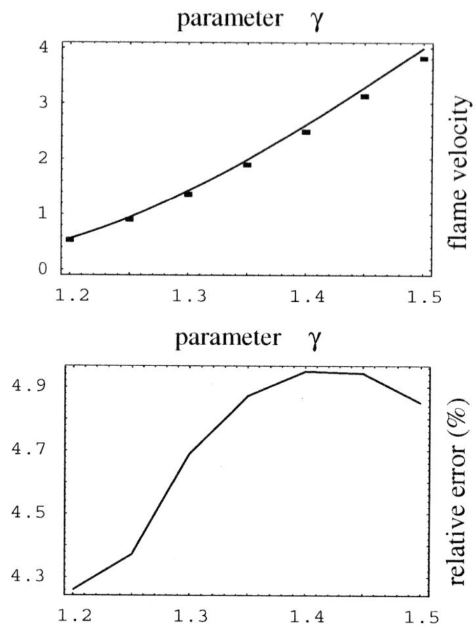

Figure 9.8. *Theoretical prediction of the flame velocity dependence on the dimensionless parameter γ versus results of numerical simulations. Upper figure (a) - flame velocity dependence on the parameter γ, solid line - formula, black squares - numerical data; Lower figure (b) - relative error of the theoretical prediction, which is calculated according to the expression (9.46). Other system parameters: q=5.66,* $\beta = 0.0295$

In what follows we present an application of the current version of the MIM to the problem described above i.e. pressure driven flames in porous media, the friction force being proportional to the square of the gas velocity, and the inertia effects are accounted for.

The corresponding dimensionless system of equations reads

$$\frac{d\theta}{d\xi} = \frac{\left(1 - \varepsilon_{inert}\frac{(\Pi-\theta)(1+\beta\theta)}{(1+\beta\Pi)^3\sigma}\right)\Lambda_F^3 H_F(\theta,\Pi) + \left(1 - \varepsilon_{inert}\frac{(1+\beta\theta)}{(1+\beta\Pi)^2\sigma\beta}\right) H_R(\theta,\Pi)}{\left(1 - \varepsilon_{inert}\frac{(1+\beta\theta)}{(1+\beta\Pi)^2\sigma\beta\gamma}\right)}, \tag{9.49}$$

$$\frac{d\Pi}{d\xi} = \frac{\left(1 - \varepsilon_{inert}\frac{(\Pi-\theta)}{(1+\beta\Pi)^2}\right)\Lambda_F^3 H_F(\theta,\Pi) - \frac{\varepsilon_{inert}}{(1+\beta\Pi)\beta} H_R(\theta,\Pi)}{\left(1 - \varepsilon_{inert}\frac{(1+\beta\theta)}{(1+\beta\Pi)^2\sigma\beta\gamma}\right)},$$

$$\frac{d\eta}{d\xi} = -\frac{\varepsilon_2}{\varepsilon_1} H_R(\theta,\Pi). \tag{9.50}$$

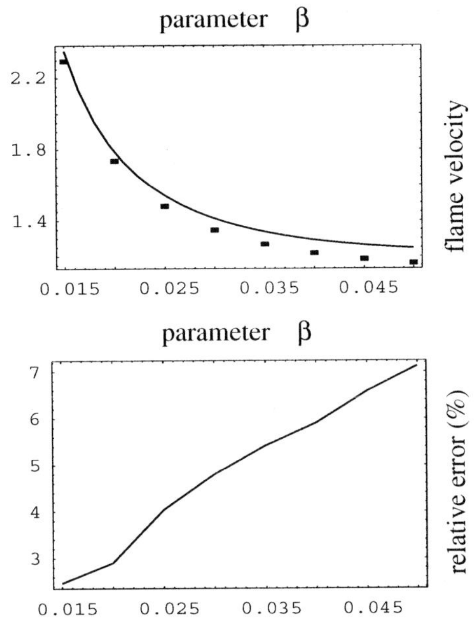

Figure 9.9. *Flame velocity Λ_F dependence on the dimensionless parameter β. Upper figure (a) - dependence of the flame velocity on the parameter β, solid line - theoretical prediction, black squares - numerical data. Lower figure (b) relative error of the theoretical prediction, which is calculated according to the expression. Other system parameters: q=5.66,* $\gamma = 1.3$

Here the functions H_F and H_R are given by (9.43)-(9.44). In addition to the dimensionless parameters (9.9) a new dimensionless factor ε_{inert} appears (responsible for the inertia effects)

$$\varepsilon_{inert}^3 = \frac{\beta D^2}{C_p T_0} \ll 1, \tag{9.51}$$

where D is defined in (9.47). The system (9.49)-(9.50) is subject to the same initial conditions (9.10).

The energy integral can be derived as a result of an appropriate combination of the equations (9.49)-(9.50) and further integration. It reads:

$$\eta - 1 + \frac{\varepsilon_2}{\varepsilon_1}(\theta - \sigma\Pi - \frac{\varepsilon_{inert}}{2}\frac{(\Pi - \theta)^2}{(1+\beta\Pi)^2}) = 0.$$

Equality between the dimensionless temperature θ_b and pressure Π_b behind the flame front and the natural assumption that the reaction has been completed in this region ($\eta_b = 0$) allow us to determine the main characteristics of the medium at the stationary point of the system:

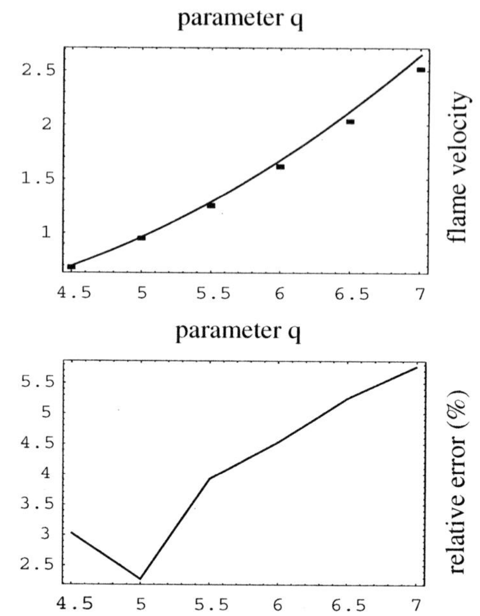

Figure 9.10. *Flame velocity Λ_F dependence on the dimensionless parameter q. Upper figure (a) - dependence of the flame velocity on the parameter q, solid line - theoretical prediction (9.46), black squares - numerical data. Lower figure (b) - relative error of the theoretical prediction (9.46), which is calculated according to the expression (9.48). Other system parameters:* $\beta = 0.0295$, $\gamma = 1.3$

$$\theta_b = \Pi_b = \frac{\varepsilon_1}{\varepsilon_2(1-\sigma)}.$$

The number of equations can be reduced due to the existence of the energy integral (say, concentration, η, can be excluded, as above). The expression for η reads

$$\eta(\theta, \Pi) = 1 - \frac{\varepsilon_2}{\varepsilon_1}\left(\theta - \sigma\Pi - \frac{\varepsilon_{inert}}{2}\frac{(\Pi-\theta)^2}{(1+\beta\Pi)^2}\right).$$

9.7.1 Singularly perturbed system - SPS

In order to reformulate the equations (9.49)-(9.50) in the form of SPS and to make them tractable to the MIM, we need to introduce a pair of new variables instead of θ and Π. The new variables should allow us to emphasize the difference in time scales.

All previous arguments could be repeated. Following the proposed algorithm for the determination of the new auxiliary variables, we define two new variables in the form:

$$v = \theta - \sigma\Pi - \frac{\varepsilon_{inert}}{2}\frac{(\Pi-\theta)^2}{(1+\beta\Pi)^2}, \tag{9.52}$$

$$u = \sigma\Pi - \frac{\varepsilon_{inert}}{\beta}\frac{(\Pi-\theta)}{(1+\beta\Pi)}. \tag{9.53}$$

Here the variable v can be interpreted as a deviation from the approximate law of energy conservation (in the absence of reaction - preheat sub-zone), whereas u can be understood as a deviation from the approximate law of momentum conservation (reaction sub-zone). Having determined the new variables (9.52), (9.53), we expect a slow variation within the relevant sub-zones.

With the introduction of variables u and v we rewrite the system (9.49)-(9.50) in the form

$$\frac{dv}{d\xi} = H_R\left(\theta(u,v), \Pi(u,v)\right), \tag{9.54}$$

$$\frac{du}{d\xi} = \Lambda_F^3 H_F\left(\theta(u,v), \Pi(u,v)\right). \tag{9.55}$$

To successfully perform an asymptotic analysis of the system (9.49)-(9.50) we divide the physically interesting region of the phase plane into the corresponding two sub-regions. The asymptotic solutions obtained separately in these sub-regions are then to be matched. We now introduce a new additional variable, w, defined by

$$w = v - u = \theta - 2\sigma\Pi - \frac{\varepsilon_{inert}}{2}\frac{(\Pi-\theta)^2}{(1+\beta\Pi)^2} + \frac{\varepsilon_{inert}}{\beta}\frac{(\Pi-\theta)}{(1+\beta\Pi)}.$$

The new variable w has a distinguishing feature - it is a fast variable within both of the sub-zones. Indeed, in the region where an input of the chemical reaction is considered as negligible (the preheat sub-zone), the term H_R is exponentially small and the inequality $H_R << H_F$ is valid. Therefore, the rate of change of the variable v is much slower than that of w. One can conclude, that the variable v can be interpreted as slow, whereas the variable w is fast. Within the reaction sub-zone the relation is the opposite. The heat release term H_R becomes dominant and the inequality $H_R >> H_F$ becomes valid. Hence, the rate of change of the variable u is much slower than that of w. A natural conclusion is that the variable u can be treated as slow, whereas the variable w is again fast.

Thus, the current system of governing equations decouples into two different singularly perturbed systems of equations within the appropriate sub-zones of the flame front. Both of the systems contain one of the pair of equations (9.54) or (9.55), and a differential equation for the new variable w

$$\frac{dw}{d\xi} = H_R - \Lambda_F^3 H_F. \tag{9.56}$$

Thus, there is good reason to consider a flame configuration within the preheat sub-zone governed by the pair of the equations (9.54), (9.56) (v - slow, w - fast), whereas the combustion wave structure within the reaction sub-zone would be better described by the pair (9.55), (9.56) (u - slow, w - fast). This decomposition permits us to rewrite the original system (9.49)-(9.50) as two separate systems for the different sub-zones of the flame. Each of these systems has the conventional form of SPS of ODEs, which makes a direct application of MIM legitimate.

9.7.2 Flame front structure - sub-zones

We now turn to the analysis of the possible dynamical scenarios of the solutions of the reduced system derived from (9.49)-(9.50). We catalogue the different possibilities according to the criteria mentioned above.

A. Preheat sub-zone. The variables (v, w) are the most suitable ones to analyze the dynamics in the preheat sub-zone. Here w is a fast variable and v is a slow variable. For these variables the reduced version of the system (9.49)-(9.50) can be rewritten as the following:

$$\frac{dw}{d\xi} = H_R(\theta(v, W), \Pi(v, w)) - \Lambda_F^3 H_F(\theta(v, w), \Pi(v, w)), \tag{9.57}$$

$$\frac{dv}{d\xi} = H_R(\theta(v, w), \Pi(v, w)).$$

The slow curve of the singularly perturbed system is derived by equating the RHS of the equation (9.57) for the fast variable (w) to zero. Hence, for the preheat sub-zone the slow curve for the current system is given by the equation:

$$\Omega_{wv} = H_R(\theta(v, W), \Pi(v, w)) - \Lambda_F^3 H_F(\theta(v, w), \Pi(v, w)). \tag{9.58}$$

We note again that the exact form and location of the slow curve depend on the value Λ_F of the flame velocity, which serves as an unknown parameter of the problem (to be determined).

The slow surface can consist of stable and unstable parts, in the sense that they attract or repel trajectories. For the initial point P_{in} the branch $R_{wv}S_{wv}$ of the slow curve, Ω_{wv} is attractive and the trajectory $P_{in}Q_{wv}$ moves towards it (Fig. 9.11). The trajectory moves in such a way, that the value of the slow variable v conserves its initial value while the value of the fast variable w changes rapidly. We can readily show that this initial stage of the system dynamics is characterized by the constant value of the concentration η, which conserves its initial value $\eta_0 = 1$. This outcome justifies our assumption that this part of the trajectory can be interpreted as the preheat zone of the flame, where the temperature and pressure change while the concentration remains constant.

B. Reaction sub-zone. The variables (u, w) are the most suitable to analyze the dynamics in the reaction sub-zone. Here w is a fast variable and u is a slow one. For these variables the reduced version of the system (9.49)-(9.50) can be rewritten as

$$\frac{dw}{d\xi} = H_R(\theta(u, w), \Pi(u, w)) - \Lambda_F^3 H_F(\theta(u, w), \Pi(u, w)), \tag{9.59}$$

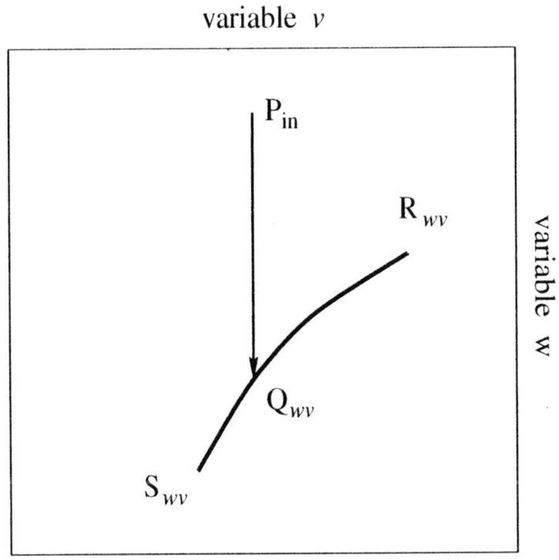

Figure 9.11. *The possible slow curves and trajectories. The slow curve $R_{wv}S_{wv}$ and the fast part $P_{in}Q_{wv}$ of the trajectory in the plane (v, w)*

$$\frac{du}{d\xi} = \Lambda_F^3 H_F\left(\theta(u, w), \Pi(u, w)\right). \tag{9.60}$$

The slow curve of the set (SPS) of equations (9.59)-(9.60) is derived by equating the RHS of the equation (9.59) for the fast variable (w) to zero. The slow curve for this zone coincides with slow curve for the preheat zone:

$$\Omega_{wu} = H_R\left(\theta(u, w), \Pi(u, w)\right) - \Lambda_F^3 H_F\left(\theta(u, w), \Pi(u, w)\right). \tag{9.61}$$

For the initial point Q of this part of trajectory the branch $R_{wu}S_{wu}$ of the slow curve Ω_{wu}, (9.61) is repulsive and the trajectory $Q_{wu}P_{fin}$ moves away from it in the direction of the final point P_{fin}, that corresponds to the burnt mixture (θ_b, Π_b) (Fig. 9.12). Using the definition of u given by (9.53), the slow variable u is constant and equal to its final value. As in the previous stage, the value of the fast variable w is fast changing.

C. Entire trajectory. We note that the same equation for the slow curves (9.58) and (9.61) represents the same curve for two different systems of coordinates: Ω_{wv} - in the (v, w) plane, and Ω_{wu} - in the (u, w) plane. Hence, the points Q_{wu} (Fig. 9.12) and Q_{wv} (Fig. 9.11) should represent the same point Q in the original (θ, Π) plane. In other words, the final point Q_{wv} of the first part $P_{in}Q_{wv}$ (Fig. 9.11) of the trajectory should serve as the initial point Q_{wu} for the second part $Q_{wu}P_{fin}$ (Fig. 9.12) of the trajectory (the points Q_{wu} and Q_{wv} coincide with the point Q).

Summarizing the previous analysis and combining the two dynamical scenarios we can give an approximation of the entire trajectory of the system in the plane (θ, Π) in the following way (Fig. 9.13). Starting from the initial point P_{in}, the approximate trajectory $P_{in}Q_i$ (i=1, 2, 3) of the system moves toward the attractive branch R_iS_i (i=1, 2, 3). This part of the approximate trajectory matches the

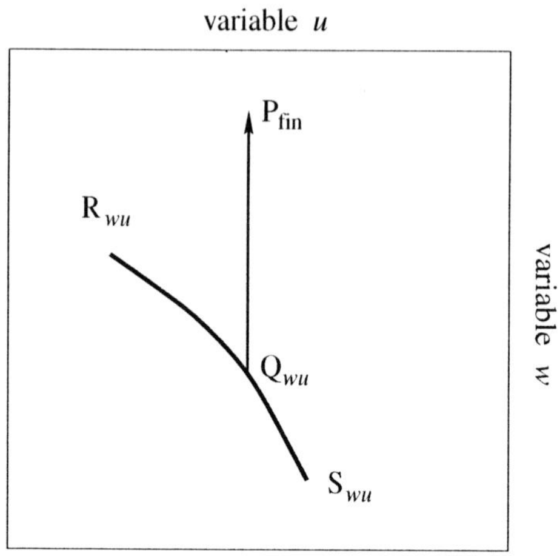

Figure 9.12. *The possible slow curves and trajectories. The slow curve $R_{wu}S_{wu}$ and the fast part $Q_{wu}P_{fin}$ of the trajectory in the plane (u, w)*

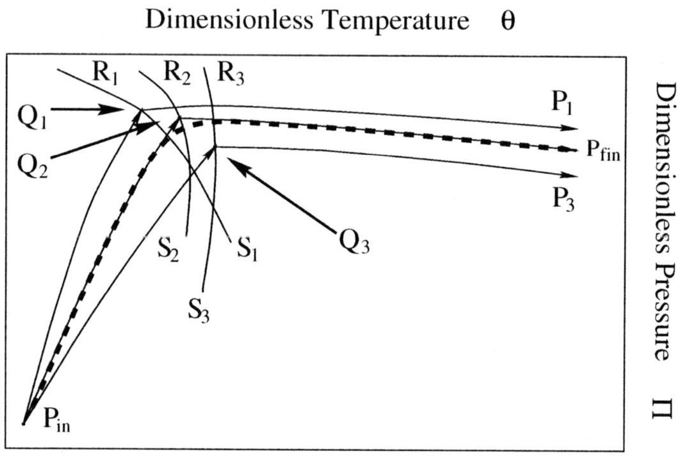

Figure 9.13. *The possible slow curves and trajectories. Presentation of various trajectories in the plane (θ, Π): $\Omega(R_iS_i)$ (i=1,2,3) - possible locations of the slow curve (due to various values of the flame velocities Λ_F), $P_{in}Q_1P_1$, $P_{in}Q_2P_{fin}$, $P_{in}Q_3P_3$ - possible trajectories, $P_{in}Q_2P_{fin}$ - the single trajectory approaching the final point P_{fin}*

part of the approximate trajectory shown in Fig. 9.11 (preheat sub-zone) and it is characterized by the constant value of the system energy (the variable v is asymptotically constant in the (v, w) plane).

We recall that the location of the slow curve, the position of the matching point Q dividing the phase plane into sub-zones and the choice of coordinate systems in sub-zones all depend on the value of the flame velocity Λ_F. In Fig. 9.13 we show the slow curve, the matching point and approximate trajectory for three different values of the flame velocity Λ_F. The existence of a number of slow curves R_iS_i (i=1, 2, 3) and matching points Q_i (i=1, 2, 3) reflects this fact (different values of the index i correspond to various values of the flame velocity).

The trajectory can reach the singular (final) point P_{fin} (with coordinates $\theta_b = \Pi_b = \frac{\varepsilon_1}{\varepsilon_2(1-\sigma)}$) only for a special value of the flame velocity Λ_F. In Fig. 9.13 this trajectory is depicted as $P_{in}Q_2P_{fin}$ (i=2). This special trajectory contains two parts. The first part has a starting point P_{in}. On approaching the point Q_2, the trajectory changes its behavior. The part of the trajectory starting at Q_2 moves to approach the singular (final) point P_{fin}. All other trajectories (such as $P_{in}Q_1P_1$ or $P_{in}Q_3P_3$) move above or below the final point P_{fin}. This situation is illustrated in Fig. 9.13 by ($P_{in}Q_1P_1$ or $P_{in}Q_3P_3$). It would make sense to say that the part Q_2P_{fin} of the second trajectory $P_{in}Q_2P_{fin}$ is compatible with the trajectory $Q_{wu}P_{fin}$ presented in Fig. 9.12 (reaction sub-zone) in the (u, w) plane. Its distinguishing feature is an approximate conservation of momentum (geometrically this means that the variable u is a slow variable in the (u, w) plane).

9.7.3 Matching point

The location of the final point P_{fin} does not depend on the flame velocity Λ_F and has a fixed position in the phase plane. Therefore, the only way to find the desired trajectory ($P_{in}Q_2P_{fin}$, Fig. 9.13) is to vary the slow curve in such a way that the matching point Q_2 belongs to it. To determine the location of the slow curve we must match the two different parts of the trajectory $P_{in}Q_2P_{fin}$, but first determine the location of the point Q_2. In fact Q_2 serves as a matching point between the two parts of the approximate trajectory that corresponds to the preheat and reaction sub-zones of the flame front. Performing the matching, we remember that at the point Q_2 the variable v has the same value that it has at the starting point P_{in} of the trajectory $P_{in}Q_{wv}$ (Fig. 9.11). Like the variable v, the variable u at Q_2 has the same value that it has at the singular (final) point P_{fin}. Hence, we have the following system of equations for evaluating the coordinates of the point Q_2:

$$\begin{aligned} H_R(\theta_Q, \Pi_Q) - \Lambda_F^3 H_F(\theta_Q, \Pi_Q) &= 0, \\ u_Q = u_0 = 0;\ v_Q = v_{fin} &= v_b = \sigma\Pi_b, \end{aligned} \tag{9.62}$$

where θ_Q and Π_Q are coordinates of the point Q_2 in the (θ, Π) plane.

We note that the first equation of the three in (9.62) guarantees that the point Q_2 belongs to the slow curve, and the last two are obtained from the asymptotic analysis of the system trajectory. The last two relations of (9.62) can be rewritten in the form

$$\left\{ \begin{array}{c} v_Q = v_{fin} = \sigma\Pi_b \\ u_Q = u_0 = 0 \end{array} \right\} \Leftrightarrow \left\{ \begin{array}{c} \sigma\Pi_Q - \frac{\varepsilon_{inert}}{\beta}\frac{(\Pi_Q - \theta_Q)}{(1+\beta\Pi_Q)} = \sigma\Pi_b \\ \theta_Q - \sigma\Pi_Q - \frac{\varepsilon_{inert}}{2}\frac{(\Pi_Q-\theta_Q)^2}{(1+\beta\Pi_Q)^2} = 0 \end{array} \right\}. \tag{9.63}$$

The system (9.63) can be solved analytically as was done in the simpler case when the impact of the inertia was neglected (see Section 6). In this model, the unknown values Π_Q and θ_Q represent explicit functions of the parameter $\theta_Q(\varepsilon_{inert})$ describing the impact of the inertia.

We do not give these functions here because we plan to use a further asymptotic expansion of $\theta_Q(\varepsilon_{inert})$ and $\Pi_Q(\varepsilon_{inert})$ as functions of the small parameter ε_{inert}.

9.7.4 Flame velocity

An analytical expression for the flame velocity Λ_F can be determined in the following manner. We consider first the functions $\theta_Q(\varepsilon_{inert})$ and $\Pi_Q(\varepsilon_{inert})$ in (9.63). As a second step we derive the relation between the flame velocity Λ_F and the inertia parameter ε_{inert}. Finally, we substitute all these relations, as functions of the inertia parameter ε_{inert}, into the first of (9.62). This equation means that the matching point Q_2 belongs to the slow curve Ω. The final equation is

$$\begin{aligned}&\Lambda_F^3(\varepsilon_{inert})\left(\Pi_Q(\varepsilon_{inert})-\theta_Q(\varepsilon_{inert})\right)^2\\&\qquad-\varepsilon_1\left(1+\beta\Pi_Q(\varepsilon_{inert})\right)^2 exp\left(\tfrac{\theta_Q(\varepsilon_{inert})}{1+\beta\theta_Q(\varepsilon_{inert})}\right)=0.\end{aligned} \tag{9.64}$$

Using the connection between ε_{inert} and D in (9.51), we get an approximate formula for the flame velocity D. We return to the functions $\theta_Q(\varepsilon_{inert})$ and $\Pi_Q(\varepsilon_{inert})$, which were derived as a solution of the system (9.63). As the flame velocity is subsonic, the value of the parameter ε_{inert} responsible for the inertia is relatively small (with respect to unity). Therefore, the functions $\theta_Q(\varepsilon_{inert})$ and $\Pi_Q(\varepsilon_{inert})$ can be expanded with respect to the small parameter ε_{inert}. For our purposes we get

$$\begin{aligned}\Pi_Q&=A_0+\varepsilon_{inert}A_1;\\\theta_Q&=B_0+\varepsilon_{inert}B_1;\\\Lambda_F^3&=C\varepsilon_{inert}^{3/2},\end{aligned} \tag{9.65}$$

with

$$\begin{aligned}A_0=\Pi_b;\ A_1&=\tfrac{(1-\sigma)\Pi_b}{\sigma\beta(1+\beta\Pi_b)};\ C=\tfrac{K_F}{A}\left(\tfrac{c_PT_0}{\beta}\exp\left(\tfrac{1}{\beta}\right)\right)^{1/2};\\B_0=\sigma\Pi_b;\ B_1&=\tfrac{(1-\sigma)\Pi_b}{\beta(1+\beta\Pi_b)}-\tfrac{1}{2}\tfrac{(1-\sigma)^2\Pi_b^2}{(1+\beta\Pi_b)^2}.\end{aligned}$$

Referring back to the parameter definitions for ε_{inert} and Λ_F, (9.47) and (9.51), we note that the parameter ε_{inert} is proportional to the square of the dimensional flame velocity D (9.51), whereas, the dimensionless flame velocity Λ_F is directly proportional to the dimensional flame velocity D (9.39). Hence, within the framework of this model there is a simple connection between Λ_F^3 and the inertia parameter ε_{inert}: $\Lambda_F^3\sim\varepsilon_{inert}^{3/2}$.

This relation and the expansions (9.65) can be substituted into the equation (9.62) (the slow curve equation). After simplification the equation (9.64) can be written as

$$D_0 + \varepsilon_{inert} D_1 + \varepsilon_{inert}^{3/2} + \varepsilon_{inert}^{5/2} D_{5/2} = 0, \qquad (9.66)$$

where the coefficients D_0, D_1, $D_{3/2}$, $D_{5/2}$ are given by

$$D_0 = \varepsilon_1 (1+\beta A_0)^2 \exp\left(\tfrac{B_0}{1+\beta B_0}\right);$$
$$D_1 = \varepsilon_1 (1+\beta A_0)\left(2\beta A_1 + \tfrac{(1+\beta A_0)}{(1+\beta B_0)^2} B_1\right) \exp\left(\tfrac{B_0}{1+\beta B_0}\right);$$
$$D_{3/2} = (A_0 - B_0)^2 C; \; D_{5/2} = 2(A_0 - B_0)(A_1 - B_1)C.$$

The equation (9.66) is an equation for the inertia parameter ε_{inert} and can be solved numerically. Exploiting the definition (9.51) of the parameter ε_{inert} responsible for the inertia impact, one can get a relation between ε_{inert} and the dimensional flame velocity D

$$\varepsilon_{inert} = \frac{\beta D^2}{c_P T_0} \Rightarrow D^2 = c_P T_0 \frac{\varepsilon_{inert}}{\beta} = \frac{\varepsilon_{inert}}{(\gamma - 1)\beta}. \qquad (9.67)$$

9.7.5 Theory vs numerics

To check the accuracy of the theoretical formulae, a number of direct numerical simulations have been performed. The Cauchy problem for the system of dimensionless equations (9.49)-(9.50) has been solved numerically for some typical combinations of the parameter values, and we present a comparative analysis of the theoretical predictions and the numerical simulations.

Figs. 9.14 - 9.15 allow us to get a visual impression of how well the approximation based on the asymptotic approach (MIM) describes a real trajectory in the phase space. The figures are a projection of the real trajectory in the three-dimensional space (θ, Π, η) onto the (θ, Π) plane (Fig. 9.14) and the (θ, η) plane (Fig. 9.15). The smooth solid curve $P_{in}WP_{fin}$ represents the result of direct numerical simulation, whereas the dashed lines $P_{in}Q_2$ and Q_2P_{fin} result from the asymptotic approximations (MIM, within preheat and reaction sub-zones). The approximation $P_{in}Q_2$ of the preheat sub-zone corresponds to the stage $P_{in}Q_{wv}$ (Fig. 9.11) and to the component $P_{in}Q_2$ in Fig. 9.13. In turn, the approximation Q_2P_{fin} of the reaction sub-zone relates to the component $Q_{wu}P_{fin}$ (Fig. 9.12) and section Q_2P_{fin} in Fig. 9.13.

One can see that within the preheat sub-zone (θ in the interval 0-35) the energy of the system is almost constant - the difference between the two curves (theoretical and numerical) hardly visible. Visually the pressure Π is strictly proportional to the temperature θ. The origin of this proportionality can be found in the expression (9.52). As mentioned above, within the subsonic flame velocity region the parameter ε_{inert} is small, and the relation (9.52) reads as $v = \theta - \sigma\Pi$. Taking into account the fact that the variable v is slow within the preheat sub-zone one can easily understand the reason of the proportionality $\theta \sim \Pi$. The coefficient of the proportionality (σ) is

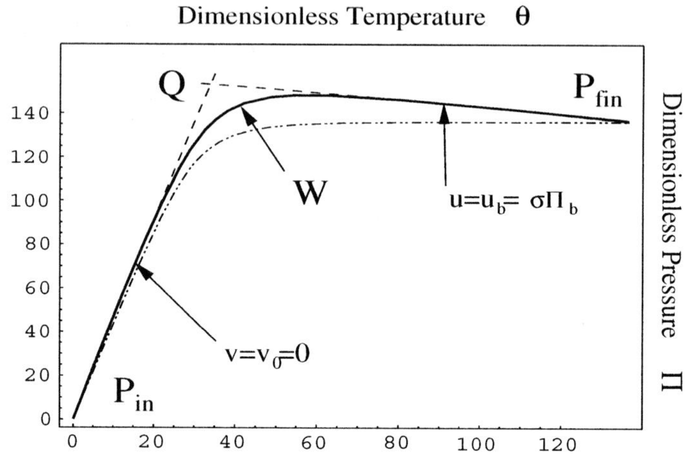

Figure 9.14. *Projection of the system trajectory on the plane (θ, Π). Solid line $P_{in}WP_{fin}$ - result of numeric simulations, dashed lines $P_{in}Q_2$ (preheat zone) and Q_2P_{fin} (reaction zone) - asymptotic approximations. For the system parameters - see caption to Fig. 9.1*

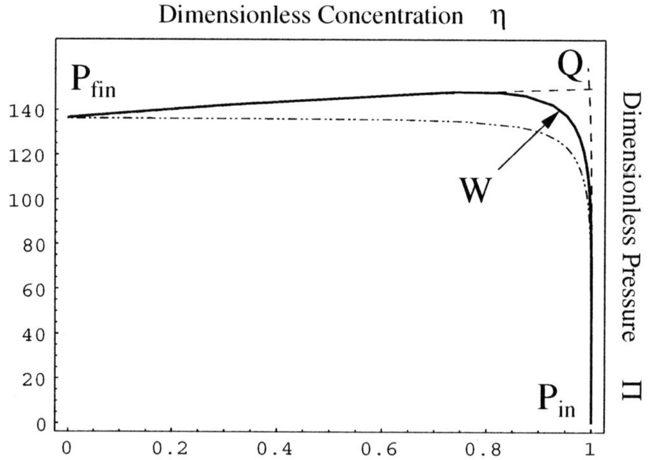

Figure 9.15. *Projection of the system trajectory on the plane (Π, η). Solid line $P_{in}WP_{fin}$ - result of numeric simulations, dashed lines $P_{in}Q_2$ (preheat zone) and Q_2P_{fin} (reaction zone) - asymptotic approximations. For the system parameters - see caption to Fig. 9.1*

more than unity, which means that the pressure rises faster than the temperature. A discrepancy begins to be observable only in the vicinity of the point Q_2 where the character of the trajectory changes.

The results computed for the temperature, pressure and fuel concentration

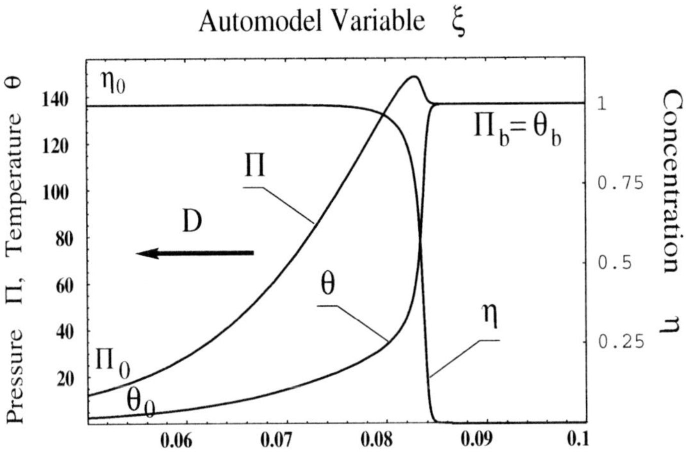

Figure 9.16. *Typical time histories of the flame front. The system parameters - see caption to Fig. 9.1*

are presented in Fig. 9.16 and should be compared with the analytical predictions. In the Fig. 9.16, one can see that the pressure Π and the temperature θ begin to change in the preheat sub-zone (the pressure Π rises faster than the temperature θ), while the value of concentration η looks constant (the energy of the system - the variable v - is asymptotically constant). Pressure rises and at some point reaches a value higher than the final one. Then the temperature θ and the concentration η begin to change quickly while the pressure Π remains constant (the momentum of the system - the variable u - is asymptotically constant within this sub-zone). In accordance with the theoretical predictions, the pressure rises faster than the temperature and it reaches its final value (pressure of the reaction products) at the point close to where the temperature jumps. This is the point Q_2, which separates the two sub-zones of the flame (reaction and preheat). Note here, a comparison between Figs. 9.1 and 9.16 (time histories) shows the inertia effects even in the region of the sub-sonic flames.

The dependence of the dimensional flame velocity D on the small parameter β for different cases and the accuracy of the theoretical predictions are presented in the Figs. 9.17 - 9.18. The solid curves in the Fig. 9.17 depict the results of numerical modeling, whereas the dashed lines give the theoretical predictions. The odd numbers correspond to the non-inertial case, whilst the even numbers (2, 4) relate to the model with inertia taken into account and compare theoretical predictions (9.46) (non-inertial approximation of the flame velocity Λ_F (9.67)) and numerical data. The speed of the flame grows with increasing β.

The accuracy of the analytic prediction can be estimated by a formula similar to (9.48):

$$\frac{\Lambda_F^{theor} - \Lambda_F^{num}}{\Lambda_F^{num}} 100 \tag{9.68}$$

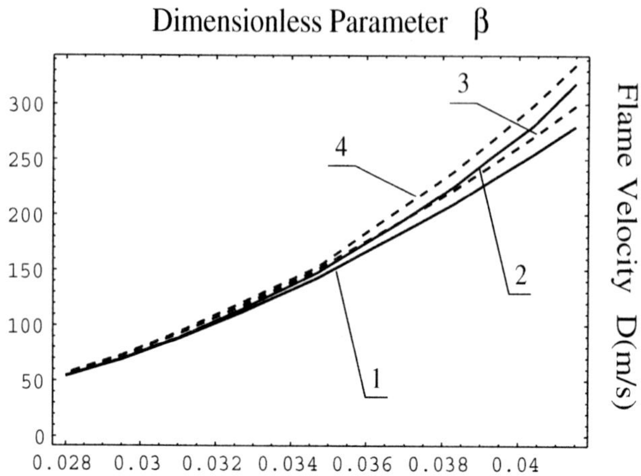

Figure 9.17. *Theoretical prediction (9.67) of the flame velocity dependence on the dimensionless parameter β versus results of numerical simulations. Dimensional flame velocity D versus parameter β. Solid lines - numerical simulations, dashed ones - theoretical predictions; the odd numbers (1, 3) - non-inertial model, the even numbers (2, 4) - inertia is accounted for*

where Λ_F^{theor} is the theoretical prediction (9.67) for the flame velocity and Λ_F^{num} represents results of an appropriate numerical simulation. In Fig. 9.18 the results of the comparison are plotted. The curve 1 presents the relative accuracy of our approximation for the non-inertial model (relative difference between the lines 1 and 3 in the Fig. 9.17), whereas the line 2 depicts the result of an application of the formula (9.68) to the model with the inertia mechanism accounted for. Fig. 9.18 illustrates how the accuracy of the asymptotic formulae (9.67) and (9.46) changes as the parameter β increases. Both calculations were performed with respect to a numerical simulation of the model with the inertia mechanism accounted for. In fact, the curve 1 in Fig. 9.18 presents a relative difference between the lines 2 and 4 in Fig. 9.17 (relative error is and remains positive), whereas the curve 2 in Fig. 9.18 depicts a relation between the curves 2 and 3 where the relative error fast becomes, and remains, negative. The analysis of the relevant numerical simulations shows that the accuracy of the theoretical predictions depends significantly on the parameter region. For the specific set of the problem parameters, the relative error of the analytical formulae is about 4-6% and grows slowly within this interval as the parameter β increases. The discrepancy between the analytical formula and the results of the numerical simulations is rather small.

9.8 Conclusion

We now summarize the basic results of the present work. The general problem of combustion in porous media has been under detailed investigation by a large number of researchers over the last decades and there are thousands of scientific

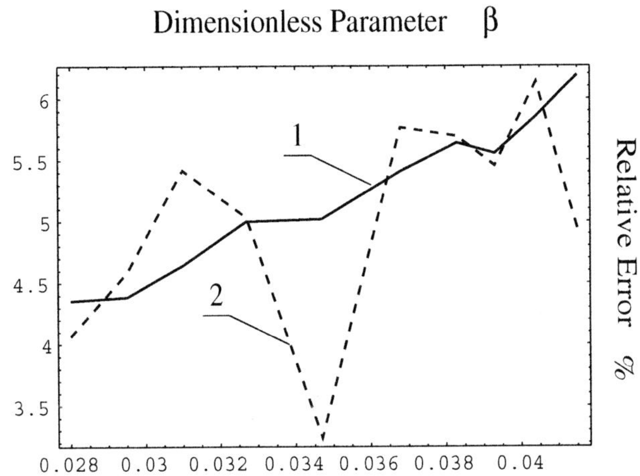

Figure 9.18. *Theoretical prediction (9.67) of the flame velocity dependence on the dimensionless parameter β versus results of numerical simulations. Relative accuracy of the theoretical formulae versus dimensionless parameter β. Digits designate: 1 - relative accuracy of the approximation (9.46) compared with numerics for the non-inertial model (relative difference between the lines 1 and 3 in the Fig. 9.17); 2 - relative accuracy of the approximation (9.67) compared with numerics for the inertial model (relative difference between the lines 2 and 4 in Fig. 9.17)*

papers devoted to the problem. Despite this the phenomenon of a combustion wave driven by a local pressure elevation through the inert solid skeleton filled with a combustible gaseous mixture represents a relatively new class of problems in the mathematical theory of combustion. The problem is new from both the conceptual and technical standpoints. The conceptual novelty lies in a conclusion that pressure diffusion through a porous medium (so-called "barodiffusion") plays the leading role in the determination of the system dynamics. Recall that the pressure does not affect open flames. The technical novelty of pressure driven flames in porous media consists in a significantly higher level of mathematical difficulty. The present paper is devoted to the detailed investigation of a series of problems of pressure driven flames in inert porous media. A step-by-step growth of the complexity level of the models of the phenomenon is illustrated and the corresponding evolution of asymptotic tools developed for their solution. The new models represented a challenge to researchers dealing with flame propagation problems. We developed an original mathematical technique for this class of combustion problems. The technique is based on two main sources:

(1) The Zel'dovich idea regarding the possible sub-division of the flame front into a number of sub-zones with significantly distinct properties,

(2) The method of integral (invariant) manifolds (MIM), in its geometrical version adapted to combustion problems.

The asymptotic tool can be briefly described in the following way. The original complex system of PDEs is reduced to the system of ODEs by introducing the automodel time-like coordinate. The introduction of new auxiliary variables transforms the original system of ODEs to a corresponding singularly perturbed system of ODEs. To determine these new variables one divides the flame front into two distinct sub-zones (preheat and reaction ones) and determines approximate conservation laws within each of the sub-zones. The new variables are defined as deviations from the laws of conservation. The dynamics of the singular perturbed system is then examined using the novel mathematical technique, namely, the geometrical version of the method of the integral (invariant) manifolds (MIM). This approach permits us to explore analytically the internal fine structure of the reaction zone, to derive explicit formulae for the parameters of the burnt products, and to obtain an explicit formula for the flame speed. To check the accuracy of the analytical results obtained from the model analysis, a series of direct numerical simulations is performed. A comparison of the asymptotic results with the direct numerical simulations was performed for each one of the models presented here. The comparison showed good agreement, suggesting that the asymptotic tools capture some of the essential physics associated with flames driven by local pressure elevation. Additional physical effects, such as supersonic flames and combustion waves in the vicinity of Chapman-Jouguet point, require further investigation. Curiously, it was found that a naive application of the Zel'dovich approach to the simplest version (Sections 4) of the generic model gives exactly the same results as the asymptotic method. In more complicated problems (Sections 6, 7) a direct application of the Zel'dovich algorithm becomes impossible and the proposed mathematical technique plays a critical role in our search for the problem solution. The authors would like to underline that, at the current stage of development of the proposed asymptotic tool, it represents an empirical algorithm which lacks a rigorous justification.

9.9 Appendix - Recent Results

The present Appendix contains our most recent results. Due to the nature of the material, the presentation will be brief.

Let us return to the original full model (9.1)-(9.8). Consider its approximation, characterized by (1) linear dependence of the friction force F_D (9.7) on the gas velocity and by (2) an absence of the inertia effect. Performing a suitable integration and substituting the dimensionless variables determined by (9.9), we get the dimensionless system (9.16)-(9.17) subject to the initial conditions (9.10). To demonstrate the essence of the new proposed approach (which is still under development), let us additionally assume the Frank-Kamenetskii [8] approximation and put $\beta = 0$ in the set (9.16)-(9.17). The simplified system reads:

$$\frac{d\theta}{d\xi} = \Lambda\left(\Pi - \theta\right) + \varepsilon_1 \eta \exp\left(\theta\right), \tag{9.69}$$

$$\sigma \frac{d\Pi}{d\xi} = \Lambda\left(\Pi - \theta\right), \tag{9.70}$$

$$\frac{d\eta}{d\xi} = -\varepsilon_2 \eta \exp\left(\theta\right). \tag{9.71}$$

The system (9.69)-(9.71) has an energy integral:

$$\eta - 1 + \frac{\varepsilon_2}{\varepsilon_1}(\theta - \sigma\Pi) = 0,$$

which permits us to determine the parameters of the burnt mixture (behind the flame front):

$$\eta_b = 0;\ \theta_b = \Pi_b = \frac{\varepsilon_1}{\varepsilon_2(1-\sigma)}.$$

In addition, the existence of the energy integral allows us to reduce the number of governing equations by one. The resulting system reads

$$\frac{d\theta}{d\xi} = \Lambda\,(\Pi - \theta) + \varepsilon_1\left(1 - \frac{\varepsilon_2}{\varepsilon_1}(\theta - \sigma\Pi)\right)\exp(\theta)\,, \tag{9.72}$$

$$\sigma\frac{d\Pi}{d\xi} = \Lambda\,(\Pi - \theta)\,. \tag{9.73}$$

This system is subject to initial conditions

$$\theta(0) = \varepsilon \ll 1;\ \Pi(0) = \frac{\varepsilon}{\sigma} \ll 1,$$

where the choice of the form (9.73) eliminates the so-called cold-boundary problem, at the initial stage, in the burned gas. The chosen boundary conditions allow us to avoid the cold-boundary problem and find the burning-rate eigenvalue Λ because the system is underdetermined. Before proceeding with the analysis it is worthwhile to estimate some characteristic values of the system parameters: $0.1 \le \sigma \le 0.3$, $10^{-6} \le \varepsilon_2 \ll \varepsilon_1 \le 10^{-4}$. Hence, the characteristic value of the adiabatic dimensional temperature and pressure has the order of magnitude $\approx 10^2$. Note also, that the singular (stationary) point of the system (9.72)-(9.73) is a saddle. This means, in particular, that this point is reachable along a single direction (separatrix) and an arbitrary trajectory of the system approaches its final point along this direction.

Consider the system (9.72)-(9.73), and construct an approximate solution in the vicinity of the stationary (adiabatic) point of the system. The trajectory behaviour of the system in the phase plane (θ, Π) is determined by the solution of the following equation:

$$\frac{d\theta}{d\Pi} = \sigma + \frac{\varepsilon_1\left(1 - \frac{\varepsilon_2}{\varepsilon_1}(\theta - \sigma\Pi)\right)}{\Lambda\,(\Pi - \theta)}\exp(\theta)\,. \tag{9.74}$$

The Eq. (9.74) can not be solved straightforwardly. Nevertheless, it is possible to simplify this equation and obtain an approximate solution. Recall that any system trajectory must approach a saddle along the separatrix. This direction can be readily written:

$$\frac{d\theta}{d\Pi}\Big|_{\substack{\theta \to \theta_b \\ \Pi \to \Pi_b}} = \sigma + A\exp(\theta_b) = 1 + \frac{\sigma\varepsilon_2}{\Lambda}\exp(\theta_b)\,,$$

where the coefficient A is evaluated in the following way (an exponentially small term is omitted):

$$A = ((1-\sigma)\exp(-\theta_b)) + \frac{\sigma\varepsilon_2}{\Lambda} \approx \frac{\sigma\varepsilon_2}{\Lambda}.$$

Thus, the simplified solution can be easily found from Eq. (9.74)

$$\int \frac{d\theta}{\left(1+\frac{\varepsilon_2}{\Lambda}\exp(\theta)\right)} = \sigma \int d\Pi,$$

and, after integration, yields

$$\theta - \ln\left(1+\frac{\varepsilon_2}{\Lambda}\exp(\theta)\right) = \sigma\Pi + Const.$$

The constant of integration is determined by the values of the variables at the stationary point. Thus

$$\theta - \theta_b - \ln\left(\frac{1+\frac{\varepsilon_2}{\Lambda}\exp(\theta)}{1+\frac{\varepsilon_2}{\Lambda}\exp(\theta_b)}\right) = \sigma(\Pi - \Pi_b).$$

Now, expanding the approximate solution up to the initial point and taking the limit $\varepsilon \to 0$ one obtains

$$-\ln\left(\frac{1+\frac{\varepsilon_2}{\Lambda}}{1+\frac{\varepsilon_2}{\Lambda}\exp(\theta_b)}\right) = \theta_b(1-\sigma).$$

Finally, an estimate of the parameter Λ is obtained as

$$\Lambda = \varepsilon_2 \exp(\sigma\theta_b) \frac{(1-\exp(-\sigma\theta_b))}{(1-\exp(-(1-\sigma)\theta_b))} \approx \varepsilon_2 \exp(\sigma\theta_b).$$

Comparison with (9.36) shows that these estimates coincide to an exponentially small order of magnitude ($\theta_b \gg 1$).

$$\Lambda = \frac{\varepsilon_1}{(1-\sigma)\theta_b}\exp(\sigma\theta_b) = \varepsilon_2 \exp(\sigma\theta_b). \tag{9.75}$$

Formula (9.75) finalizes the solution of the problem (9.69)-(9.71). We underline that the relation (9.75) was derived without subdivision of the flame front into two distinct sub-zones, as we did in this paper. It is our intention to explore this aspect further.

9.10 Acknowledgments

Authors are greatly thankful to Gregory Sivashinsky and Peter Gordon for fruitful discussions. This research was partly supported by the Grant G-695-15.10/2001 from the G.I.F., the German-Israeli Foundation for Scientific Research and Development.

Bibliography

[1] V. S. BABKIN, *Filtration combustion of gases. Present state of affairs and prospects*, Pure and Applied Chemistry, 65 (1993), pp. 335–344.

[2] N. N. BOGOLYUBOV AND YU. A. MITROPOLSKY, *Asymptotic Methods in the Theory of Nonlinear Oscillations*, Gordon and Breach, New York, 1961.

[3] I. BRAILOVSKY, V. GOLDSHTEIN, I. SHREIBER, AND G. SIVASHINSKY, *On combustion wave driven by diffusion of pressure*, Combustion Science and Technology, 124 (1996), pp. 145–165.

[4] I. BRAILOVSKY, M. FRANKEL, AND G. SIVASHINSKY, *Galloping and spinning modes of subsonic detonation*, Combustion Theory and Modelling, 4 (2000), pp. 47–60.

[5] I. BRAILOVSKY AND G. SIVASHINSKY, *Hydraulic resistance as a mechanism for deflagration-to- detonation transition*, Combustion and Flame, 122 (2000), pp. 492–499.

[6] V. BYKOV, I. GOLDFARB, AND V. GOL'DSHTEIN, *On an asymptotic approach to pressure driven flames in porous media*, Int. J. of Pure and Appl. Math., 9(4) (2003), pp. 403–418.

[7] N. FENICHEL, *Geometric singular perturbation theory for ordinary differential equations*, J. Differential Equations, 31 (1979), pp. 53–98.

[8] D. A. FRANK-KAMENETSKII, *Diffusion and Heat Exchange in Chemical Kinetics*, Plenum Press, New York, 1969.

[9] I. GOLDFARB, V. GOL'DSHTEIN, AND G. KUZMENKO, *Pressure driven flame in porous media*, Physics Letters A, 251 (1999), pp. 394–403.

[10] I. GOLDFARB, V. GOL'DSHTEIN, G. KUZMENKO, AND J. B. GREENBERG, *Monodisperse spray effects on thermal explosion in a gas* HTD-352, Proc. of ASME Heat Transfer Division, ASME International Congress and Exposition, Dallas, 2, pp. 199–206, 1997.

[11] ———, *On thermal explosion of a cool spray in a hot gas*, in Proc. of Twenty-Seventh Symposium (International) on Combustion The Combustion Institute, Pittsburgh, PA, pp. 2367–2371, 1998.

[12] I. GOLDFARB, V. GOL'DSHTEIN, I. SHREIBER, AND A. ZINOVIEV, *Liquid drop effects on self-ignition of combustible gas*, in Proc. of the Twenty-Sixth Symposium (International) on Combustion, The Combustion Institute Pittsburgh PA, pp. 1557–1563, 1996.

[13] I. GOLDFARB, V. GOL'DSHTEIN, AND A. ZINOVIEV, *Delayed thermal explosion in porous media: method of integral manifolds*, IMA J. of Applied Mathematics, 67 (2002), pp. 263–280.

[14] V. M. GOL'DSHTEIN , I. R. SHREIBER, AND G. A. SIVASHINSKY, *On creeping detonation in filtration combustion*, Shock Waves, 4(1) (1994), pp. 109–112.

[15] V. M. GOL'DSHTEIN AND V. SOBOLEV, *Integral manifolds in chemical kinetics and combustion*, in Singularity Theory and Some Problems of Functional Analysis, AMS Translations, Series 2, 153, pp. 73–92, 1992.

[16] P. V. GORDON, L. S. KAGAN, AND G. I. SIVASHINSKY, *Fast subsonic combustion as a free-interface problem*, Interfaces and Free Boundaries, 5 (2003), pp. 47–62.

[17] J. HALE, *Ordinary Differential Equations*, Wiley, New York, 1969.

[18] YU. A. MITROPOLSKIY AND O. B. LYKOVA, *Lectures on the Methods of Integral Manifolds*, Inst. Mathematics, Ukrainian Academy of Science, Kiev, 1968 (in Russian, MR 40 #1655)).

[19] N. N. SEMENOV, *Zur theorie des verbrennungsprozesses*, Z. Phys, 48 (1928), pp. 571–581.

[20] V. SOBOLEV, *Integral manifolds and decomposition of singularly perturbed systems*, System and Control Letters, 5 (1984), pp. 169–179.

[21] ——, *Integral manifolds, singular perturbations and optimal control*, Ukrainian Mathematical J., 39(1) (1987), pp. 111-116.

[22] V. STRYGIN AND V. SOBOLEV, *Separation of Motions by the Integral Manifolds Method*, Nauka, Moscow, 1988 (in Russian, MR 89k:34071).

[23] YA. B. ZEL'DOVICH, G. I. BARENBLATT, V. B. LIBROVICH, AND G. M. MAKHVILADZE, *Mathematical Theory of Combustion and Explosions*, Plenum, New York, 1985.

Chapter 10

Split-Hyperbolicity and the Analysis of Systems with Hysteresis and Lang-Kobayashi Equations

A. Pokrovskii, O. Rasskazov, and R. Studdert

We explain the opportunities which are provided in the areas of analysis of systems with hysteresis and computer-aided analysis of chaotic behavior by the machinery of split-hyperbolicity, as suggested in [26]. In particular we present some sufficient conditions for the robustness of the unstable oscillations in nonlinear ODEs with respect to small hysteresis perturbations. We also apply split-hyperbolicity to study the dynamics of the truncated Lang-Kobayashi equations that describe the behavior of semiconductor lasers with feedback. For particular values of the parameters we rigorously prove that the system has chaotic behavior in the Smale sense: some power of a suitable first return map is topologically conjugate to the left shift operator in the set of symbolic sequences. The proof is computer assisted, with all errors estimated and taken into account.

10.1 Introduction

In the recent years there has been rapidly growing interest in the influence of small hysteresis perturbations on the dynamics of physical systems, see [8, 9, 10] and references therein. In [3] it was shown that the properties of exponential asymptotic stability are robust with respect to small hysteresis perturbations under natural technical assumptions. To analyze the robustness of complicated, quasi-chaotic behaviour one needs similar properties concerning hyperbolic, but not exponentially stable, solutions of the ODE without hysteresis. This work uses the notion of split-hyperbolicity to obtain one of the first results of this kind.

The machinery of split-hyperbolicity was suggested in [26] as a refined version of semi-hyperbolicity [6]. This concept is applicable to Lipschitz continuous maps in metric spaces and ensures that a map under discussion has all the main properties of hyperbolic maps, including topological conjugacy to the appropriate symbolic dynamics. It should be noted that the usual hyperbolicity theory cannot be applied here, as the hysteresis operators are non-smooth and must be considered as operators in metric spaces. General definitions are given in Section 10.4.

In Section 10.4, we apply split-hyperbolicity to the general area of computer-aided rigorous analysis of chaotic behavior. The most important point is that the verification of the split-hyperbolicity is reduced to establishing a few simple inequalities; thus it can be done numerically.

In Section 10.5, we apply this technique to the rigorous analysis of the chaotic behaviour in the Lang-Kobayashi model of lasers with an external cavity, a topic commenced in [31]. Put simply, a laser with an external cavity is a device as shown in Fig. 10.1. The dynamics of a semiconductor laser with an external cavity can be reasonably well described by the so-called reduced Lang-Kobayashi equations, which are a system of three ODEs including some physical parameters. For particular values of the parameters the reduced LK system demonstrates some features of chaotic behavior and this behavior is qualitatively and quantitatively consistent with experimental results. Further details are given in Section 10.5.1.

Finally, Appendix 10.6 is devoted to the details of the computer-aided proof. In particular, we estimate the global integration error for a 4th order Runge-Kutta method applied to a system of twelve ODEs. A simple algorithm is offered to establish split-hyperbolicity of the numerically constructed Poincaré map.

10.2 Split-Hyperbolicity in Product-Spaces

Let $J \in \mathcal{Z}$ be a set, possibly infinite, of consecutive integers. Let M_n^s, M_n^u where $n \in Z$ be complete metric spaces with metrics ρ_n^s, ρ_n^u. Suppose that the non-empty balls in M_n^s and in M_n^u are connected, i.e. they cannot be represented as a disjoint union of two nonempty sets each of which is both relatively open and relatively closed.

Elements from the Cartesian product $M_n = M_n^s \times M_n^u$ are treated as pairs $x = (x^s, x^u)$. The spaces M_n are endowed with the usual metric

$$\rho_n(x, y) \triangleq \max\left\{\rho_n^s\left(x^s, y^s\right),\ \rho_n^u(x^u, y^u)\right\}.$$

Let $\mathbf{f}$ denote the bi-infinite sequence $\{f_i\}_{i\in Z}$ of continuous maps, which may be partially defined, from M_n to M_{n+1}: $f_n(x^s, x^u) = (f_n^s(x^s, x^u),\ f_n^u(x^s, x^u))$ where $f_n^s : M_n^s \times M_n^u \mapsto M_{n+1}^s$ and $f_n^u : M_n^s \times M_n^u \mapsto M_{n+1}^u$.

Let the sequence $X = \{x_n\}_{n\in J}$ of the elements $x_n \in M_n$ be fixed, such that the images $f_n(x_n)$ are defined for all $n \in J$. Denote by $B_n^s[r]$ the closed r-ball in M_n^s centered at $x_n^s \in M_n^s$; the balls $B_n^u[r]$ for $x_n^u \in M_n^u$, $r > 0$, are defined in a similar way. Let δ^s, δ^u be some positive constants. Denote $U_n \triangleq B_n^s[\delta^s] \times B_n^u[\delta^u]$. Let $\mathcal{D}_n$ be the set of those $y \in U_n$ which satisfy $f_n(y) \in U_{n+1}$.

A four-tuple $\mathbf{s} = (\lambda^s, \lambda^u, \mu^s, \mu^u)$ of non-negative real numbers is called *split* if $\lambda^s < 1 < \lambda^u$ and $\Delta(\mathbf{s}) \triangleq (1 - \lambda^s)(\lambda^u - 1) - \mu^s\mu^u > 0$.

Definition: The sequence $\mathbf{f}$ of the maps f_n is said to be *split* ($\mathbf{s}$-*) hyperbolic in the* (δ^s, δ^u)*-neighborhood of the sequence* X if it satisfies the following three conditions:

C0 $\mathcal{D}_n$ is closed for all $n \in J$, and, for each boundary point y of $\mathcal{D}_n$, either y belongs to the boundary of U_n or $f(y)$ belongs to the boundary of U_{n+1} (whenever $n+1 \in J$).

C1 For all $n, n+1 \in J$ and all $y, z \in \mathcal{D}_n$ the following inequalities hold

$$\rho^s_{n+1}(f^s_n(y), f^s_n(z)) \leq \lambda^s \rho^s_n(y^s, z^s) + \mu^s \rho^u_n(y^u, z^u),$$
$$\rho^u_{n+1}(f^u_n(y), f^u_n(z)) \geq -\mu^u \rho^s_n(y^s, z^s) + \lambda^u \rho^u_n(y^u, z^u).$$

C2 The map $w \mapsto f^u_n(v, w)$ is open as a map from $B^u_n[\delta^u]$ to $B^u_{n+1}[\delta^u]$ for each $v \in B^s_n[\delta^s]$, in the sense that the image $f^u_n(v, U)$ of an open subset U of $B^u_n[\delta^u]$ is relatively open in $B^u_{n+1}[\delta^u]$ (whenever $n, n+1 \in J$).

Conditions C0 and C2 are technical and intended for use in some complicated situations. We offer the following simple lemma without proof.

Lemma 10.2.1. *Let f_n be defined on U_n and let $M^u_n = \mathbb{R}^d_u$. Then conditions C0 and C2 from the definition of split-hyperbolicity hold.*

For a given sequence $\bar{X} = \{x_n\}_{n \in J}$ and a given $\mathbf{f}$ the *discrepancy* $D(\bar{X}; \mathbf{f})$ is defined by the formula: $D(\bar{X}; \mathbf{f}) \triangleq \max_{n,n+1 \in J} \rho_{n+1}(f_n(x_n), x_{n+1})$. We introduce the numbers

$$a^s \triangleq \frac{\lambda^u - 1 + \mu^s}{\Delta(\mathrm{s})}, \qquad a^u \triangleq \frac{1 - \lambda^s + \mu^u}{\Delta(\mathrm{s})}.$$

One of the main tools in the investigation of split-hyperbolic systems is the Shadowing Theorem given below, see [26].

Theorem 10.2.1. *(Shadowing Theorem) Let the sequence $\mathbf{f}$ be split-hyperbolic in the (δ^s, δ^u)-neighborhood of the sequence $\bar{X} = \{\bar{x}_n\}_{n \in J}$ and let the discrepancy $D(\bar{X}; \mathbf{f})$ satisfy the inequality $D(\bar{X}; \mathbf{f}) < \min\left\{\frac{\delta^s}{a^s}, \frac{\delta^u}{a^u}, \delta^u\right\}$. Then there exists a trajectory $X = \{x_n\}_{n \in J}$ of $\mathbf{f}$ satisfying*

$$\rho^s_n(x^s_n, \bar{x}^s_n) \leq a^s\, D(\bar{X}; \mathbf{f}) \text{ and } \rho^u_n(x^u_n, \bar{x}^u_n) \leq a^u\, D(\bar{X}, \mathbf{f}).$$

Corollary 10.2.1. *Assume that $\bar{X}$ is a sequence satisfying Theorem 10.2.1 and X a corresponding trajectory. If n is an initial time, then $\bar{x}^s_n = x^s_n$. If n is a final time point, then $\bar{x}^u_n = x^u_n$.*

Various properties of split-hyperbolicity were investigated in [30].

10.3 Main Result

Let $f(t, x) : \mathbb{R}_+ \times \mathbb{R}^d$ be T-periodic in time t and Lipschitz continuous in x: $|f(t,x) - f(t,y)| < l_f |x - y|$. Consider an ordinary differential equation

$$x' = f(t, x),$$

with $x \in \mathbb{R}^d$. Suppose that this equation has a T-periodic solution with $x(0) = x_T$. Let $\xi(t) : \mathbb{R} \to \mathbb{R}^d$ be continuous and let $F(x, \xi(\cdot))$ be a time T shift operator along the trajectories of the perturbed equation

$$x' = f(t, x) + \xi(t)$$

with the initial condition $x(0) = x$.

Let $\mathbb{R}^d = \mathbb{R}^{d_s} \times \mathbb{R}^{d_u}$ and let the coordinate decomposition of $F = (F^s, F^u)$; $F^s : \mathbb{R}^d \mapsto \mathbb{R}^{d_s}$, $F^u : \mathbb{R}^d \mapsto \mathbb{R}^{d_u}$ satisfy the inequalities

$$|F^u(x_1, \xi(\cdot)) - F^u(x_2, \eta(\cdot))| \\ \geq \lambda^u |x_1^u - x_2^u| - \gamma_c \max_{0 \leq t \leq T} |\xi(t) - \eta(t)|, \tag{10.1}$$

$$|F^s(x_1, \xi(\cdot)) - F^s(x_2, \eta(\cdot))| \\ \leq \lambda^s |x_1^s - x_2^s| + \gamma_c \max_{0 \leq t \leq T} |\xi(t) - \eta(t)|, \tag{10.2}$$

for $0 < \lambda^s < 1 < \lambda^u$, for x_1, x_2 such that $|x_1 - x_T|, |x_2 - x_T| < \delta_c$ and for small perturbations of the original system: $|\xi(t)|, |\eta(t)| \leq \varepsilon_c$. Note that $F(x) = F(x, 0)$ is split-hyperbolic as a map from some neighborhood of x_T to $\mathbb{R}^{d_s} \times \mathbb{R}^{d_u}$ with the split $(\lambda^s, \lambda^u, 0, 0)$, see Lemma 10.2.1.

In particular, inequalities (10.1) and (10.2) hold for a hyperbolic flow after an appropriate change of variables.

Now we consider an extended system that involves the output of a hysteresis nonlinearity:

$$x' = f(t, x), \qquad z(t) = (\Gamma[z_0]Lx)(t). \tag{10.3}$$

Here $L : \mathbb{R}^d \mapsto \mathbb{R}^m$ is a linear map with the Euclidean norm estimate $||L|| \leq l$; $\Gamma[z_0]$ is an operator which transforms functions $u : \mathbb{R}_+ \mapsto \mathbb{R}^m$, $\mathbb{R}_+ = [0, +\infty)$, to functions $z : \mathbb{R}_+ \mapsto \mathbf{Z}$, where $\mathbf{Z}$ is a complete metric space equipped with a metric ρ_z; the argument z_0 represents initial memory. As usual, the notation $(\Gamma[z_0]u)(t)$ refers to the value of the function $z = \Gamma[z_0]u$ at the time t. We require Γ to be a normal nonlinearity (as introduced in [3]).

Let $W_t^{1,1}$ be the Banach space of absolutely continuous functions $u : [0, t] \mapsto \mathbb{R}^m$, equipped with the standard norm

$$\|u\|_{W_t^{1,1}} = |u(0)| + \int_0^t |u'(s)| ds.$$

We define $W_{loc}^{1,1} = \{u : \mathbb{R}^+ \mapsto \mathbb{R}^m, u_{|[0,t]} \in W_t^{1,1}\}$.

A nonlinearity $\Gamma : W_{loc}^{1,1} \times \mathbf{Z} \mapsto C(\mathbb{R}_+; \mathbb{R}^m)$ is called a *normal nonlinearity with threshold* $h > 0$ if it satisfies the Volterra property

$$u(s) = v(s), \quad 0 \leq s \leq t, \quad \text{implies} \\ (\Gamma[z_0]u)(t) = (\Gamma[z_0]v)(t), \quad \text{for all} \quad t \geq 0,$$

the semigroup property

$$(\Gamma[(\Gamma[z_0]u)(t_1)]v)(t_2 - t_1) \equiv (\Gamma[z_0]u)(t_2),$$

where $v(t) = u(t - t_1)$, the Lipschitz condition (N1) and the contraction property (N2) below:

(N1) There exists a constant $\gamma_u > 0$ such that for every $z_0 \in \mathbf{Z}$, every $t \geq s \geq 0$ and every $u, v \in W_t^{1,1}$ the inequality $\rho_z((\Gamma[z_0]u)(s), (\Gamma[z_0]v)(s)) \leq \gamma_u \|u - v\|_{W_t^{1,1}}$ holds.

(N2) There exists a continuous and bounded function $q : \mathbb{R}_+ \mapsto \mathbb{R}_+$ with $q(\alpha) < 1$ for $\alpha > h$ such that

$$\rho_z((\Gamma[z_0]u)(t), (\Gamma[z_1]u)(t)) \leq q(osc_t u)\rho_z(z_0, z_1)$$

holds for all $t \geq 0$, all $z_0, z_1 \in \mathbf{Z}$ and all $u \in W_t^{1,1}$. Here $osc_t u$ is defined as $\sup_{0 \leq \tau, \sigma \leq t} |u(\tau) - u(\sigma)|$.

Although the definition seems to be complicated, numerous hysteresis nonlinearities are in fact normal nonlinearities. As examples we have the so-called *stop* or *von Mises* nonlinearity, see [17, 19], and certain modifications of the Preisach nonlinearity, see [29].

It is possible to show that system (10.3), where Γ is a normal nonlinearity, satisfies the conditions of split hyperbolicity. Let $M^s = \mathbb{R}^{d_s} \times \mathbf{Z}$ and $M^u = \mathbb{R}^{d_u}$. Introduce new metrics

$$\rho_s((x_1^s, z_1), (x_1^s, z_1)) = K|x_1^s - x_2^s| + \rho_z(z_1, z_2),$$
$$\rho_u(x_1^u, x_2^u) = |x_1^u - x_2^u|,$$

where $K = \frac{2\gamma_u l e^{l_f T}}{1-\lambda^s}$. Let $x(t)$ with $x(0) = x_T$ be the periodic orbit with the period T and let the oscillation satisfy $osc_T Lx > h$ where h is a threshold of the normal nonlinearity Γ.

Consider (10.3) with the initial conditions x_T, z_0 where $z_0 \in \mathbf{Z}$ and denote $z_k = z(kT) = (\Gamma[z_{k-1}]Lx)(T)$ for all integer $k \geq 1$. Condition (N2) implies that

$$\rho_z(z_{k+1}, z_k) \leq q(osc_T x)\rho(z_k, z_{k-1}) \leq q^k(osc_T x)\rho(z_1, z_0).$$

Since $q(osc_T Lx) < 1$ and $\mathbf{Z}$ is a complete metric space there exists z_f such that $z_f = (\Gamma[z_f]Lx)(T)$.

Let us choose the neighborhood of $((x_T^s, z_f), x_T^u)$ such that the T-shift is defined for $((x^s, z), x^u) \in U = B^s[(x_T^s, z_f); \delta^s] \times B^u[x_T^u; \delta^u]$ and such that $osc_T(Lx) > h$ for all $x(t)$ with the initial condition $x(0) \in B^s[x_T^s; \delta^s] \times B^u[x_T^u; \delta^u]$, $0 \leq t \leq T$.

Proposition 10.3.1. *A T-shift along the trajectories of (10.3) is split-hyperbolic in the (δ^s, δ^u) neighborhood of $((x_T^s, z_f), x_T^u)$.*

Let $g : \mathbb{R}^d \times \mathbf{Z} \to \mathbb{R}^d$ be a global Lipschitz continuous function with the Lipschitz constant λ_g: $|g(x_1, z_1) - g(x_2, z_2)| \leq \lambda_g|x_1 - x_2| + \lambda_g\rho(z_1, z_2)$. Consider a small perturbation of (10.3) in the form

$$x' = f(t, x) + \varepsilon g(x, z), \quad z(t) = (\Gamma[z_0]Lx)(t). \tag{10.4}$$

Theorem 10.3.1. *There exists $\varepsilon_0, \delta^u, \delta^s > 0$ such that for every $0 < \varepsilon < \varepsilon_0$, the shift operator along the trajectories of the perturbed system (10.4) for a time T is split-hyperbolic in $U = B^s[(x_T^s, z_f); \delta^s] \times B^u[x_T^u; \delta^u]$.*

Corollary 10.3.1. *There exists ε_1 such that for all $0 < \varepsilon < \varepsilon_1$, the time T shift operator along (10.4) has a unique fixed point in U.*

Proofs of Proposition 10.3.1, Theorem 10.3.1 and Corollary 10.3.1 can be found in [29].

10.4 Split-Hyperbolicity in the Analysis of Chaotic Behavior

10.4.1 Maps strongly compatible with topological Markov chains

For any positive integer m, let us denote by $\Omega(m)$ the totality of all bi-infinite sequences $\omega = \{\omega_i\}_{i=-\infty}^{\infty}$ with $\omega_i \in \{0, \dots, m\}$ for $i = 0, \pm 1, \pm 2, \dots$, and denote by $\sigma = \sigma_m$ the *(left) shift* on $\Omega(m)$ given by $\sigma_m(\omega) = \omega' = (\dots, \omega'_{-1}, \omega'_0, \omega'_1, \dots)$ where $\omega'_i = \omega_{i+1}$. Let $A = (a_{i,j})$, $i, j = 0, \dots, m$, be a square $(m+1)\times(m+1)$-matrix whose entries are either zeros or ones, and introduce the set

$$\Omega_A = \left\{\omega \in \Omega(m) : a_{\omega_i, \omega_{i+1}} = 1, i = 0, \pm 1, \pm 2, \dots\right\}.$$

The set Ω_A is shift invariant and the restriction σ_A of σ_m to Ω_A is a *topological Markov chain* with the matrix A.

Let $f : M \mapsto M$ be a continuous map in a complete metric space M with the metric ρ. A *trajectory* of f (or, to be more precise, of the dynamical system generated by f) is a sequence $\mathbf{x} = \{x_i\}_{i=-i_-}^{\infty}$ satisfying $x_{i+1} = f(x_i)$, for $i = -i_-$, $\dots$, 0, 1, 2, $\dots$, where $0 \le i_- < \infty$ (note that $i_- = i_-(\mathbf{x})$ depends on the particular trajectory $\mathbf{x}$). Let σ_f be the left shift map naturally defined on the set $\mathrm{Tr}(f)$ of bi-infinite trajectories of f. It is convenient to endow the sets Ω_A and $\mathrm{Tr}(f)$ with the usual metrics:

$$d(\omega, \omega') = \sum_{i=-\infty}^{\infty} \frac{|\omega_i - \omega'_i|}{2^{|i|}}$$

and

$$d(\mathbf{x}, \mathbf{x}') = \sum_{i=-\infty}^{\infty} \frac{\rho(x_i, x'_i)}{2^{|i|}}. \tag{10.5}$$

Let $\mathcal{X} = (X_0, \dots, X_m)$ be a finite family of compact connected subsets of M. A continuous map f is *strongly* $(\mathcal{X}, \sigma_A)$*–compatible* if for any $\omega \in \Omega_A$ there exists a unique trajectory $\mathbf{x} = \varphi(\omega) \in \mathrm{Tr}(f)$ satisfying $x_i \in X_{\omega_i}$.

Lemma 10.4.1. *Let f be strongly $(\mathcal{X}, \sigma_A)$-compatible. Then the map φ has the following properties*

(i) *A shift of $\omega \in \Omega_A$ induces a shift of the trajectory $\varphi(\omega)$: $\varphi\sigma_A = \sigma_f\varphi$;*

(ii) *if* $\omega \in \Omega_A$ *is p-periodic, then the trajectory* $\mathbf{x} = \varphi(\omega)$ *is also p-periodic;*

(iii) *the map* φ *is continuous.*

Proof: This assertion is well known. We supply a proof only for the reader's convenience.

(i) Let $\mathbf{x} = \varphi\omega$. It will suffice to prove that

$$\sigma_f \mathbf{x} = \varphi \sigma_A \omega. \tag{10.6}$$

Then obviously $\mathbf{y} = \sigma_f \mathbf{x}$ satisfies $y_i \in X_{\omega_{i+1}} = X_{\sigma_A \omega_i}$. On the other hand $\mathbf{z} = \varphi(\sigma_A \omega)$ is the only trajectory satisfying $\mathbf{z}_i \in X_{\sigma_A \omega_i}$ and (10.6) follows.

(ii) If $\mathbf{x}$ is not p periodic, then $\sigma_f^p \mathbf{x}$ satisfies $\sigma_f^p \mathbf{x} = \varphi \sigma_A^p \omega = \varphi\omega$, and is different from $\mathbf{x}$; this contradicts the definition of strong compatibility.

(iii) Suppose that $\omega^{(k)}$ converges to ω, but $\varphi(\omega^{(k)})$ does not converge to $\varphi(\omega)$. Then, since the set of all bi-infinite trajectories $\mathbf{x}$ satisfying $x_i \in X_i$ is compact in the metric (10.5), there exists a limit point $\mathbf{y} \neq \varphi(\omega)$ of the sequence $\varphi(\omega^{(k)})$. By construction, this limit point is a bi-infinite trajectory of f and satisfies $y_i \in X_{\omega_i}$. This contradicts the definition of strong compatibility. The lemma is proved. □

By this lemma, a strongly $(\mathcal{X}, \sigma_A)$–compatible map f is $(\mathcal{X}, \sigma_A)$–compatible in the terminology of [27].

The $(\mathcal{X}, \sigma_A)$-compatible maps have some features of chaotic behavior if A has many ones, and if sufficiently many sub-families of $\mathcal{X}$ have empty intersections. For instance, from the definitions we have:

Proposition 10.4.1. *Let* $\bigcap_{i=0}^m X_i = \emptyset$, *and the matrix* A *be* k*-transitive, in the sense that its power* A^k *has all positive entries, see Definition 1.9.6 [16]. Then the topological entropy* $\mathcal{E}^{top}(f)$ *is positive, and moreover,*

$$\mathcal{E}^{top}(f) \geq \frac{\ln\left(1 + \frac{1}{m}\right)}{k},$$

where $\ln(\cdot)$ *denotes the natural logarithm.*

Proof: We prove the inequality

$$\mathcal{E}^{top}(f^k) \geq \ln\left(1 + \frac{1}{m}\right). \tag{10.7}$$

Let I be some subset of $\{0, \ldots, m\}$; we say that this subset is *maximal* if $X_I = \bigcap_{i \in I} X_i \neq \emptyset$, but $X_I \bigcap X_j = \emptyset$ for any $j \in \{0, \ldots, m\} \setminus I$. The family of all maximal sets will be denoted by $\mathcal{I}$. By the compactness of the sets X_i, $i \in \{0, \ldots, m\}$, there exists a positive ε such that the inclusions

$$x \in X_I, y \in X_j$$

imply $\rho(x, y) > \varepsilon$ for any $I \in \mathcal{I}$, $j \notin I$.

Now, for a given n we will construct a set $\mathrm{Tr}(n)$ of n-trajectories

$$\mathbf{x} = \Big(x_0,\ x_1 = f^k(x_0),\ x_2 = f^{2k}(x_0),\ \ldots,\ x_{n-1} = f^{k(n-1)}(x_0)\Big)$$

of the system f^k such that

$$\max_{i=0,\ldots,n-1} \rho(x_i, y_i) \geq \varepsilon, \quad \mathbf{x}, \mathbf{y} \in \mathrm{Tr}(n).$$

We prove the following estimate for the number of elements in $\mathrm{Tr}(n)$:

$$\mathrm{card}(\mathrm{Tr}(n)) \geq \left(1 + \frac{1}{m}\right)^n. \tag{10.8}$$

The inequality (10.7) will then follow from the definition of topological entropy [16] and (10.8).

Let Ω_m denote the set of all sequences $\omega = \{\omega_i\}_{i=0}^{n-1}$ with $\omega_i \in \{0, \ldots, m\}$. Let us choose any sequence $\omega^{(0)} \in \Omega_m$. Then, since the matrix A is k-transitive, there exists a sequence $\mathbf{x}^0 \in \mathrm{Tr}(n)$ satisfying

$$x_i^0 \in X_{\omega_i^{(0)}}, \quad i \in \{0, \ldots, n-1\}.$$

We now choose any sequence

$$\mathbf{I}^{(0)} = (I_0^{(0)}, \ldots, I_{n-1}^{(0)}), \quad I_i^{(0)} \in \mathcal{I},$$

satisfying the additional property $\omega_i^{(0)} \in I_i^{(0)}$ for $i = 0, \ldots, n-1$. Finally, denote by $\Omega^{(0)}$ the set of sequences $\omega \in \Omega_m$ satisfying

$$\omega_i \in I_i^{(0)}, \quad i = 0, \ldots, n-1.$$

Suppose now that the sequences $\mathbf{x}^\iota$ and the sets $\Omega^{(\iota)}$ are constructed for $0 \leq \iota \leq \kappa$, for a non-negative integer κ, and that

$$\bigcup_{\iota=0}^{\kappa} \Omega^{(\iota)} \neq \Omega_m.$$

Now we choose a sequence $\omega^{(\kappa+1)} \in \Omega_m$, which satisfies

$$\omega^{(k+1)} \notin \bigcup_{\iota=0}^{\kappa} \Omega^{(\iota)},$$

and then construct the sequences $\mathbf{x}^{\kappa+1} \in \mathrm{Tr}(n)$, $\mathbf{I}^{(\kappa+1)}$ and $\Omega^{(k+1)}$ as before. We proceed as long as possible.

Since

$$\bigcap_{i=0}^{m} X_i = \emptyset$$

by supposition, each set $\Omega^{(\iota)}$ contains not more than m^n elements. On the other hand, the size of the set of all sequences ω of the length n is $(m+1)^n$. Therefore, we can construct at least $(1+1/m)^n$ sets $\Omega^{(\iota)}$. The sequence has the necessary properties and (10.8) follows. Thus (10.7) is proved.

To complete the proof of the proposition it remains to prove the formula $\mathcal{E}^{top}(f^k) = k\mathcal{E}^{top}(f)$; see Property (3) in Proposition 3.1.7 [16]. The proposition is proved. □

We note in passing that if the matrix A is k-transitive for some positive integer k then it is also (m^2+1)-transitive; this is a reformulation of the well known Wielandt theorem, see further references and improvements in [12, 24]. Thus, under the conditions of the previous proposition, the universal estimate

$$\mathcal{E}^{top}(f) \geq \frac{\ln\left(1+\frac{1}{m}\right)}{m^2+1}$$

is also valid.

The topological conjugacy of two maps $f : A \mapsto A$ and $g : B \mapsto B$ means the existence of the homeomorphism $h : A \mapsto B$ such that $g = h^{-1}fh$ ([16], p.68). (Loosely speaking, f and g can be treated as 'the same map in different coordinates'.)

Under simple technical conditions a suitable restriction of an $(\mathcal{X}, \sigma_A)$-compatible map is topologically conjugate to σ_A. For instance,

Proposition 10.4.2. *Suppose that the map f is invertible in the sense that the simultaneous relationships $f(x) = f(y)$, $x, y, f(x) \in \mathcal{X}$ imply $x = y$. Suppose that for any two different symbolic trajectories $\omega^{(1)}, \omega^{(2)} \in \Omega_A$ the equality $X_{\omega_i^{(1)}} \bigcap X_{\omega_i^{(2)}} = \emptyset$ holds at least for one i. Then there exists a set $K \subset \mathcal{U} = \bigcup X_i$ such that the restriction $f|_K$ is topologically conjugate to the topological Markov chain σ_A.*

Proof: It is sufficient to define the homeomorphism h by

$$h(\omega) = \varphi(\omega)_0.$$

□

We also formulate an analogue of the above proposition for non-invertible maps. Recall that the one-sided topological Markov chain $\sigma_A^R \Omega_m^R \mapsto \Omega_m^R$ is defined by

$$\sigma_m^R(\omega_0, \omega_1, \omega_2, \ldots) = (\omega_1, \omega_2, \omega_3, \ldots),$$

where Ω_A^R is the one-sided space of symbolic sequences

$$\Omega_A^R = \{\omega = (\omega_0, \omega_1, \omega_2, \ldots) : \omega_i \in \{0, 1, \ldots, m\} \text{ for } i = 0, 1, 2, \ldots\},$$

with the metric

$$d(\omega, \omega') = \sum_{i=0}^{\infty} \frac{|\omega_i - \omega_i'|}{2^{|i|}}.$$

Proposition 10.4.3. *Suppose that for any different symbolic trajectories $\omega^{(1)}, \omega^{(2)} \in \Omega_A$, for at least one i the equality $X_{\omega_i^{(1)}} \bigcap X_{\omega_i^{(2)}} = \emptyset$ holds. Suppose also that in each column of the matrix A there is at least one 1. Then the restriction of the map f to a suitable f-invariant set $K \subset \mathcal{U} = \bigcup X_i$ is topologically conjugate to the one-sided left shift σ_m^R. Moreover, the inclusions*

$$f^i(\varphi(\omega)_0) \in X_{\omega_i}, \quad \omega \in \Omega_A^R$$

are satisfied.

Consider specifically the case when the union set $\mathcal{U} = \bigcup X_i$ has $\ell > 1$ connected components $U_0, \dots, U_{\ell-1}$.

Proposition 10.4.4. *Let the union set $\mathcal{U} = \bigcup X_i$ have $\ell > 1$ connected components and let the matrix A be k-transitive. Then the restriction $f^k|_K$ of the iterated map f^k to a suitable f invariant Cantor set $K \subset \mathcal{U}$ is topologically conjugate to the one-sided left shift $\sigma_{\ell-1}^R$. Moreover, the inclusions*

$$f^{ki}(\varphi(\omega)_0) \in U_{\omega_i}, \quad \omega \in \Omega_{\ell-1}^R$$

can be satisfied and the estimate of topological entropy $\mathcal{E}^{top}(f)$

$$\mathcal{E}^{top}(f) \geq \ln(\ell)/k \tag{10.9}$$

holds.

If additionally the map f is invertible, in the sense that the simultaneous relationships $f(x) = f(y)$, $x, y, f(x) \in \mathcal{U}$ imply $x = y$, then there exists (another) Cantor set $K_1 \subset \mathcal{U}$ such that the restriction $f^k|_{K_1}$ is topologically conjugate to the shift $\sigma_{\ell-1}$.

Proof: Again this statement is well known. First we consider the case when the map f is invertible.

Let there correspond to any i, $0 \leq i \leq \ell-1$ a number $I(i)$ satisfying $X_{I(i)} \subseteq U_i$.

With any pair (i,j), $0 \leq i,j \leq \ell - 1$ we associate a sequence $s^{(i,j)}$ of the form $s_0^{(i,j)} = I(i), s_1^{(i,j)}, s_2^{(i,j)}, \dots, s_k^{(i,j)} = I(j)$ such that $a_{s_n^{(i,j)}, s_{n+1}^{(i,j)}} = 1$ for $n = 0, \dots, k-1$. Such a sequence exists because the entries of $B = A^k$ are strictly positive, and (by induction) an entry $b_{i,j}$ of B coincides with the number of possible sequences $s^{(i,j)}$.

Finally, there corresponds to a sequence $\omega \in \Omega(\ell - 1)$ the extended sequence $R(\omega) \in \Omega_A$, which is defined by $R(\omega)_{k \cdot i+j} = s_j^{(\omega_i, \omega_{i+1})}$ for an integer i and $0 \leq j \leq k-1$.

This new sequence, by definition, belongs to Ω_A. Thus the element $\varphi(R(\omega))_0$ is well defined for each $\omega \in \Omega(\ell - 1)$. Now we define

$$K = \{\varphi(R(\omega))_0 : \omega \in \Omega(\ell - 1)\} \tag{10.10}$$

and

$$h(\omega) = \varphi(R(\omega))_0 \text{ for } \omega \in \Omega(\ell - 1). \tag{10.11}$$

This map is an homeomorphism, because the simultaneous relationships

$$f(x) = f(y),\ x, y, f(x) \in \mathcal{U}$$

imply $x = y$, and K is compact. Thus the proposition is proved for an invertible f.

In the general case it is sufficient that there corresponds to a sequence $\omega^R \in \Omega^R(\ell - 1)$ the sequence $R(\omega) \in \Omega_A$ where $\omega_i = \omega_i^R$ for $i \geq 0$ and $\omega_i = 0$ for $i < 0$. Then we can again use the formulas (10.10) and (10.11).

It remains to establish (10.9). Indeed, by the invariance of the topological entropy with respect to the topological conjugacy, $\mathcal{E}^{top}(f^k) \geq \ln(\ell)$ (see Proposition 3.2.5 [16]). We also note the formula $\mathcal{E}^{top}(f^k) = k\mathcal{E}^{top}(f)$ from Proposition 3.1.7 [16]. The proposition is proved. □

10.4.2 Split-hyperbolicity in the analysis of strong compatibility

To simplify applications we introduce a new notation and reformulate the results presented in Section 10.2 in a more convenient form. Let $\hat{M}^u, \hat{M}^s, \check{M}^u, \check{M}^s$ be complete metric spaces with metrics $\hat{\rho}^u$, $\hat{\rho}^s$, $\check{\rho}^u$, $\check{\rho}^s$ respectively. The product space $\hat{M} = \hat{M}^u \times \hat{M}^s$ is endowed with the usual metric

$$\hat{\rho}(x, y) \triangleq \max\left\{\hat{\rho}^s\left(x^s, y^s\right),\ \hat{\rho}^u(x^u, y^u)\right\}.$$

The product space $\check{M}$ with the metric $\check{\rho}$ is similarly introduced.

Let $\hat{V} = \hat{V}^u \times \hat{V}^s, \check{V} = \check{V}^u \times \check{V}^s$ be some closed product sets in the corresponding spaces. Recall, that a *split* is a four-tuple $\mathbf{s} = (\lambda^u, \lambda^s, \mu^u, \mu^s)$ of nonnegative real numbers for which $\lambda^s < 1 < \lambda^u$ and $\Delta(\mathbf{s}) \triangleq (1 - \lambda^s)(\lambda^u - 1) - \mu^s\mu^u > 0$.

The map $g : \hat{V} \mapsto \check{M}$ is said to be $\mathbf{s}$-*hyperbolic over* $(\hat{V}, \check{V})$ if it satisfies the following two conditions.

C1. The inequalities

$$\check{\rho}^s\left(g^s(z), g^s(y)\right) \leq \lambda^s\hat{\rho}^s(z^s, y^s) + \mu^s\hat{\rho}^u(z^u, y^u) \tag{10.12}$$

and

$$\check{\rho}^u\left(g^u(z), g^u(y)\right) \geq -\mu^u\hat{\rho}^s(z^s, y^s) + \lambda^u\hat{\rho}^u(z^u, y^u) \tag{10.13}$$

hold for $y, z \in \hat{V}$, $g(y), g(z) \in \check{V}$.

C2. The map $v \mapsto g^u(v, w)$ is open as a map from $\hat{V}^u$ to $\check{V}^u$ for each $w \in \hat{V}^s$, in the sense that the image $g^u(U, w)$ of an open subset U of $\hat{V}^u$ is relatively open in $\check{V}^u$.

Let M be a metric space, let $f : M \mapsto f(M) \subset M$, and let $x_0, \ldots, x_m \in M$. Let us choose $m+1$ metric spaces $M_i = M_i^u \times M_i^s$ and homeomorphisms $h_i : M_i \mapsto M$. Suppose that the elements $y_i = (y_i^u, y_i^s) = h_i^{-1}(x_i)$ are well defined. Denote by $B_i^u[\delta]$ and $B_i^s[\delta]$ closed δ-balls centered at y_i^u and y_i^s respectively; denote also $B_i[\delta^u, \delta^s] = B_i^u[\delta^u] \times B_i^s[\delta^s]$.

We introduce the numbers

$$a^u \triangleq \frac{1 - \lambda^s + \mu^u}{\Delta(\mathbf{s})} = \frac{1 - \lambda^s + \mu^u}{(1 - \lambda^s)(\lambda^u - 1) - \mu^s\mu^u}$$

and

$$a^s \triangleq \frac{\lambda^u - 1 + \mu^s}{\Delta(\mathbf{s})} = \frac{\lambda^u - 1 + \mu^s}{(1 - \lambda^s)(\lambda^u - 1) - \mu^s \mu^u}.$$

Let δ^s, δ^u be some positive numbers. Let A be a square $m+1 \times m+1$-matrix whose entries are either zeros or ones.

Theorem 10.4.1. *Suppose that the nonempty balls in the metric spaces M_i^s, M_i^u are connected, i.e. cannot be represented as a disjoint union of two nonempty sets each of which is both relatively open and relatively closed. Suppose that $g_{i,j} = h_j^{-1} f h_i$ is* s*-hyperbolic on $B_i[\delta^u, \delta^s], B_j[\delta^u, \delta^s]$ whenever $a_{i,j} = 1$. Suppose also that*

$$\rho_j(g_{i,j}(\tilde{y}_i), \tilde{y}_j) < \min \left\{ \frac{\delta^s}{a^s}, \frac{\delta^u}{a^u}, \delta^u \right\} \tag{10.14}$$

whenever $a_{i,j} = 1$. Then f is strongly $(\mathcal{X}, \sigma_A)$-compatible, where

$$\mathcal{X} = \{X_i = h_i(B_i[\delta^u, \delta^s]);\ i = 0, \dots, m\}.$$

Proof: Let

$$\omega = (\dots, \omega_{-1}, \omega_0, \omega_1, \dots) \in \Omega_A$$

be a given sequence. To prove the proposition we show that there exists a unique sequence $\mathbf{x}$ satisfying

$$x_{i+1} = f(x_i), \quad x_i \in X_{\omega_i} \text{ for } i = 0, \pm 1, \pm 2, \dots. \tag{10.15}$$

Consider the sequence of maps

$$g_i = h_{\omega_{i+1}}^{-1} f h_{\omega_i} : M_{\omega_i} \mapsto M_{\omega_{i+1}}.$$

The sequence

$$\mathbf{y} = (\dots, y_{-1}, y_0, y_1, \dots), \quad y_i \in M_i$$

satisfies

$$y_{i+1} = g_i(y_i),$$

if and only if the sequence $x_i = h_i(y_i)$ satisfies $x_{i+1} = f(x_i)$, which is the first of the relationships (10.15). Similarly, the inclusion $y_i \in B_{\omega_i}(\delta^u, \delta^s)$ holds if and only if $x_i = h_i(y_i)$ belongs to X_{ω_i}. Thus, it remains to prove the following assertion

Lemma 10.4.2. *There exists a unique sequence* $\mathbf{y}$ *satisfying*

$$y_{i+1} = g_i(y_i), \quad y_i \in B_{\omega_i}[\delta^u, \delta^s] \text{ for } i = 0, \pm 1, \pm 2, \dots.$$

Proof: Let us fix some ω. The sequence $\mathbf{g}$ is $\mathbf{s}$-hyperbolic in the (δ^u, δ^s)-neighborhood of the sequence

$$\mathbf{z} = \{z_i\}_{i=-\infty}^{\infty} = \{y_{\omega_i}\}_{i=-\infty}^{\infty}$$

in terms of [26]: the conditions C1, C2 from [26] for the sequence **g** follow from C1, C2 above, whereas C0 from [26] holds because each g_i is defined over the whole (δ^u, δ^s)-neighborhood of z_i. Also the discrepancy

$$D(\mathbf{z}, \mathbf{g}) = \sup_{-\infty<i<\infty} \rho_{i+1}\left(g_i(z_i), z_{i+1}\right)$$

satisfies the inequality

$$D(\mathbf{z}, \mathbf{g}) < \min\left\{\frac{\delta^s}{a^s}, \frac{\delta^u}{a^u}, \delta^u\right\},$$

by (10.14).

The result now follows from the shadowing theorem for split-hyperbolic maps, see Theorem 4.3 in [26].

Theorem 10.4.2. *(Shadowing theorem)*
Suppose that the nonempty balls in the metric spaces M_i^u, M_i^s are connected. Let the sequence of functions **g** *be* **s***-hyperbolic in the (δ^s, δ^u)-neighborhood of the sequence $\mathbf{z} = \{z_n\}_{n=-\infty}^{\infty}$ and let the discrepancy $D(\mathbf{z}, \mathbf{g})$ satisfy the inequality*

$$D(\mathbf{z}, \mathbf{g}) < \min\left\{\frac{\delta^s}{a^s}, \frac{\delta^u}{a^u}, \delta^u\right\}.$$

Then in the (δ^s, δ^u)-neighborhood of **z** *there exists a unique trajectory $\mathbf{x} = \{x_n\}_{n=-\infty}^{\infty}$ of* **g** *satisfying*

$$\rho_n^s(x_n^s, \bar{x}_n^s) \le a^s\, D(\mathbf{z}, \mathbf{g}) \quad \text{and} \quad \rho_n^u(x_n^u, \bar{x}_n^u) \le a^u\, D(\mathbf{z}, \mathbf{g}).$$

□

10.5 Application to the Truncated Lang-Kobayashi Equations

10.5.1 Lang-Kobayashi equations

External cavity semiconductor lasers present many interesting features for both technological applications and fundamental non-linear science. Their dynamics has been the subject of numerous studies over the last twenty years, see for example [15] and references therein. Motivations for these studies vary from the need for stable tunable laser sources, for laser cooling or multiplexing, to the general understanding of their complex stability and chaotic behavior. A typical experiment is usually described by a set of delay differential equations introduced by Lang and Kobayashi [20]:

$$\begin{aligned} \dot{E} &= \kappa\,(1 + i\alpha)\,(N - 1)\,E + \gamma e^{-i\varphi_0} E\,(t - \tau)\,, \\ \dot{N} &= -\gamma_{\|}\left(N - J + |E|^2 N\right). \end{aligned} \tag{10.16}$$

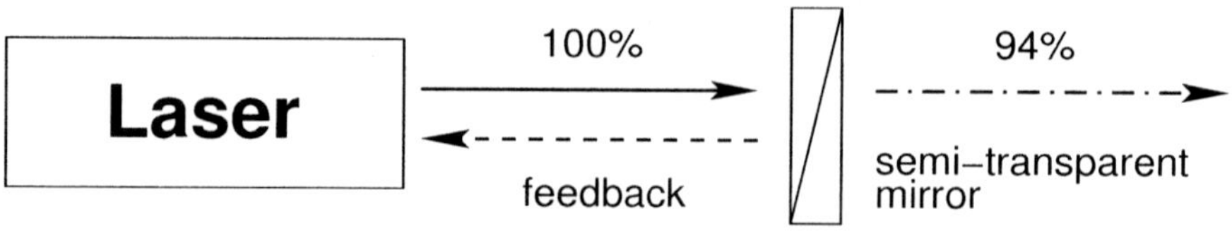

Figure 10.1. *Laser with a semi-transparent mirror*

Here E is the complex amplitude of the electric field, N is the carrier density, J is pumping current, κ is the field decay rate, $1/\gamma_{\|}$ is the spontaneous time scale, α is the linewidth enhancement factor, γ represents the feedback level, φ_0 is the phase of the feedback if the laser emits at the solitary laser frequency and τ is the external cavity round trip time. A typical setup is shown on Fig. 10.1. Numerical simulations of these equations have successfully reproduced many experimental observations, such as mode hopping between external cavity modes [23] and a period doubling route to chaos [22]. However there are few analytical results since delay equations are nonlocal.

This model was recently reduced to a 3-D dynamical system describing the temporal evolution of the laser power $P = |E|^2$, carrier density N and phase difference $\eta(t) = \varphi(t) - \varphi(t-\tau)$. This was achieved by assuming $P(t-\tau) = P(t)$ together with the approximation $\dot{\varphi} = \eta/\tau + \dot{\eta}/2$. This expression remains valid when the phase fluctuates on a time scale much shorter than the re-injection time τ. Under these approximations the Lang-Kobayashi equations (10.16) reduce to:

$$\begin{aligned}
\dot{P} &= 2\left(\kappa\left(N-1\right)+\gamma\cos\left(\eta+\varphi_0\right)\right)P,\\
\dot{N} &= -\gamma_{\|}\left(N-J+PN\right),\\
\dot{\eta} &= -\frac{1}{\tau_s}\eta + 2\kappa\alpha\left(N-1\right) - 2\gamma\sin\left(\eta+\varphi_0\right).
\end{aligned}$$

This model was successfully used to describe low frequency fluctuations commonly observed in semiconductor lasers with optical feedback, but its behavior for a low feedback level has not yet been investigated.

10.5.2 Poincaré map

To improve some estimates crucial for the numerical computations and to somehow center the graph of trajectories while keeping all parameters as rational numbers it is convenient to introduce new scaled variables:

$$x^{(1)} = \ln P, \quad x^{(2)} = 5N - 97/20, \quad x^{(3)} = 5(\eta+2)/16.$$

Choosing numerical values for the parameters in line with [31] we can rewrite the system under investigation as

$$\frac{dx^{(1)}}{dt} = -\frac{3}{50} + \frac{2}{5}x^{(2)} + \frac{3}{25}\cos\left(1 - \frac{16}{5}x^{(3)}\right),$$

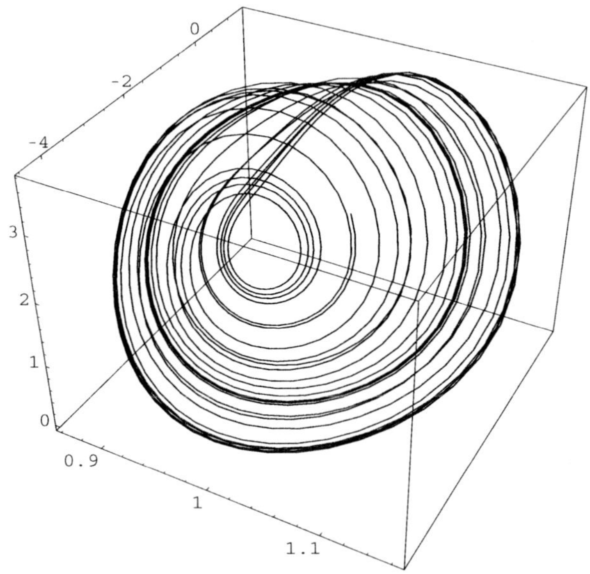

Figure 10.2. $\gamma_{||} = 0.03, \quad \gamma = 0.06$

$$\frac{dx^{(2)}}{dt} = \frac{609}{2000} - \frac{3}{100}x^{(2)} - \frac{3}{100}e^{x^{(1)}}\left(\frac{97}{20} + x^{(2)}\right), \tag{10.17}$$
$$\frac{dx^{(3)}}{dt} = -\frac{7}{160} + \frac{3}{8}x^{(2)} - \frac{1}{50}x^{(3)} + \frac{3}{80}\sin\left(1 - \frac{16}{5}x^{(3)}\right).$$

We will use the vector notation $\dot{x} = \mathcal{F}(x)$ for the system (10.17) (where $x = \left(x^{(1)}, x^{(2)}, x^{(3)}\right)$). For the elements of the plane $x^{(3)} = 0$ we will use the notation $x = (x^{(1)}, x^{(2)})$ as a synonym for $x = (x^{(1)}, x^{(2)}, 0)$. Note that the condition of transverse intersection with W, $\dot{x}^{(3)} \neq 0$, is broken only for the set of W defined by

$$x^{(2)} = \frac{7}{60} - \frac{3}{10}\sin(1) = x_T^{(2)}.$$

Let us denote by S the half-plane $x^{(3)} = 0$, $x^{(2)} > x_T^{(2)}$. Since S is, by definition, transversal to trajectories of the system, we can introduce the Poincaré map Φ, which is defined as a (partial) map from S to itself, obtained by following trajectories from one intersection with S to the next.

Let us denote by Π the union set of two parallelograms $R_1, R_2 \subset S$ given by

$$R_1 = \{(0.32, 0.43) + (-0.3675, 0.3255)\alpha + (0.039, 0.039)\beta : \alpha, \beta \leq 1\};$$
$$R_2 = \{(-0.04, 0.83) + (0.075, 0)\alpha + (0, 0.1)\beta : \alpha, \beta \leq 1\}.$$

Theorem 10.5.1. *The Poincaré map Φ is defined for all $y \in \Pi$ and $\Phi(\Pi) \subset \Pi$.*

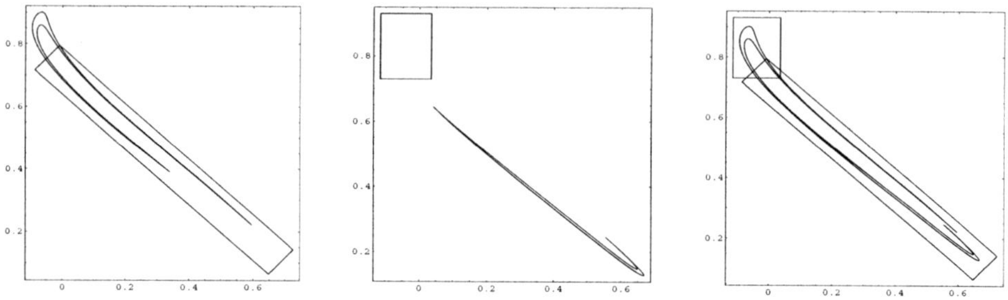

(a) Rectangle R_1 and its image

(b) Rectangle R_2 and its image

(c) Previous figures combined

Figure 10.3. *Geometry of the Poincaré map on the set* $\Pi = R_1 \bigcup R_2$

The geometry of the map Φ on Π is illustrated by the numerically constructed Figs. 10.3(a), 10.3(b) and 10.3(c). On the naive level, Fig. 10.3(c) provides a "proof" for the proposition above. However, the detailed justification of this figure is cumbersome. It requires a rigorous proof of continuity of the Poincaré map Φ as well as estimates of the accuracy of discrete computer arithmetic and estimates of the precision of the numerical method. A rigorous computer assisted proof of Theorem 10.5.1 was given in [31].

10.5.3 Main theorem

Consider the set $\mathcal{X}^*$ consisting of nine parallelograms X_i^*, $i = 0, \ldots, 19$ given by

$$X_i^* = \{x_i + p_i\alpha + q_i\beta : \ |\alpha|, |\beta| \leq 0.00002\}. \tag{10.18}$$

Here the coordinates of the center points $x_i = \left(x_i^{(1)}, x_i^{(2)}\right)$ are given in Table 10.1 and the coordinates of $p_i = \left(p_i^{(1)}, p_i^{(2)}\right)$, $q_i = \left(q_i^{(1)}, q_i^{(2)}\right)$, are given in Table 10.2. Fig. 10.4 graphs these parallelograms together with R_1, R_2.

We introduce a 20×20 matrix $A_* = (a_{i,j})$ $i, j = 0, \ldots, 19$ given by

$$\begin{array}{ll} a_{i,i+1} = 1; & \text{for all } 0 \leq i \leq 18, \\ a_{19,0} = 1; \ a_{2,2} = 1, & \\ a_{i,j} = 0; & \text{if pair } (i,j) \text{ is none of above.} \end{array} \tag{10.19}$$

Theorem 10.5.2. *The Poincaré map Φ is strongly $(\mathcal{X}^*, \sigma_{A_*})$ compatible.*

By definition the union set $\mathcal{U}^* = \bigcup X_i^*$ has 15 connected components $U_0^*, \ldots, U_{14}^*$ (see Fig. 10.4). Since Φ is an homeomorphism, Theorem 10.5.2 together with Proposition 10.4.4 imply the following corollary.

Table 10.1. *Vectors x_i*

i	0	1	2	3	4	5	6	7
$x_i^{(1)}$	0.019587	0.020515	0.020479	0.020479	0.020482	0.020475	0.020492	0.020453
$x_i^{(2)}$	0.66699	0.66915	0.66906	0.66907	0.66907	0.66907	0.66906	0.66910

i	8	9	10	11	12	13	14	15
$x_i^{(1)}$	0.020545	0.020328	0.020842	0.019627	0.022516	0.015729	0.032140	-0.0048159
$x_i^{(2)}$	0.66900	0.66923	0.66869	0.66996	0.66695	0.67407	0.65706	0.69655

i	16	17	18	19
$x_i^{(1)}$	0.092305	-0.074542	0.44510	0.042707
$x_i^{(2)}$	0.59933	0.79970	0.30273	0.72057

Table 10.2. *Vectors p_i, q_i*

i	0	1	2	3	4	5	6	7
$p_i^{(1)}$	0.69432	0.69432	0.69432	0.69432	0.69432	0.69432	0.69432	0.69432
$p_i^{(2)}$	-0.71967	-0.71967	-0.71967	-0.71967	-0.71967	-0.71967	-0.71967	-0.71967
$q_i^{(1)}$	0.39613	0.39613	0.39613	0.39613	0.39613	0.39613	0.39613	0.39613
$q_i^{(2)}$	0.91820	0.91820	0.91820	0.91820	0.91820	0.91820	0.91820	0.91820

i	8	9	10	11	12	13	14	15
$p_i^{(1)}$	0.69432	0.69432	0.69432	0.69432	-0.75835	0.79844	-0.92879	0.84009
$p_i^{(2)}$	-0.71967	-0.71967	-0.71967	-0.71967	0.78856	-0.84560	0.94270	-0.95290
$q_i^{(1)}$	0.39613	0.39613	0.39613	0.39613	-0.36270	0.35310	-0.29803	0.34887
$q_i^{(2)}$	0.91820	0.91820	0.91820	0.91820	-0.85146	0.80491	-0.73433	0.74206

i	16	17	18	19
$p_i^{(1)}$	-1.2827	0.29435	-1.0545	-0.58329
$p_i^{(2)}$	1.1753	-0.73198	0.86816	0.65993
$q_i^{(1)}$	-0.13304	0.72214	-1.0288	-0.38881
$q_i^{(2)}$	-0.67436	1.3124	-0.51327	-0.89935

Corollary 10.5.1. *The restriction of Φ^{38} to a suitable Φ-invariant Cantor set $K \subset \mathcal{U}^*$ is topologically conjugate to the left shift σ_{14}. Moreover, the inclusion*

$$\Phi^{38i}(\varphi(\omega)_0) \in U_{\omega_i}, \quad \omega \in \Omega_{14}$$

can be satisfied, and the estimate

$$\mathcal{E}^{top}(\Phi) \geq \ln(15)/38 > 0.07$$

holds.

Proof: We need only mention that the matrix A_* is 38-transitive. □

10.5.4 Idea of the proof of Theorem 10.5.2

The theorem will be obtained as a special case of Theorem 10.4.1 above. Let us specify $M = \mathbb{R}^2$, $M_i^s = M_i^u = \mathbb{R}^1$, $i = 0, \ldots, 19$. We denote by x the points from M and by y the points from $M_i^u \times M_i^s$. Let the points x_i be given by Table 10.1

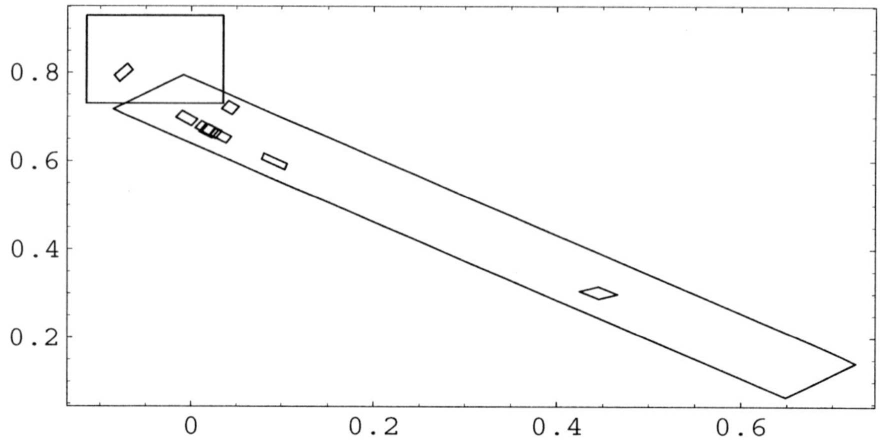

Figure 10.4. *Parallelograms* R_1,R_2

above. Introduce homeomorphisms $h_i : M_i^u \times M_i^s \mapsto M$ by

$$h_i(y^u, y^s) = \begin{pmatrix} x_i^{(1)} \\ x_i^{(2)} \end{pmatrix} + \mathcal{H}_i \begin{pmatrix} y^u \\ y^s \end{pmatrix} \quad \text{where } \mathcal{H}_i = \begin{pmatrix} p_i^{(1)} & p_i^{(2)} \\ q_i^{(1)} & q_i^{(2)} \end{pmatrix}.$$

In particular, all elements $y_i = h_i^{-1}(x_i)$ are zeros and all closed balls $B_i[\delta^u, \delta^s]$ are centered at $(0,0)$. We finally choose $\delta^u = \delta^s = 0.00002$ and fix the split

$$\mathbf{s}_* = (\lambda_*^u, \lambda_*^s, \mu_*^u, \mu_*^s) = (1.8, 0.35, 0.32, 0.32).$$

Lemma 10.5.1.

(a) *Sets* X_i^**, given by (10.18), satisfy* $X_i^* = h_i(B_i[\delta^u, \delta^s])$, $i = 0, \ldots, 19$.

(b) *For all pairs* i, j *such that* $a_{i,j} = 1$*, the inequality*

$$|g_{i,j}(0)| < \min\left\{\frac{\delta^s}{a^s}, \frac{\delta^u}{a^u}, \delta^u\right\} \qquad \textit{holds}\ .$$

(c) *The map* $g_{i,j} = h_j^{-1}\Phi h_i$ *is* $\mathbf{s}_*$*-hyperbolic over* $(B_i[\delta^u, \delta^s], B_j[\delta^u, \delta^s])$ *whenever* $a_{i,j} = 1$.

Proof:

(a) Follows from the definitions.

(b) Straightforward calculations, within the precision $4 \cdot 10^{-7}$, give us

$$\max_{i,j:\ a_{i,j}=1} |g_{i,j}(0)| < 4.2 \cdot 10^{-6} \quad \text{and} \quad 7 \cdot 10^{-6} < \min\left\{\frac{\delta^s}{a^s}, \frac{\delta^u}{a^u}, \delta^u\right\},$$

where a^s, a^u are calculated from $\mathbf{s}_*$ and the assertion (b) follows.

(c) We consider C1. The inequalities (10.12), (10.13) can be rewritten as

$$|g^u_{i,j}(z) - g^u_{i,j}(y)| \geq \lambda^u|z^u - y^u| - \mu^u|z^s - y^s| \tag{10.20}$$

and

$$|g^s_{i,j}(z) - g^s_{i,j}(y)| \leq \lambda^s|z^s - y^s| + \mu^s|z^u - y^u|. \tag{10.21}$$

Thus it remains to establish the inequalities above for all i, j satisfying $a_{i,j} = 1$ and all $y, z \in B_i[\delta^u, \delta^s]$.

Let $S^{(1)}, S^{(2)}$ be 2×2 matrices with non-negative entries. We say that

$$S^{(1)} \succ S^{(2)}$$

if and only if

$$s^{(1)}_{11} < s^{(2)}_{11} \quad s^{(1)}_{12} > s^{(2)}_{12} \quad s^{(1)}_{21} > s^{(2)}_{21} \quad s^{(1)}_{22} > s^{(2)}_{22}.$$

If the four-tuple $\mathbf{s}^{(1)} = (s^{(1)}_{11}, s^{(1)}_{22}, s^{(1)}_{12}, s^{(1)}_{21})$ is a split, then the four-tuple $\mathbf{s}^{(2)} = (s^{(2)}_{11}, s^{(2)}_{22}, s^{(2)}_{12}, s^{(2)}_{21})$ is also a split; moreover if, under certain conditions, the function f is split-hyperbolic with the split $\mathbf{s}^{(2)}$ over $(\check{V}, \hat{V})$, then it is also split-hyperbolic with the split $\mathbf{s}^{(1)}$.

Lemma 10.5.2. *The relationship*

$$\begin{pmatrix} \lambda^u_* & \mu^u_* \\ \mu^s_* & \lambda^s_* \end{pmatrix} \succ g'_{i,j}(y)$$

holds for all i, j such that $a_{i,j} = 1$ and for all y from $B_i[\delta^u, \delta^s]$.

The proof of the lemma is computer assisted and presented in Section 10.6. The result of the lemma can be restated as four inequalities

$$\begin{aligned}
\left|\frac{\partial}{\partial y^s} g^s_{i,j}(y^u, y^s)\right| &< 0.35 = \lambda^s_*, \\
\left|\frac{\partial}{\partial y^u} g^s_{i,j}(y^u, y^s)\right| &< 0.32 = \mu^s_*, \\
\left|\frac{\partial}{\partial y^s} g^u_{i,j}(y^u, y^s)\right| &< 0.32 = \mu^u_*, \\
\frac{\partial}{\partial y^u} g^u_{i,j}(y^u, y^s) &> 1.8 = \lambda^u_*,
\end{aligned}$$

that hold for all i, j satisfying $a_{i,j} = 1$ and all $(y^u, y^s) \in B_i[\delta^u, \delta^s]$.

Now Condition C1 follows from the Cauchy formulas:

$$\begin{aligned}
g^s_{i,j}(y_1) - g^s_{i,j}(y_2) = (y^s_1 - y^s_2) &\int_0^1 \frac{\partial}{\partial y^s} g^s_{i,j}(\lambda y_1 + (1-\lambda)y_2)\, d\lambda \\
+(y^u_1 - y^u_2) &\int_0^1 \frac{\partial}{\partial y^u} g^s_{i,j}(\lambda y_1 + (1-\lambda)y_2)\, d\lambda,
\end{aligned}$$

and

$$g^u_{i,j}(y_1) - g^u_{i,j}(y_2) = (y^s_1 - y^s_2)\int_0^1 \frac{\partial}{\partial y^s} g^u_{i,j}(\lambda y_1 + (1-\lambda)y_2)\,d\lambda$$
$$+(y^u_1 - y^u_2)\int_0^1 \frac{\partial}{\partial y^u} g^u_{i,j}(\lambda y_1 + (1-\lambda)y_2)\,d\lambda.$$

Taking the absolute value and using the triangle inequality we obtain

$$|g^s_{i,j}(y_1) - g^s_{i,j}(y_2)| \le |y^s_1 - y^s_2| \max_z \left|\frac{\partial}{\partial y^s} g^s_{i,j}(z)\right| + |y^u_1 - y^u_2| \max_z \left|\frac{\partial}{\partial y^u} g^s_{i,j}(z)\right|,$$

and

$$|g^u_{i,j}(y_1) - g^u_{i,j}(y_2)| \ge |y^u_1 - y^u_2| \min_z \left|\frac{\partial}{\partial y^u} g^u_{i,j}(z)\right| - |y^s_1 - y^s_2| \max_z \left|\frac{\partial}{\partial y^s} g^u_{i,j}(z)\right|,$$

where min, max are taken over all $z \in B_i[\delta^u, \delta^s]$.

The inequalities (10.21), (10.20) follow from the formulas above and the estimates given in Lemma 10.5.2. Thus the condition C1 is verified.

Finally, since the Poincaré map Φ is continuous on Π, inequality (10.20) implies strict monotonicity of $g_{i,j}(y^u, y^s)$ in the second variable for all i, j satisfying $a_{i,j} = 1$ and all $(z^u, z^s) \in B_i[\delta^u, \delta^s]$, and the condition C2 follows. Thus, the assertion (c) is proved, and so is the lemma.

□

Theorem 10.5.2 follows from Theorem 10.4.1 and Lemma 10.5.2.

To conclude this section we explain how the maps h_i were chosen. Firstly, we considered a pseudo orbit of length 20 that has elements in a small vicinity of a pseudo fixed point and elements that are far away from it. Both orbit and fixed point were constructed using the broken orbits method [28]. Thus we obtained the sequence of x_i, $i = 0, \dots, 19$ (see Table 10.1).

Secondly, we found the approximate eigenvectors for the Φ^{20} shift at the points x_i and used them as the first approximations for p_i, q_i. Then the vectors p_i,q_i were stretched or compressed to make the eigenvalues of $g_{i,j} = h_i^{-1}\Phi h_j$ approximately the same for all i, j such that $a_{i,j} = 1$.

Finally, we chose δ^u, δ^s to satisfy the inequality for the discrepancy. Since the total computational time is proportional to $\delta^u \cdot \delta^s$, those values had to be chosen as small as possible.

10.6 Proof of Lemma 10.5.2

10.6.1 Auxiliary translation operator

Let $\varphi(t, x)$ denote the translation operation for time t along the trajectory of (10.17) starting at x. Recall that the Poincaré map $\Phi(x_0) : \Pi \mapsto \Pi$ is defined as $\Phi(x_0) =$

$\varphi(T(x_0), x_0)$ where $T(x_0)$ is the time of the first intersection with Π. Thus

$$\frac{\partial \Phi}{\partial x} = P_{x_0} \frac{\partial \varphi}{\partial x}(T(x_0), x_0) \tag{10.22}$$

and P_{x_0} is the matrix of the projector to Π along the vector $f(\varphi(T(x_0), x_0))$. Note that the matrix $\frac{\partial \varphi}{\partial x}(t, x_0) = \mathcal{J}(t, x_0)$ satisfies the matrix differential equation

$$\dot{\mathcal{J}} = \frac{\partial \mathcal{F}}{\partial x}(\varphi(t, x_0))\mathcal{J},$$

with the initial condition $J(0) = I$. Thus, $\mathcal{J}$ is a part of the solution of the 12-dimensional system

$$\begin{aligned} \dot{x} &= \mathcal{F}(x), \\ \dot{\mathcal{J}} &= \frac{\partial \mathcal{F}}{\partial x}(x)\mathcal{J}, \end{aligned} \tag{10.23}$$

with initial conditions

$$\begin{aligned} x(0) &= x_0, \\ \mathcal{J}(0) &= I, \end{aligned} \tag{10.24}$$

where I stands for the identity matrix. We rewrite (10.23) in vector form. Introduce the vector

$$\tilde{y} = (x^{(1)}, x^{(2)}, x^{(3)}, j^{(1,1)}, j^{(2,1)}, j^{(3,1)}, j^{(1,2)}, j^{(2,2)}, j^{(3,2)}, j^{(1,3)}, j^{(2,3)}, j^{(3,3)}),$$

where $j^{(i,k)}$, $i, k = 1, 2, 3$ are the elements of the matrix $\mathcal{J}$. Sometimes we will use the equivalent notation for the components of $\tilde{y}$:

$$\tilde{y} = (\tilde{y}^{(1)}, \tilde{y}^{(2)}, \dots, \tilde{y}^{(12)}),$$

and for the equation (10.23):

$$\frac{d\tilde{y}}{dt} = \tilde{\mathcal{F}}(\tilde{y}) \tag{10.25}$$

Denote by $\tilde{\varphi}(t, \tilde{y})$ the translation operator along trajectories of the differential equation (10.25). Since we are interested in initial conditions of type (10.24), where $\mathcal{J}(0) = I$, sometimes we will denote the translation operator by $\tilde{\varphi}(t, x_0)$ assuming that $x_0 \in S$.

To conclude this subsection we define an area of the phase space that is of interest for further consideration. Let Ω be an open rectangular area given by

$$\Omega = (-0.6, 1.34) \times (-0.6, 1.02) \times (-0.92, 1.14).$$

Introduce $\tilde{\Omega} = \Omega \times \{J\}$ where $J = (j^{(k,l)})$ satisfies $|j^{(k,l)}| < j_*^{(k,l)}$. The matrix $J_* = (j_*^{(k,l)})$ is given as:

$$J_* = \begin{pmatrix} 0.24 & 0.12 & 0.2 \\ 0.2 & 0.1 & 0.16 \\ 0.16 & 0.08 & 0.14 \end{pmatrix}.$$

Define the ε-inflation and the ε-deflation of Ω as

$$\begin{aligned}\tilde{\Omega}(\varepsilon) &= \{\tilde{z} = (\tilde{z}_i) : \max_{1\le i\le 12} |\tilde{z}_i - \tilde{y}_i| < \varepsilon \text{ for some } \tilde{y} = (\tilde{y}_i) \in \tilde{\Omega}\},\\ \tilde{\Omega}(-\varepsilon) &= \{\tilde{z} = (\tilde{z}_i) \in \tilde{\Omega} : \max_{1\le i\le 12} |\tilde{z}_i - \tilde{y}_i| > \varepsilon \text{ for all } \tilde{y} = (\tilde{y}_i) \in \partial\tilde{\Omega}\}.\end{aligned} \tag{10.26}$$

For convenience we introduce a projector $J(\tilde{y})$ as

$$J(\tilde{y}) = \begin{pmatrix} \tilde{y}^{(4)} & \tilde{y}^{(7)} & \tilde{y}^{(10)} \\ \tilde{y}^{(5)} & \tilde{y}^{(8)} & \tilde{y}^{(11)} \\ \tilde{y}^{(6)} & \tilde{y}^{(9)} & \tilde{y}^{(12)} \end{pmatrix}.$$

10.6.2 Lipschitz continuity of the translation operator

In this subsection we will estimate the Lipschitz constant of $\tilde{\varphi}(t, \tilde{y})$ in $\tilde{y}$ provided that the corresponding solutions belong to $\tilde{\Omega}$.

Proposition 10.6.1. *Let $\tilde{\varphi}(t, \tilde{y}_1), \tilde{\varphi}(t, \tilde{y}_2) \in \tilde{\Omega}, \quad 0 \le t \le T$. Then*

$$|\tilde{\varphi}(T, \tilde{y}_1) - \tilde{\varphi}(T, \tilde{y}_2)| < e^{0.51*T} |\tilde{y}_1 - \tilde{y}_2|.$$

Proof: First we introduce an auxiliary definition. Recall that the logarithmic norm (with respect to Euclidean metric) $\mu(Q)$ of a square matrix Q is defined by

$$\mu(Q) = \lim_{h\to 0,\ h>0} \frac{||I + hQ|| - 1}{h},$$

where I is the identity matrix and $||\cdot||$ is the operator norm of the matrix.

To prove the proposition we reformulate in our notation the well-known theorem that can be found, for example, in [11], p. 60, Theorem 10.6.

Theorem 10.6.1. *Let $\varphi(t, x)$ denote the translation operator along the trajectories of the differential equation $\dot{x} = \mathcal{F}(x)$. Suppose that we have the estimate $\mu\left(\frac{\partial \mathcal{F}}{\partial x}(x)\right) \le L(t)$ for all x from the coordinate interval $[\varphi(t, x_0), y(t)]$, and*

$$|y'(t) - \mathcal{F}(t, y(t))| \le \delta(t), \quad |y_0 - x_0| \le \rho.$$

Then for $t > 0$ we have

$$|\varphi(t, x_0) - y(t)| \le e^{L_0(t)} \left(\rho + \int_0^t e^{-L_0(s)} \delta(s) ds\right), \tag{10.27}$$

with $L_0(t) = \int_0^t L(s) ds$.

To obtain the statement of the proposition we assume that $y(t) = \tilde{\varphi}(t, \tilde{y}_2)$ which leads to $\delta(t) = 0$. Now (10.27) can be rewritten as

$$|\tilde{\varphi}(t, \tilde{y}_1) - \tilde{\varphi}(t, \tilde{y}_2)| \le \rho e^{\int_0^t L(s) ds} \le e^{Lt} |\tilde{y}_1 - \tilde{y}_2|,$$

where $L = \max_{0 \le s \le t} L(s)$. This inequality proves the proposition if

$$\sup_{\tilde{z} \in \tilde{\Omega}} \mu \left(\frac{\partial \tilde{\mathcal{F}}}{\partial \tilde{y}}(\tilde{z}) \right) < 0.51.$$

This estimate can be obtained using the same technique as in Lemma 3.1 in [31], c.f. Mathematica program from

http://phys.ucc.ie/~oll/lksplit/A

□

10.6.3 Numerical integration

Let $\mathcal{J} = (j^{(i,k)})$ where $i, k = 1, 2, 3$. To integrate the system (10.25) we use a Runge-Kutta method of 4th order given below. Let $\tilde{\nu}_n(x_0)$ denote the numerical solution of the equation (10.25) with initial conditions $x(0) = x_0, \mathcal{J}(0) = I$. Recall ([11], p. 137) that this method for autonomous ODEs is given by

$$\begin{aligned} \tilde{\nu}_{n+1}(x_0) = \tilde{\nu}_n(x_0) + \frac{h}{6}(&\mathbf{k}_1(\tilde{\nu}_n(x_0), h) + 2\mathbf{k}_2(\tilde{\nu}_n(x_0), h) + \\ &2\mathbf{k}_3(\tilde{\nu}_n(x_0), h) + \mathbf{k}_4(\tilde{\nu}_n(x_0), h)) + \omega_n. \end{aligned} \tag{10.28}$$

Here

$$\begin{aligned} \mathbf{k}_1(\tilde{y}, \tau) &= \tilde{\mathcal{F}}(\tilde{y}), \\ \mathbf{k}_2(\tilde{y}, \tau) &= \tilde{\mathcal{F}}(\tilde{y} + \tfrac{\tau}{2}\mathbf{k}_1(\tilde{y}, \tau)), \\ \mathbf{k}_3(\tilde{y}, \tau) &= \tilde{\mathcal{F}}(\tilde{y} + \tfrac{\tau}{2}\mathbf{k}_2(\tilde{y}, \tau)), \\ \mathbf{k}_4(\tilde{y}, \tau) &= \tilde{\mathcal{F}}(\tilde{y} + \tau\mathbf{k}_3(\tilde{y}, \tau)); \end{aligned}$$

and ω_n is the error in the IEEE implementation of the method. The values $\tilde{\nu}_n(x_0)$ are approximations for the values $\tilde{\varphi}(nh, x_0)$ of the translation operator $\tilde{\varphi}$ on the time grid. We will refer to this method as RK4.

The RK4 method provides an immediate approximation to the solution of the system (10.17) only at the discrete times nh. To approximate a solution between these discrete times we will use linear interpolation. In other words, our approximate solution is the function $\tilde{\nu}(t, x)$, $t \ge 0$, given by

$$\tilde{\nu}(t, x) = (1 - \frac{\theta}{h})\tilde{\nu}_k(x) + \frac{\theta}{h}\tilde{\nu}_{k+1}(x)$$

for $t \in [kh, (k+1)h]$, $\theta = t - kh$.

Now we can formulate a convenient proposition about the accuracy of the RK4 method. We fix the time step $h = 0.001$.

Proposition 10.6.2. *Let $x_0 \in \Pi$ and $\tilde{\nu}_n(x_0) \in \tilde{\Omega}(-4 \cdot 10^{-7})$ for all $nh \le 22$. Then the inequality*

$$|\tilde{\varphi}(t, x_0) - \tilde{\nu}(t, x_0)| < 4 \cdot 10^{-7}, \qquad 0 \le t \le 22,$$

is valid and $\tilde{\varphi}(t, x_0) \in \tilde{\Omega}$ for all $t \le 22$.

The rest of this section is devoted to the proof of this proposition.

Local error

The local error $e(h, \tilde{\Omega})$ for the RK4 method in $\tilde{\Omega}$ is given by

$$e(h, \tilde{\Omega}) = \sup_{\tilde{y} \in \tilde{\Omega}} \{ |\tilde{\varphi}(h, \tilde{y}) - \tilde{\nu}_1(\tilde{y})| : \ \tilde{y}, \tilde{\nu}_1(\tilde{y}) \in \tilde{\Omega} \},$$

where h is the time step, and $\tilde{\nu}_1(\tilde{y})$ is the point calculated with the RK4 method.

Lemma 10.6.1. *Let $\tilde{y} \in \tilde{\Omega}$, $0 < h \leq 0.001$ and $\tilde{\nu}_1(\tilde{y}) \in \tilde{\Omega}$. Then the local error satisfies*

$$e(h, \tilde{\Omega}) \leq 0.42h^5 + \omega, \tag{10.29}$$

where $\omega < 10^{-13}$.

Proof: General estimates of the local error for the RK4 method can be found for example in Theorem 3.1 from [11], p. 157, and we reformulate them for our particular case:

$$e(h, \tilde{\Omega}) \leq \frac{h^5}{(5)!} \sup_{\tau \in [0,h]} |\tilde{\varphi}_t^{(5)}(\tau, \tilde{y})| + \frac{h^5}{(4)!} \sum_{i=1}^{4} |b_i| \sup_{\tau \in [0,h]} |\mathbf{k}_i^{(4)}(\tilde{y}, \tau)| + \omega. \tag{10.30}$$

Here b_i are coefficients of $h\mathbf{k}_i$ in (10.28): $b_1 = \frac{1}{6}$, $b_2 = \frac{1}{3}$, $b_3 = \frac{1}{3}$, $b_4 = \frac{1}{6}$; ω is the error in the computer implementation of the RK4 method; all derivatives are taken with respect to the system (10.17). For example

$$\mathbf{k}_1^{(1)}(\tilde{y}, \tau) = \frac{\partial \tilde{\mathcal{F}}}{\partial \tilde{y}}(\tilde{y}) \tilde{\mathcal{F}}(\tilde{y}),$$

$$\mathbf{k}_2^{(1)}(\tilde{y}, \tau) = \frac{\partial \tilde{\mathcal{F}}}{\partial \tilde{y}} \left(\tilde{y} + \frac{\tau}{2} \tilde{\mathcal{F}}(\tilde{y}) \right) \left(\tilde{\mathcal{F}}(\tilde{y}) + \frac{\tau}{2} \frac{\partial \tilde{\mathcal{F}}}{\partial \tilde{y}}(\tilde{y}) \tilde{\mathcal{F}}(\tilde{y}) \right).$$

The estimate (10.29) follows immediately from (10.30) and the inequalities

$$|\mathbf{k}_1^{(4)}(\tilde{y}, \tau)| < 8, \quad \tilde{y} \in \tilde{\Omega}, \ 0 \leq \tau \leq h, \tag{10.31}$$

$$|\mathbf{k}_2^{(4)}(\tilde{y}, \tau)| < 8, \quad \tilde{y} \in \tilde{\Omega}, \ 0 \leq \tau \leq h, \tag{10.32}$$

$$|\mathbf{k}_3^{(4)}(\tilde{y}, \tau)| < 8, \quad \tilde{y} \in \tilde{\Omega}, \ 0 \leq \tau \leq h, \tag{10.33}$$

$$|\mathbf{k}_4^{(4)}(\tilde{y}, \tau)| < 8, \quad \tilde{y} \in \tilde{\Omega}, \ 0 \leq \tau \leq h, \tag{10.34}$$

$$|\tilde{\varphi}_t^{(5)}(\tau, \tilde{y})| < 10, \quad \tilde{y} \in \tilde{\Omega}, \ 0 \leq \tau \leq h, \tag{10.35}$$

$$\omega \leq 10^{-13}. \tag{10.36}$$

It simply remains to justify the estimates (10.31) – (10.36). The inequalities (10.31) – (10.34) were obtained via the following scheme. Using Mathematica we obtain the explicit analytical formulas for $\mathbf{k}_i^4$, and then use interval arithmetic to estimate the range of change for the variables. Programs can be found in

http://phys.ucc.ie/˜oll/lksplit/A

To estimate $\tilde{\varphi}_t^{(5)}(\tau, \tilde{y})$ we need some additional constructions. Consider the auxiliary set $\tilde{\Omega}(0.08)$, see (10.26). Let us prove that

$$\tilde{\varphi}(\tau, \tilde{y}) \in \tilde{\Omega}(0.08), \qquad \tilde{y} \in \tilde{\Omega},\ 0 \le \tau \le h. \tag{10.37}$$

Suppose that it is not true. Introduce the distance between the point $\tilde{y}$ and the set Y as $d(\tilde{y}, Y) = inf_{y \in Y}|\tilde{y} - y|$. Then there is $\tau_0 \le h$ such that $\tilde{\varphi}(\tau, \tilde{y}) \in \tilde{\Omega}(0.08)$ for $0 \le \tau < \tau_0$ and $\tilde{\varphi}(\tau_0, \tilde{y}) \in \partial\tilde{\Omega}(0.08)$. The last inclusion implies that $d(\tilde{\Omega}, \tilde{\varphi}(\tau_0, \tilde{y})) \ge 0.08$. On the other hand, since $\tilde{y} \in \tilde{\Omega}$,

$$d(\tilde{\varphi}(\tau, \tilde{y}), \tilde{\Omega}) \le |\tilde{\varphi}(\tau, \tilde{y}) - \tilde{y}| = |\tilde{\varphi}'(\tau_0, \tilde{y})\tau| < \tilde{\mathcal{F}}_{max} h,$$

where $\tilde{\mathcal{F}}_{max} = \sup_{\tilde{y} \in \tilde{\Omega}(0.08)} |\tilde{\mathcal{F}}(\tilde{y})|$. Computations in Mathematica show that $\tilde{\mathcal{F}}_{max} < 15.9$ and the last formula is reduced to

$$d(\tilde{\varphi}(\tau, \tilde{y}), \tilde{\Omega}) < 0.08.$$

We have arrived at a contradiction, and (10.37) is proved.

By the definition of $\mathbf{k}_1$

$$\tilde{\varphi}^{(5)}(\tau, \tilde{y}) \equiv \mathbf{k}_1^{(4)}(\tilde{\varphi}(\tau, \tilde{y}), 0),$$

which together with (10.37) implies

$$|\tilde{\varphi}^{(5)}(\tau, \tilde{y})| \le \sup_{\tilde{y}^* \in \tilde{\Omega}(0.08)} |\mathbf{k}_1^{(4)}(\tilde{y}^*, 0)|.$$

Finally, the estimate

$$\sup_{\tilde{y}^* \in \tilde{\Omega}(0.08)} |\mathbf{k}_1^{(4)}(\tilde{y}^*, 0)| < 10$$

is proved in a similar way to (10.31).

The inequality (10.36) follows from the description of IEEE arithmetic, and how trigonometric and exponential functions are realized in PC architecture.

The lemma is proved. □

Global Error

Now we can estimate the global error $E(nh, x_0)$ of the RK4 method, which is defined for a given x_0 and n as

$$E(nh, \tilde{y}_0) = |\tilde{\varphi}(nh, \tilde{y}_0) - \tilde{\nu}_n(\tilde{y}_0)|,$$

that is, as the absolute value of the difference between the exact solution and the one obtained by numerical integration at the points of the discrete time grid. Define

$$E(nh) = \left(0.83h^4 + 2\frac{\omega}{h}\right)\left(e^{0.51nh} - 1\right).$$

Lemma 10.6.2. *Let $\tilde{\varphi}(t,\tilde{y}_0), \tilde{\nu}_k(\tilde{y}_0) \in \tilde{\Omega}$ for all $t \in [0, nh]$ and all $k \le n$. Let $0 < h \le 0.001$. Then*

$$E(nh, \tilde{y}_0) < E(nh).$$

Proof: We can use Theorem 3.4 from [11], p. 160, to obtain a global error estimate $E(nh)$ for the RK4 method. Here we reformulate it for our situation.

Theorem 10.6.2. *Let $y(t)$ be an exact solution belonging to $\tilde{\Omega}$ and $[0, nh]$ be a numerical integration time. Assume that h is small enough so that the numerical solution remains in $\tilde{\Omega}$. If the local error is estimated as $e(h) \le Ch^5 + \omega$ in $\tilde{\Omega}$ and the logarithmic norm satisfies $\mu(\frac{\partial F}{\partial x}) \le L$ in $\tilde{\Omega}$, then*

$$E(nh, y(0)) \le \left(\frac{Ch^4}{L} + \frac{\omega}{Lh}\right)\left(e^{Lnh} - 1\right).$$

Supplying the values $C = 0.42$ and $L = 0.51$ from Lemma 10.6.1 and Proposition 10.6.1 we obtain the statement of the lemma. □

Now we finalize the proof of Proposition 10.6.2. Let us assume that $\tilde{\varphi}(t, x_0) \in \tilde{\Omega}$ for $t \le 22$. Let k satisfy $t = kh + \theta$ where $0 \le \theta \le h$. Then we define

$$\tilde{y}(\theta) = (1 - \theta/h)\tilde{\varphi}(kh, x_0) + (\theta/h)\tilde{\varphi}(kh + h, x_0).$$

By the triangle inequality:

$$|\tilde{\varphi}(t, x_0) - \tilde{\nu}(t, x_0)| \le |\tilde{y}(\theta) - \tilde{\nu}(t, x_0)| + |\tilde{\varphi}(t, x_0) - \tilde{y}(\theta)|. \tag{10.38}$$

Now we estimate the terms on the right-hand side. For the first term, recalling the definitions for $\tilde{y}(\theta)$ and $\tilde{\nu}(t, x_0)$, we obtain

$$\begin{aligned} |\tilde{y}(\theta) - \tilde{\nu}(t, x_0)| &\le |(1 - \theta/h)(\tilde{\varphi}(kh, x_0) - \tilde{\nu}_k(x_0))| \\ &\quad + |(\tilde{\varphi}(kh + h, x_0) - \tilde{\nu}_{k+1}(x_0))\theta/h| \\ &\le (1 - \theta/h)E(kh) + (\theta/h)E(kh + h) \\ &< E(kh + h), \end{aligned} \tag{10.39}$$

since the global error estimate $E(t)$ increases in time. The second term can be rewritten as

$$|\tilde{\varphi}(t, x_0) - \tilde{y}(\theta)| = |(\tilde{\varphi}(t, x_0) - \tilde{\varphi}(nh, x_0)) + (\tilde{\varphi}(nh, x_0) - \tilde{\varphi}(nh + h, x_0))\theta/h|. \tag{10.40}$$

We use Taylor expansions of $\tilde{\varphi}(t, x_0)$, $\tilde{\varphi}(nh, x_0)$ and $\tilde{\varphi}(nh + h, x_0)$ at the point $kh + h/2$. They are similar, so we only write that for $\tilde{\varphi}(t, x_0)$:

$$\begin{aligned} \tilde{\varphi}(t, x_0) \;=\; & \tilde{\varphi}\,(kh + h/2, x_0) + \tilde{\varphi}'\,(kh + h/2, x_0)\,(\theta - h/2) \\ & + \tilde{\varphi}''(kh + \theta_1, x_0)\,(\theta - h/2)^2/2. \end{aligned}$$

Substituting the expansions into (10.40) we get

$$|\tilde{\varphi}(t,x_0) - \tilde{y}(\theta)| = |(\tilde{\varphi}'(h/2)\theta - \tilde{\varphi}''(\theta_2)h^2/8 + \tilde{\varphi}''(\theta_1)(\theta - h/2)^2/2) \\ +(-\tilde{\varphi}'(h/2)h + \tilde{\varphi}''(\theta_2)h^2/8 - \tilde{\varphi}''(\theta_3)h^2/8)\theta/h|$$

or

$$|\tilde{\varphi}(t,x_0) - \tilde{y}(\theta)| \le |\tilde{\varphi}''(\theta_2)|h^2/4 + |\tilde{\varphi}''(\theta_1)|h^2/8 + |\tilde{\varphi}''(\theta_3)|h^2/8 \\ \le \tilde{\varphi}''_{max}h^2/2, \qquad (10.41)$$

where $\tilde{\varphi}''_{max} = \sup_{\tilde{\Omega}} |\tilde{\varphi}''(t,x_0)|$. Substituting (10.39) and (10.41) into (10.38), we get

$$|\tilde{\varphi}(t,x_0) - \tilde{\nu}(t,x_0)| < E(kh+h) + \tilde{\varphi}''_{max}\frac{h^2}{2}.$$

As in the case of the fifth derivative of $\tilde{\varphi}$, with the help of Mathematica we get the estimate $\tilde{\varphi}''_{max} < 12$. From the conditions of the proposition we know that $t \le 22$, and by the monotonicity of $E(nh)$

$$E(kh+h) \le E(22) = \left(0.82h^4 + \frac{2\cdot 10^{-13}}{h}\right)(e^{0.51*22} - 1).$$

With the $h = 0.001$ we get the inequality stated in the proposition for the situation when $\tilde{\varphi}(t,x_0) \in \tilde{\Omega}$ and

$$|\tilde{\varphi}(t,x_0) - \tilde{\nu}(t,x_0)| < 4\cdot 10^{-7}.$$

Now we assume that there exists $t < n_2 h$ such that $\tilde{\varphi}(t,x_0)$ does not belong to $\tilde{\Omega}$. Then from the continuity of $\tilde{\varphi}$ there is a $t_0 < 22$ such that $\tilde{\varphi}(t_0,x_0) \in \partial\tilde{\Omega}$ and $\tilde{\varphi}(t,x_0) \in \tilde{\Omega}$ for all $t < t_0$. Let k_0 satisfy $k_0 h \le t_0 \le k_0 h + h$. By the conditions of the lemma $\tilde{\nu}_{k_0}(x_0)$ and $\tilde{\nu}_{k_0+1}(x_0)$ belong to $\tilde{\Omega}(-4\cdot 10^{-7})$ and so does $\tilde{\nu}(t_0,x_0)$, because $\tilde{\Omega}(-4\cdot 10^{-7})$ is a convex set. Thus $d(\tilde{\nu}(t_0,x_0),\tilde{\varphi}(t_0,x_0)) > 4\cdot 10^{-7}$.

On the other hand for all $0 \le t < t_0$, $\tilde{\varphi}(t,x_0) \in \tilde{\Omega}$ and $d(\tilde{\nu}(t,x_0),\tilde{\varphi}(t,x_0)) < 4\cdot 10^{-7}$. By continuity of $\tilde{\varphi}(t,x_0)$ we get $d(\tilde{\nu}(t_0,x_0),\tilde{\varphi}(t_0,x_0)) \le 4\cdot 10^{-7}$ and arrive at a contradiction. This means that $\tilde{\varphi}(t,x_0) \in \tilde{\Omega}$ for all $t \le 22$ and the proposition is proved. □

10.6.4 Proof of s-hyperbolicity

Preliminary constructions

Suppose we have a dynamical system, a plane in the phase space of the dynamical system and numerically calculated trajectories that somehow approximate the original system. What kind of properties should the numerical trajectory have to allow us to state a strict proposition about the Poincaré map of the original system? We will address part of this question.

We introduce the notation

$$
\begin{aligned}
W &= \{\tilde{y} : \tilde{y}^{(3)} = 0\}, \\
W^- &= \{\tilde{y} : \tilde{y}^{(3)} < 0\}, \\
W^+ &= \{\tilde{y} : \tilde{y}^{(3)} > 0\}, \\
\tilde{S} &= \{\tilde{y} : \tilde{y}^{(3)} = 0, \tilde{y}^{(2)} > x_T^{(2)}\}, \\
\tilde{\ell} &= \{\tilde{y} : \tilde{y}^{(2)} = x_T^{(2)}, \tilde{y}^{(3)} = 0\}.
\end{aligned}
$$

Note that $\tilde{\ell}$ is a boundary of $\tilde{S}$ in the plane W and that for all $\tilde{y} \in \tilde{S}$ the inclusion $\tilde{\mathcal{F}}(\tilde{y}) \in W^+$ holds, which means that $\tilde{S}$ is pierced by trajectories of the system (10.23) only in the direction from W^- to W^+.
Let $\mathcal{A}$ be a convex set in $\tilde{S}$, and P_* be a parallel projector to W along some vector $\tilde{y}_* \notin W$. Introduce

$$
\begin{aligned}
\mathcal{A}^- &= \{\tilde{y} : P_*\tilde{y} \in \mathcal{A}, \tilde{y}^{(3)} < 0\}, \\
\mathcal{A}^+ &= \{\tilde{y} : P_*\tilde{y} \in \mathcal{A}, \tilde{y}^{(3)} > 0\}, \\
\mathcal{A}^\pm &= \{\tilde{y} : P_*\tilde{y} \in \mathcal{A}\}.
\end{aligned}
$$

We define the distance between the point $\tilde{z}$ and the set Y by $d(\tilde{z}, Y) = \inf_{\tilde{y} \in Y} |\tilde{z} - \tilde{y}|$. Let $\mathcal{B}(\tilde{y}; r)$ denote an open ball in $\mathbb{R}^{12}$ of radius r centered in $\tilde{y}$.

Let $\tilde{y} \in \tilde{S}$, and suppose that positive integers n_1, n_2 and the real $\varepsilon > 0$ meet the following requirements:

(r_1) The inequality $|\tilde{\varphi}(t, \tilde{y}) - \tilde{\nu}(t, \tilde{y})| < \delta$ holds for $0 \le t \le n_2 h$.

(r_2) The neighborhood of the numerical solution $\tilde{\nu}_n(\tilde{y})$ stays in $\tilde{\Omega}$ for all $0 \le n \le n_2$: $\mathcal{B}(\tilde{\nu}_n(\tilde{y}); \varepsilon + \delta) \subset \tilde{\Omega}$.

(r_3) The numerical solution $\tilde{\nu}_n(\tilde{y})$ satisfies

$$\mathcal{B}(\tilde{\nu}_{n_1}(\tilde{y}); \varepsilon + \delta) \subset \mathcal{A}^-, \qquad \mathcal{B}(\tilde{\nu}_{n_2}(\tilde{y}); \varepsilon + \delta) \subset \mathcal{A}^+$$

and

$$\mathcal{B}(\tilde{\nu}_n(\tilde{y}); \varepsilon + \delta) \subset \mathcal{A}^\pm \text{ for all } n_1 \le n \le n_2.$$

(r_4) There is no $0 < t_0 < n_1 h$ such that $\tilde{\nu}_n(t_0, \tilde{y}) \in \tilde{S}$.

(r_5) The inequality $d(\tilde{\nu}(t, \tilde{y}), \tilde{\ell}) \ge \varepsilon + \delta$ holds for all $0 \le t \le n_1 h$.

We will then say that the triplet (n_1, n_2, ε) is $(\mathcal{A}, \delta, \tilde{y})$-fortunate.

Lemma 10.6.3. *Let (n_1, n_2, ε) be $(\mathcal{A}, \delta, \tilde{y})$-fortunate and let $\tilde{y} \in \tilde{S}$. Then for all $\tilde{y}_1 \in \tilde{S}$ such that $|\tilde{y} - \tilde{y}_1| \le \varepsilon e^{-Ln_2h}$*

- *there is the time $n_1 h < T(\tilde{y}_1) < n_2 h$ for which $\tilde{\varphi}(T(\tilde{y}_1); y_1) \in \tilde{S}$;*
- *there is no $t \in (0; T(\tilde{y}_1)) \cup (T(\tilde{y}_1); n_2 h)$ for which $\varphi(t; y_1) \in \tilde{S}$;*

- *the Poincaré map $\tilde{\Phi}$ is defined and continuous and $\tilde{\Phi}(\tilde{y}_1) \in \mathcal{A}$.*

Proof: We show first that the map $\tilde{\Phi}$ is correctly defined.

Firstly we show that $\tilde{\varphi}(t, \tilde{y}_1) \subset \tilde{\Omega}$ for all $0 \le t \le n_2 h$ and all $\tilde{y}_1 \in \mathcal{B}(\tilde{y}; \varepsilon e^{-Ln_2h})$. Let us assume the contrary. Then there is $t_0 < n_2 h$ such that $\tilde{\varphi}(t, \tilde{y}_1) \subset \tilde{\Omega}$ for all $t < t_0$ and $\tilde{\varphi}(t_0, \tilde{y}_1) \subset \partial\tilde{\Omega}$. We estimate the distance between the points of the actual trajectory and its numerical counterpart at the time t_0

$$|\tilde{\varphi}(t_0, \tilde{y}_1) - \tilde{\nu}(t_0, \tilde{y})| \le |\tilde{\varphi}(t_0, \tilde{y}) - \tilde{\varphi}(t_0, \tilde{y}_1)| + |\tilde{\varphi}(t_0, \tilde{y}) - \tilde{\nu}(t_0, \tilde{y})|. \tag{10.42}$$

From Theorem 10.6.1

$$|\tilde{\varphi}(t_0, \tilde{y}) - \tilde{\varphi}(t_0, \tilde{y}_1)| \le e^{Lt_0}|\tilde{y} - \tilde{y}_1| \le \varepsilon e^{L(t_0 - n_2)h} < \varepsilon, \tag{10.43}$$

and, by (r_1),

$$|\tilde{\varphi}(t_0, \tilde{y}) - \tilde{\nu}(t_0, \tilde{y})| < \delta. \tag{10.44}$$

Substituting (10.43),(10.44) into (10.42) we obtain

$$|\tilde{\varphi}(t_0, \tilde{y}_1) - \tilde{\nu}(t_0, \tilde{y})| < \delta + \varepsilon,$$

which, together with (r_2), implies $\tilde{\varphi}(t_0, \tilde{y}_1) \in \tilde{\Omega}$. We have arrived at a contradiction.

Secondly, let us show that $\tilde{\Phi}(\mathcal{B}(\tilde{y}; \varepsilon e^{-Ln_2h})) \subset \mathcal{A}$. Let $\tilde{y}_1$ be a point in the set $\mathcal{B}(\tilde{y}; \varepsilon e^{-Ln_2h}) \cap \tilde{S}$. From the third inclusion of (r_3), $\mathcal{B}(\tilde{\nu}_n(\tilde{y}); \varepsilon + \delta) \subset \mathcal{A}^{\pm}$ for all $n_1 \le n \le n_2$. Since $\mathcal{A}^{\pm}$ is convex and $\tilde{\nu}(t, \tilde{y})$ is defined as a linear approximation between points of $\tilde{\nu}_n(\tilde{y})$, $\mathcal{B}(\tilde{\nu}(t, \tilde{y}); \delta + \varepsilon) \subset \mathcal{A}^{\pm}$ for all $n_1 h \le t \le n_2 h$. Theorem 10.6.1 with (r_1) implies that

$$\tilde{\varphi}(t, \tilde{y}_1) \in \mathcal{B}(\tilde{\varphi}(t, \tilde{y}); \varepsilon) \subset \mathcal{B}(\tilde{\nu}(t, \tilde{y}); \delta + \varepsilon) \subset \mathcal{A}^{\pm}. \tag{10.45}$$

From the first part of (r_3), $\tilde{\varphi}(n_1 h, \tilde{y}_1) \in \mathcal{A}^{-}$. From the second part of (r_3), $\tilde{\varphi}(n_2 h, \tilde{y}_1) \in \mathcal{A}^{+}$ Combining those facts with (10.45), we obtain that there is only one moment in time $n_1 h < T(\tilde{y}_1) < n_2 h$ such that $\tilde{\varphi}(T(\tilde{y}_1), \tilde{y}_1) \in \mathcal{A} \subset \tilde{S}$.

Finally we show that $T(\tilde{y}_1)$ is indeed the time of the first intersection with $\tilde{S}$, i.e.

$$\tilde{\varphi}(t, \tilde{y}_1) \notin \tilde{S} \text{ for all } t \le n_1 h.$$

Let us assume the contrary. Then either $\tilde{\varphi}(t, \tilde{y}_1)$ intersects the line $\tilde{\ell}$, or the trajectory $\tilde{\varphi}(t, \tilde{y}_1)$ makes a loop around the line $\tilde{\ell}$ (this follows from the piercing properties of plane W and definition of $\tilde{S}$).

The former case is impossible because $\tilde{\varphi}(t, \tilde{y}_1) \in \mathcal{B}(\tilde{\nu}(t, \tilde{y}); \varepsilon + \delta)$ and according to (r_5) this ball can not intersect the line $\tilde{\ell}$.

We consider the latter case. In this situation, $\tilde{\varphi}(t, \tilde{y}_1)$ must contain the points

$$\tilde{y}_{1,1} = \tilde{\varphi}(t_1, \tilde{y}_1) : \tilde{y}_{1,1}^{(2)} = x_T^{(2)} \text{ and } \tilde{y}_{1,1}^{(3)} < 0,$$

$$\tilde{y}_{1,2} = \tilde{\varphi}(t_2, \tilde{y}_1) : \tilde{y}_{1,2}^{(2)} = x_T^{(2)} \text{ and } \tilde{y}_{1,2}^{(3)} > 0.$$

In turn, $\tilde{y}_{1,1} \in \mathcal{B}(\tilde{\nu}(t_1, \tilde{y}); \varepsilon + \delta) = \mathcal{B}_1$ and $\tilde{y}_{1,2} \in \mathcal{B}(\tilde{\nu}(t_2, \tilde{y}); \varepsilon + \delta) = \mathcal{B}_2$.

Now we will examine the possible positions of $\tilde{\nu}(t_1, \tilde{y})$ and $\tilde{\nu}(t_2, \tilde{y})$ that are the centers of $\mathcal{B}_1$ and $\mathcal{B}_2$, respectively. Taking into account (r_4), two situations must be considered.

Case 1: Either $\tilde{\nu}(t_1, \tilde{y}) \in W^+$ or $\tilde{\nu}(t_2, \tilde{y}) \in W^-$, or both. We consider only the case when $\tilde{\nu}(t_1, \tilde{y}) \in W^+$, since the others are similar. We will show that, under these conditions, $\mathcal{B}_1$ intersects $\tilde{\ell}$. Now, for Euclidean metrics we have:

$$d^2(\tilde{y}_{1,1}, \tilde{\nu}(t_1, \tilde{y})) - d^2(\tilde{\nu}(t_1, \tilde{y}), \tilde{\ell}) \geq (y_{1,1}^{(3)} - \tilde{\nu}(t_1, \tilde{y})^{(3)})^2 - (\tilde{\nu}(t_1, \tilde{y})^{(3)})^2.$$

Recalling that $y_{1,1}^{(3)} < 0$, $\tilde{\nu}(t_1, \tilde{y})^{(3)} > 0$ we get $d(\tilde{\nu}(t_1, \tilde{y}), \tilde{\ell}) < d(\tilde{\nu}(t_1, \tilde{y}), \tilde{y}_{1,1}) < \varepsilon + \delta$, which implies that $\tilde{y}_* \in \mathcal{B}(\tilde{\nu}(t_1, \tilde{y}); \varepsilon + \delta)$ for $\tilde{y}_* \in \tilde{\ell}$. The last inclusion contradicts (r_5).

Case 2: Now let the center points of $\mathcal{B}_1$ and $\mathcal{B}_2$ satisfy $\tilde{\nu}(t_1, \tilde{y}) \in W^-$ and $\tilde{\nu}(t_2, \tilde{y}) \in W^+$. In this case the numerical solution $\tilde{\nu}(t, \tilde{y})$ crossed the plane W before the moment t_1 (since it began in W^+), and crosses the plane W at least once in the time interval (t_1, t_2). From (r_5) the numerical solution is far enough from $\tilde{\ell}$ and one of intersections must happen with the halfplane $\tilde{S}$. This contradicts (r_4).

Thus $n_1 h < T(\tilde{y}_1) < n_2 h$ is indeed the time of the first intersection with $\tilde{S}$ and the Poincaré map for $\tilde{y} \in \tilde{S}$ is defined as

$$\tilde{\Phi}(P_W(\tilde{y}_1)) = P_W \tilde{\varphi}(T(\tilde{y}_1), \tilde{y}_1) \in \tilde{S},$$

where P_W is the orthogonal projector to W. The inclusion $\tilde{\Phi}(\tilde{y}_1) \in \mathcal{A}$ was shown earlier and continuity of Φ on $\mathcal{B}(\tilde{y}_1; \varepsilon e^{-Ln_2 h}) \cap \tilde{S}$ easily follows from the method of construction. The lemma is proved. □

Linearization of the Poincaré map

The first step is to compute the approximation $\tilde{y}_{ref}$ of $\tilde{\Phi}(h_i(y))$. The second step is to construct the linearization $B_{ref} = (b_{k,l})_{k,l=1}^2$ of $h_j^{-1} \Phi h_i$. Now if

$$(\lambda_*^u, \lambda_*^s, \mu_*^u, \mu_*^s) = s_* \succ s_{ref} = (b_{1,1}, b_{2,2}, b_{1,2}, b_{2,1}),$$

and

$$\Delta(s_*, s_{ref}) = \min\{|b_{1,1}| - \lambda_*^u, \lambda_*^s - |b_{2,2}|, \mu_*^u - |b_{1,2}|, \mu_*^s - |b_{2,1}|\}$$

is big enough compared with the integration error, then the linearization of $h_j^{-1} \Phi h_i$ is s_*-hyperbolic in a neighborhood of y. The last step is to estimate the Lipschitz constants of the maps involved, and to specify a split-hyperbolic neighborhood of y.

For a given $\tilde{y}$ introduce the number

$$d_1(\tilde{y}) = \min_k \{\tilde{y}^{(k)} - \tilde{y}_{min}^{(k)}, \tilde{y}_{max}^{(k)} - \tilde{y}^{(k)}\},$$

where $\tilde{y}_{min}^{(k)}, \tilde{y}_{max}^{(k)}$ stand for the minimum and maximum values of $\tilde{y}^{(k)}$ over $\tilde{\Omega}$. Since $\tilde{\Omega}$ is a box with the sides parallel to the coordinate planes, $d_1(\tilde{y})$ is positive if and

only if $\tilde{y} \in \tilde{\Omega}$ and in this case it measures the distance from $\tilde{y}$ to $\partial\tilde{\Omega}$.
We introduce the number

$$d_2(\tilde{y}_1, \tilde{y}_2) = \begin{cases} \sqrt{(\tilde{y}_1^{(2)} - \tilde{y}_T^{(2)})^2 + (\tilde{y}_1^{(3)})^2} & \text{if } \alpha < 0, \\ \sqrt{(\tilde{y}_2^{(2)} - \tilde{y}_T^{(2)})^2 + (\tilde{y}_2^{(3)})^2} & \text{if } \alpha > 1, \\ \sqrt{(\tilde{y}_1^{(2)} + \alpha(\tilde{y}_2^{(2)} - \tilde{y}_1^{(2)}) - \tilde{y}_T^{(2)})^2 + (\tilde{y}_1^{(3)} + \alpha(\tilde{y}_2^{(3)} - \tilde{y}_1^{(3)}))^2} & \\ & \text{if } 0 \le \alpha \le 1, \end{cases} \tag{10.46}$$

where

$$\alpha = \frac{(\tilde{y}_T^{(2)} - \tilde{y}_1^{(2)})(\tilde{y}_2^{(2)} - \tilde{y}_1^{(2)}) - \tilde{y}_1^{(3)}(\tilde{y}_2^{(3)} - \tilde{y}_1^{(3)})}{(\tilde{y}_2^{(2)} - \tilde{y}_1^{(2)})^2 + (\tilde{y}_2^{(3)} - \tilde{y}_1^{(3)})^2}.$$

Formula (10.46) gives the distance between the line segment with the endpoints $\tilde{y}_1, \tilde{y}_2$ and the line $\tilde{\ell}$.
Introduce the map $Lin(\tilde{y}) : \tilde{y} \in \tilde{S} \mapsto$ 2x2 matrices as

$$Lin(\tilde{y}) = \begin{pmatrix} 1 & 0 & -\tilde{\mathcal{F}}(\tilde{y})^{(1)}/\tilde{\mathcal{F}}(\tilde{y})^{(3)} \\ 0 & 1 & -\tilde{\mathcal{F}}(\tilde{y})^{(2)}/\tilde{\mathcal{F}}(\tilde{y})^{(3)} \end{pmatrix} \begin{pmatrix} \tilde{y}^{(4)} & \tilde{y}^{(7)} \\ \tilde{y}^{(5)} & \tilde{y}^{(8)} \\ \tilde{y}^{(6)} & \tilde{y}^{(9)} \end{pmatrix}.$$

Note that for $\tilde{y} = \tilde{\Phi}(x) = \tilde{\varphi}(T(x), x)$, the matrix $Lin(\tilde{y})$ is the linearization of Φ at $x \in S$ since the first matrix in the formula above represents the projection along the vector $\mathcal{F}(\varphi(T(x), x))$, while the second matrix is the appropriate part of $\frac{\partial}{\partial x}\varphi(T(x), x)$; see (10.22), (10.23).
Finally, introduce the number $d_3(\tilde{y}) : \tilde{y} \in \tilde{S}$ by the formula

$$d_3(\tilde{y}) = \min_{k=1\ldots12,\, k\neq3} \{\tilde{y}^{(k)} - y_{min}^{(k)}, y_{max}^{(k)} - \tilde{y}^{(k)}\},$$

where $y_{min}^{(k)}, y_{max}^{(k)}$ are taken from the column of Tables 10.3– 10.5 identified by a pair (i, j). For example, pair $(3, 3)$ marks the first one in the Table 10.3. If $d_3(\tilde{y})$ is positive, then it measures the distance from $\tilde{y} \in \tilde{S}$ to the boundary of the rectangular set $\Xi_i = \Pi_{k=1}^{12}[y_{min}^{(k)}, y_{max}^{(k)}] \subset \tilde{S}$ with $y_{min}^{(3)} = y_{max}^{(3)} = 0$.

For the given 2x2 matrix $\mathfrak{f} = \{\mathfrak{f}_{k,l}\}_{k,l=1,2}$ introduce the notation

$$\|\mathfrak{f}\|_{max} = \max_{k,l=1,2} |\mathfrak{f}_{k,l}|.$$

Lemma 10.6.4. *In the max norm the Lipschitz constant of $\mathcal{H}_j^{-1}Lin(\tilde{y})\mathcal{H}_i$ on the set Ξ_i is estimated by L_i given in the Tables 10.3, 10.4.*

Proof: Let $\tilde{y}_1$, $\tilde{y}_2$ belong to Ξ_i and let $\mathfrak{f}(\tilde{y}) = \{\mathfrak{f}_{k,l}\}_{k,l=1}^2 = \mathcal{H}_j^{-1}Lin(\tilde{y})\mathcal{H}_i$. By the definition of the $\|\cdot\|_{max}$ norm for the matrices,

$$\|\mathfrak{f}(\tilde{y}_1) - \mathfrak{f}(\tilde{y}_2)\|_{max} = \max_{k,l=1,2} |\mathfrak{f}_{k,l}(\tilde{y}_1) - \mathfrak{f}_{k,l}(\tilde{y}_2)|.$$

By the Cauchy formulas,

$$\mathfrak{f}_{k,l}(\tilde{y}_1) - \mathfrak{f}_{k,l}(\tilde{y}_2) = \sum_{m=1}^{12} (\tilde{y}_1^{(m)} - \tilde{y}_2^{(m)}) \int_0^1 \frac{\partial}{\partial \tilde{y}^{(m)}} \mathfrak{f}_{k,l}(\lambda \tilde{y}_1^{(m)} + (1-\lambda)\tilde{y}_2^{(m)})\, d\lambda.$$

Taking the absolute value and using the triangle inequality we obtain

$$|\mathfrak{f}_{k,l}(\tilde{y}_1) - \mathfrak{f}_{k,l}(\tilde{y}_2)| \le |\tilde{y}_1 - \tilde{y}_2|_{max} \sum_{m=1}^{12} \max_{\tilde{y}\in\Xi_i} \left| \frac{\partial \mathfrak{f}_{k,l}(\tilde{y})}{\partial \tilde{y}^{(m)}} \right|.$$

Finally, the estimates

$$\max_{k,l=1,2} \sum_{m=1}^{12} \max_{\tilde{y}\in\Xi_i} \left| \frac{\partial \mathfrak{f}_{k,l}(\tilde{y})}{\partial \tilde{y}^{(m)}} \right| < L_i$$

are proved with help of a Mathematica package, see the webpage

http://phys.ucc.ie/~oll/lksplit/A

□

Coordinate transformation in (10.23)

The original coordinates $\tilde{y}(t)$ provide us with poor estimates for the constant L of the shift operator. To improve the estimates we introduce a coordinate change. Let

$$\tilde{y} = G\bar{y},$$

where G is a 12 by 12 diagonal matrix with the elements $\{g_{i,i} = 1,\ i = 1, 2, 3 \text{ and } g_{i,i} = 25,\ i = 4, \ldots, 12\}$.

It is easy to check that in the new coordinates the equation (10.23) with the initial conditions (10.24) is equivalent to the equation (10.23) with the initial conditions

$$\begin{aligned} x(0) &= x_0, \\ \mathcal{J}(0) &= 0.04 \cdot I, \end{aligned}$$

where I stands for the identity matrix. Thus

$$\tilde{\varphi}(T, \tilde{y}_1) = G \cdot \bar{\varphi}(T, \bar{y}_1), \qquad \bar{y}_1 = G^{-1} \cdot \tilde{y}_1.$$

Here we note that $\tilde{\Omega}$ was chosen with the change of variables already in mind and all the results hold.

Let $|\cdot|_{max}$ stand for $\max_{i=1,\ldots,12} |\tilde{y}^{(i)}|$. Introduce the function

$$d_5(\tilde{y}, \tilde{y}_{ref}, r_{ref}) = r_{ref} - |G \cdot \tilde{y}_{ref} - P_* G \cdot \tilde{y}|_{max}$$

which, if positive, measures the distance between $P_*\tilde{y}$ and $\mathbb{R}^{12} \backslash G^{-1} \cdot \mathcal{B}(G \cdot \tilde{y}_{ref}; r_{ref})$.

Local split-hyperbolicity algorithm

Here we discuss informally the algorithm that establishes the split-hyperbolicity of $g_{i,j}$ in a neighborhood of a given point $y = (y^u, y^s) \in M_i^u \times M_i^s$. Consider the algorithm $Alg_1(\tilde{y}, i, j)$ defined by the following 9 steps:

Step 1. Assign $\tilde{y}_0 = G^{-1}\tilde{y}$, $\varepsilon = d_1(\tilde{y}_0)$, $n = 1$, $n_1 = n_2 = 0$, $hs = 1$, $wcross = 0$.

Step 2. Compute the point $\tilde{y}_n = \tilde{\nu}(\tilde{y}_{n-1}, h)$. If ($hs \cdot \tilde{y}_n^{(3)} < 0$), then assign $hs = -hs$, $wcross = wcross + 1$.

Step 3. Compute new ε: $\varepsilon = \min\{\varepsilon, d_1(\tilde{y}_n), d_2(\tilde{y}_n, \tilde{y}_{n-1})\}$.

Step 4. If ($wcross = 2$ and $\tilde{y}_{n-1}^{(3)} \leq 0$), then Go to 6. Otherwise the algorithm terminates unsuccessfully.

Step 5. Let $n = n+1$. If ($nh \geq 22$), then the algorithm terminates unsuccessfully, otherwise Go to 2.

Step 6. Assign $n_1 = n - 1$. Calculate approximate intersection point $\tilde{y}_{ref}$ as a reference point: let $\alpha = \tilde{y}_{n-1}^{(3)}/(\tilde{y}_{n-1}^{(3)} - \tilde{y}_n^{(3)})$ and $\tilde{y}_{ref} = \alpha\tilde{y}_n + (1-\alpha)\tilde{y}_{n-1}$. Let P_* be a linear projector to W along $\tilde{\mathcal{F}}(\tilde{y})$. Calculate $B_{ref} = \mathcal{H}_j^{-1} Lin(G \cdot \tilde{y}_{ref})\mathcal{H}_i = (b_{k,l})_{k,l=1}^2$ and

$$r_{ref} = \min\{|b_{1,1}| - \lambda_*^u,\ \lambda_*^s - |b_{2,2}|,\ \mu_*^u - |b_{1,2}|,\ \mu_*^s - |b_{2,1}|\}/L_i.$$

Step 7. If ($nh \geq 22$), then the algorithm terminated unsuccessfully. Calculate a new ε:

$$\begin{aligned}\varepsilon = \min\{\ & \varepsilon,\ d_1(\tilde{y}_n),\ d_2(\tilde{y}_n, \tilde{y}_{n-1}),\ d_3(P_*\tilde{y}_n),\ d_3(P_*\tilde{y}_{n_1}),\\ & d_5(y_{n_1}, y_{ref}, r_{ref}), d_5(y_n, y_{ref}, r_{ref})\}.\end{aligned}$$

Step 8. If ($\varepsilon \geq \min\{-\tilde{y}_{n_1}^{(3)}, \tilde{y}_n^{(3)}\}$), then Go to 9. Compute $\varepsilon = \varepsilon - 4 \cdot 10^{-7} - 10^{-10}$. Let $n_2 = n$. If ($\varepsilon \leq 0$), then the algorithm terminates unsuccessfully otherwise the algorithm terminates successfully.

Step 9. Let $n_1 = n_1 - 1$ and let $n = n + 1$. Calculate $\tilde{y}_n = \tilde{\nu}(\tilde{y}_{n-1}, h)$. Go to 7.

Lemma 10.6.5. *If $Alg_1(\tilde{y}, i, j)$ is successfully terminated then the triplet (n_1, n_2, ε) is $(G^{-1} \cdot \mathcal{B}(G \cdot \tilde{y}_{ref}; r_{ref}), 4 \cdot 10^{-7}, G^{-1} \cdot \tilde{y})$ -fortunate. Moreover, if $x \in \mathcal{B}(G^{-1} \cdot \tilde{y}, \varepsilon e^{-Ln_2h}) \cap S$, then $\tilde{\Phi}(x) \in \Xi_i$.*

Proof: To simplify notation in the proof below we enumerate values taken by ε just before the increment of n as ε_n (i.e. at the end of either Step 3 or Step 7). Assume that the algorithm successfully terminated and denote by ε_* the final value of ε.

(r_2): By Steps 3 and 7, $\varepsilon_n < d_1(\tilde{y}_n)$ holds, which means $\mathcal{B}(\tilde{\nu}_n(\tilde{y}_0); \varepsilon_n) \subset \tilde{\Omega}$. Step 8 ensures that $0 < \varepsilon_* < \varepsilon_{n_2} - 4 \cdot 10^{-7}$ and takes into account numerical errors. Thus $\mathcal{B}(\tilde{\nu}_n(\tilde{y}_0); \varepsilon_* + \delta) \subset \tilde{\Omega}$, and ($r_2$) is proved.

Table 10.3. *Data for projector error estimates, part 1*

L_i	(3,3) 8.135	(1,2) 8.122	(2,3) 8.135	(3,4) 8.135	(4,5) 8.135	(5,6) 8.134	(6,7) 8.135
$\tilde{y}^{(1)}_{min}$	0.008826	0.008867	0.008826	0.008826	0.008830	0.008823	0.008839
$\tilde{y}^{(2)}_{min}$	0.6573	0.6574	0.6573	0.6573	0.6573	0.6573	0.6573
$\tilde{y}^{(4)}_{min}$	-0.4573	-0.4500	-0.4576	-0.4573	-0.4573	-0.4573	-0.4574
$\tilde{y}^{(5)}_{min}$	1.024	1.022	1.024	1.024	1.024	1.024	1.024
$\tilde{y}^{(6)}_{min}$	1.120	1.114	1.121	1.120	1.120	1.120	1.120
$\tilde{y}^{(7)}_{min}$	2.172	2.169	2.172	2.172	2.172	2.172	2.172
$\tilde{y}^{(8)}_{min}$	-0.6323	-0.6300	-0.6324	-0.6323	-0.6323	-0.6323	-0.6324
$\tilde{y}^{(9)}_{min}$	-1.745	-1.740	-1.745	-1.745	-1.745	-1.745	-1.745
$\tilde{y}^{(10)}_{min}$	1.045	1.047	1.045	1.045	1.045	1.045	1.045
$\tilde{y}^{(11)}_{min}$	0.2907	0.2911	0.2906	0.2907	0.2907	0.2907	0.2906
$\tilde{y}^{(12)}_{min}$	-0.3763	-0.3767	-0.3762	-0.3763	-0.3763	-0.3762	-0.3763
$\tilde{y}^{(1)}_{max}$	0.03214	0.03218	0.03214	0.03214	0.03215	0.03214	0.03216
$\tilde{y}^{(2)}_{max}$	0.6808	0.6809	0.6808	0.6808	0.6808	0.6808	0.6808
$\tilde{y}^{(4)}_{max}$	-0.4144	-0.4072	-0.4147	-0.4144	-0.4145	-0.4144	-0.4145
$\tilde{y}^{(5)}_{max}$	1.054	1.052	1.054	1.054	1.054	1.054	1.054
$\tilde{y}^{(6)}_{max}$	1.153	1.146	1.153	1.153	1.153	1.153	1.153
$\tilde{y}^{(7)}_{max}$	2.215	2.212	2.215	2.215	2.215	2.215	2.215
$\tilde{y}^{(8)}_{max}$	-0.5966	-0.5943	-0.5967	-0.5966	-0.5966	-0.5965	-0.5966
$\tilde{y}^{(9)}_{max}$	-1.701	-1.696	-1.701	-1.701	-1.701	-1.701	-1.701
$\tilde{y}^{(10)}_{max}$	1.077	1.079	1.077	1.077	1.077	1.077	1.077
$\tilde{y}^{(11)}_{max}$	0.3184	0.3188	0.3184	0.3184	0.3184	0.3184	0.3184
$\tilde{y}^{(12)}_{max}$	-0.3405	-0.3410	-0.3405	-0.3405	-0.3405	-0.3405	-0.3406

(r_1): By Steps 5 and 7 $n, n_2 < 22/h$ hold, and (r_1) follows from (r_2) and Proposition 10.6.2.

(r_3): Step 7 guarantees that for all $0.5(n_1 + n_2 + 1) < n \leq n_2$

$$\mathcal{B}(\tilde{y}_n; \varepsilon_n),\ \mathcal{B}(\tilde{y}_{n_1+n_2+1-n}; \varepsilon_n) \subset G^{-1} \cdot \mathcal{B}(G \cdot \tilde{y}_{ref}; r_{ref}). \tag{10.47}$$

The first inclusion in (r_3) follows from $\varepsilon_* \leq \varepsilon_{n_1}$ and (10.47), which implies $\mathcal{B}(\tilde{y}_{n_1}; \varepsilon_*) \subset \mathcal{B}(\tilde{y}_{ref}; r_{ref})$, and Step 8, from which $\varepsilon_* < -y^{(3)}_{n_1}$. The same logic works for the second inclusion of (r_3). The third one follows from (10.47) and fact that $\varepsilon_* < \varepsilon_n - 4 \cdot 10^{-7}$. Thus (r_3) is proved.

(r_5) This property holds, due to Steps 3 and 7 from which

$$\varepsilon_n \leq \min_{t \in [nh-h, nh]} d(\nu(t, \tilde{y}_0),\ \ell),$$

and Step 8 that states $\varepsilon_* < \varepsilon_n - 4 \cdot 10^{-7}$.

(r_4) By Steps 2-4 $\tilde{y}^{(3)}_{n_1} < 0$ for all $n \leq n_1$ and there was only one intersection of $\nu(t, \tilde{y}_0)$ with W for $0 < t \leq n_1 h$, and by $\tilde{y}_0 \in \tilde{S}$ it could be only an intersection from W^+ to W^-.
Assume that $\nu(t, \tilde{y}_0), 0 < t \leq n_1 h$, intersects $\tilde{S}$. Now if $\varphi(t, \tilde{y}_0)$ intersects W then by Proposition 10.6.2 and (r_5) it intersects $\tilde{S}$ from W^+ to W^- which is impossible. If $\varphi(t, \tilde{y}_0)$ is in W^+ then the continuation of the algorithm

Table 10.4. *Data for projector error estimates, part 2*

L_i	(7,8) 8.134	(8,9) 8.136	(9,10) 8.130	(10,11) 8.144	(11,12) 8.111	(12,13) 8.868	(13,14) 8.655
$\tilde{y}^{(1)}_{min}$	0.008801	0.008891	0.008676	0.009186	0.007980	0.01085	0.003978
$\tilde{y}^{(2)}_{min}$	0.6573	0.6572	0.6575	0.6569	0.6582	0.6552	0.6622
$\tilde{y}^{(4)}_{min}$	-0.4572	-0.4576	-0.4565	-0.4591	-0.4530	-0.4674	-0.4336
$\tilde{y}^{(5)}_{min}$	1.024	1.024	1.023	1.025	1.021	1.029	1.010
$\tilde{y}^{(6)}_{min}$	1.120	1.121	1.120	1.121	1.119	1.125	1.110
$\tilde{y}^{(7)}_{min}$	2.171	2.172	2.171	2.174	2.166	2.185	2.140
$\tilde{y}^{(8)}_{min}$	-0.6322	-0.6326	-0.6317	-0.6338	-0.6287	-0.6409	-0.6127
$\tilde{y}^{(9)}_{min}$	-1.745	-1.746	-1.744	-1.747	-1.740	-1.758	-1.717
$\tilde{y}^{(10)}_{min}$	1.045	1.045	1.045	1.046	1.042	1.051	1.030
$\tilde{y}^{(11)}_{min}$	0.2907	0.2905	0.2910	0.2898	0.2926	0.2861	0.3010
$\tilde{y}^{(12)}_{min}$	-0.3761	-0.3766	-0.3755	-0.3780	-0.3723	-0.3858	-0.3545
$\tilde{y}^{(1)}_{max}$	0.03212	0.03221	0.03199	0.03251	0.03128	0.03420	0.02749
$\tilde{y}^{(2)}_{max}$	0.6808	0.6807	0.6810	0.6804	0.6817	0.6787	0.6859
$\tilde{y}^{(4)}_{max}$	-0.4143	-0.4148	-0.4137	-0.4163	-0.4101	-0.4246	-0.3894
$\tilde{y}^{(5)}_{max}$	1.054	1.054	1.054	1.055	1.052	1.059	1.041
$\tilde{y}^{(6)}_{max}$	1.153	1.153	1.152	1.153	1.151	1.157	1.143
$\tilde{y}^{(7)}_{max}$	2.215	2.216	2.214	2.218	2.210	2.228	2.185
$\tilde{y}^{(8)}_{max}$	-0.5965	-0.5968	-0.5959	-0.5981	-0.5930	-0.6050	-0.5760
$\tilde{y}^{(9)}_{max}$	-1.701	-1.701	-1.700	-1.703	-1.696	-1.713	-1.671
$\tilde{y}^{(10)}_{max}$	1.077	1.077	1.077	1.078	1.074	1.083	1.063
$\tilde{y}^{(11)}_{max}$	0.3184	0.3182	0.3187	0.3176	0.3203	0.3139	0.3291
$\tilde{y}^{(12)}_{max}$	-0.3404	-0.3408	-0.3398	-0.3422	-0.3366	-0.3499	-0.3177

will result in negative ε and unsuccessful termination. We have arrived at a contradiction. Therefore (r_4) is proved.

Finally, the estimation of the number of double-precision operations and IEEE standard show that the computational error for ε is less than 10^{-10} and this fact is addressed in Step 8.

To prove that, if $x \in \mathcal{B}(\tilde{y}_0, \varepsilon e^{-Ln_2h}) \cap S$, then $\tilde{\Phi}(x) \in \Xi_i$, it is enough to mention that by Step 7, $P_*(\mathcal{B}(\tilde{y}_n, \varepsilon+\delta)) \subset \Xi_i$ for all $n_1 \leq n \leq n_2$. The last inclusion implies that the set (n_1, n_2, ε) is $(\Xi_i, \delta, \tilde{y}_0)$-fortunate and $\tilde{\Phi}(x) \in \Xi_i$ follows. The lemma is proved. □

Lemma 10.6.6. *Let i, j satisfy $a_{i,j} = 1$ where $a_{i,j}$ is an element of (10.19). If $Alg_1(\tilde{y}_0, i, j)$ is successfully terminated with the triplet (n_1, n_2, ε) then the relationship*

$$\begin{pmatrix} \lambda^u_* & \mu^u_* \\ \mu^s_* & \lambda^s_* \end{pmatrix} \succ \begin{pmatrix} \left|\dfrac{\partial g^{(1)}_{i,j}}{\partial y^{(1)}}\right| & \left|\dfrac{\partial g^{(1)}_{i,j}}{\partial y^{(2)}}\right| \\ \left|\dfrac{\partial g^{(2)}_{i,j}}{\partial y^{(1)}}\right| & \left|\dfrac{\partial g^{(2)}_{i,j}}{\partial y^{(2)}}\right| \end{pmatrix}$$

holds for all $y = h_i^{-1}(x)$ where x satisfies $\left|x - (\tilde{y}^{(1)}_0, \tilde{y}^{(2)}_0)\right| < \varepsilon e^{-Ln_2h}$.

Proof: Let us fix $\tilde{y}_0 \in \tilde{S}, J(\tilde{y}_0) = I$ and choose any y such that $x = h_i(y)$ satisfies $\left|x - (\tilde{y}^{(1)}_0, \tilde{y}^{(2)}_0)\right| < \varepsilon e^{-Ln_2h}$. Since we are looking for a linearization of the Poincaré

Table 10.5. *Data for projector error estimates, part 3*

L_i	(14,15) 10.06	(15,16) 7.681	(16,17) 15.71	(17,18) 2.701	(18,19) 44.46	(19,20) 6.159	(20,1) 8.416
$\tilde{y}^{(1)}_{min}$	0.02011	-0.01665	0.07951	-0.08521	0.4328	0.03143	0.008054
$\tilde{y}^{(2)}_{min}$	0.6450	0.6844	0.5867	0.7880	0.2908	0.7091	0.6554
$\tilde{y}^{(4)}_{min}$	-0.5134	-0.3171	-0.7214	0.4288	-0.2570	2.405	-0.6424
$\tilde{y}^{(5)}_{min}$	1.053	0.9461	1.161	0.5495	0.7840	-0.8635	1.071
$\tilde{y}^{(6)}_{min}$	1.143	1.058	1.210	0.6982	0.3107	0.3209	1.286
$\tilde{y}^{(7)}_{min}$	2.243	1.998	2.586	1.317	4.025	0.2184	2.248
$\tilde{y}^{(8)}_{min}$	-0.6817	-0.5230	-0.9126	-0.09787	-1.947	1.066	-0.6851
$\tilde{y}^{(9)}_{min}$	-1.816	-1.586	-2.146	-0.9440	-3.565	0.1904	-1.869
$\tilde{y}^{(10)}_{min}$	1.080	0.9679	1.265	0.8003	2.480	1.711	0.9868
$\tilde{y}^{(11)}_{min}$	0.2635	0.3455	0.1209	0.4515	-0.8219	0.03324	0.2837
$\tilde{y}^{(12)}_{min}$	-0.4315	-0.2562	-0.6954	0.1705	-2.157	0.3669	-0.3590
$\tilde{y}^{(1)}_{max}$	0.04419	0.007044	0.1051	-0.06386	0.4574	0.05400	0.03113
$\tilde{y}^{(2)}_{max}$	0.6691	0.7087	0.6119	0.8113	0.3146	0.7320	0.6786
$\tilde{y}^{(4)}_{max}$	-0.4695	-0.2685	-0.6817	0.4832	-0.1903	2.434	-0.6009
$\tilde{y}^{(5)}_{max}$	1.084	0.9797	1.190	0.5857	0.8298	-0.8195	1.101
$\tilde{y}^{(6)}_{max}$	1.175	1.093	1.238	0.7351	0.3785	0.3487	1.319
$\tilde{y}^{(7)}_{max}$	2.291	2.047	2.637	1.363	4.057	0.2456	2.290
$\tilde{y}^{(8)}_{max}$	-0.6435	-0.4844	-0.8711	-0.06221	-1.916	1.097	-0.6511
$\tilde{y}^{(9)}_{max}$	-1.769	-1.538	-2.093	-0.8994	-3.526	0.2175	-1.826
$\tilde{y}^{(10)}_{max}$	1.114	1.001	1.304	0.8230	2.531	1.732	1.018
$\tilde{y}^{(11)}_{max}$	0.2930	0.3736	0.1544	0.4726	-0.7813	0.06881	0.3103
$\tilde{y}^{(12)}_{max}$	-0.3927	-0.2176	-0.6517	0.2025	-2.108	0.3921	-0.3241

map, there is only one $\tilde{y} \in \tilde{S}$ that interests us, namely

$$\tilde{y} = (x^{(1)}, x^{(2)}, 0,\ 1, 0, 0,\ 0, 1, 0,\ 0, 0, 1).$$

By the successful termination of Alg_1 the inclusions $\Phi(\tilde{y}), \tilde{y}_{ref} \in \Xi_i$ hold, and by Lemma 10.6.4

$$\|g'_{i,j}(y) - B_{ref}\|_{max} \leq L_i \cdot |\tilde{\varphi}(T(\tilde{y}), \tilde{y}) - \tilde{y}_{ref}|_{max}.$$

By Steps 7 and 8 of Alg_1, $\varepsilon_* < r_{ref} - |\tilde{y}_{ref} - P_*\tilde{y}_n| - 4 \cdot 10^{-7}$ for all $n_1 \leq n \leq n_2$ and that guarantees

$$|\tilde{\varphi}(T(\tilde{y}), \tilde{y}) - \tilde{y}_{ref}| < r_{ref},$$

for all $\tilde{y} \in \mathcal{B}(\tilde{y}_0; \varepsilon e^{-Ln_2 h})$. Combining the last two inequalities we get

$$\|g'_{i,j}(y) - B_{ref}\|_{max} < L_i r_{ref}.$$

By Step 6 of the algorithm

$$\begin{pmatrix} \lambda^u_* + L_i r_{ref} & \mu^u_* - L_i r_{ref} \\ \mu^s_* - L_i r_{ref} & \lambda^s_* - L_i r_{ref} \end{pmatrix} \succ B_{ref},$$

which implies

$$\frac{\partial g^u_{i,j}}{\partial y^u}(y) > \lambda^u, \quad \left|\frac{\partial g^s_{i,j}}{\partial y^s}(y)\right| < \lambda^s, \quad \left|\frac{\partial g^u_{i,j}}{\partial y^s}(y)\right| < \mu^u, \quad \left|\frac{\partial g^s_{i,j}}{\partial y^u}(y)\right| < \mu^s.$$

The lemma is proved. □

10.6.5 Computations

Let x, p, q be given vectors in S, then the parallelogram $R(p, q, x) \subset S$ is given by

$$R = \{\alpha p + \beta q + x : \; |\alpha| \le 1, |\beta| \le 1\}.$$

Denote an open ball in the halfplane S with the radius r, centered at ξ, by $\mathcal{B}(\xi; r)$.

To prove the statement of Lemma 10.5.2 for all $\tilde{y} : (\tilde{y}^{(1)}, \tilde{y}^{(2)}) \in R$, $\tilde{y}^{(3)} = 0$, $J(\tilde{y}) = I$ we cover R with a finite number of balls such that their radii are not less than some number and the linearization of the Poincaré map satisfies the required properties on each ball.

We define the parallelogram grid with the step h_g

$$\mathcal{G}_R = \{\xi(k, l) = kh_g\tilde{y} + lh_g z + c \in S\}, \; k \in [-N_1, N_1], \; l \in [-N_2, N_2],$$

where $N_1 = \lfloor |\tilde{y}|/h \rfloor + 1$, $N_2 = \lfloor |z|/h \rfloor + 1$ and the operation $\lfloor \cdot \rfloor$ stands for the maximum integer that is not greater than number in the "floor brackets".

Consider the algorithm $Alg_2(i, j, h_g)$:

Step 0. Define $R = X_i^*$ and introduce an appropriate grid with step h_g on it. Assign $m_{k,l} = 0$ for all $k \in [-N_1, N_1]$, $l \in [-N_2, N_2]$.

Step 1. Choose k and l such that $m_{k,l} = 0$. If there are no such k, l, then the algorithm is successfully terminated.

Step 2. Calculate $Alg_1(\xi(k, l), i, j) = (n_1, n_2, \varepsilon)$. If $Alg_1(\xi(k, l), i, j)$ terminates unsuccessfully then so should Alg_2. Otherwise calculate $r = \varepsilon e^{-Ln_2 h}/h_g - \sqrt{2}/2$.

Step 3. If $r < 0$, then the algorithm terminated unsuccessfully. Otherwise assign $m_{k_1,l_1} = 1$ for all indices k_1, l_1 satisfying

$$|(k_1 - k)\tilde{y}/|y| + (l_1 - l)z/|z|| \le r.$$

Step 4. Go to 1.

Lemma 10.6.7. *The successful termination of the algorithm $Alg_2(i, j, h_g)$ implies that*

$$\begin{pmatrix} \lambda^u & \mu^u \\ \mu^s & \lambda^s \end{pmatrix} \succ g'_{i,j}(x) \tag{10.48}$$

holds for all $x \in B_i[\delta^u, \delta^s]$.

Proof: Let us consider any 'successful loop' of the algorithm above. Let n_1, n_2, ε, r denote values of the corresponding variables at the beginning of Step 3 of the loop. Note that $\tilde{\Phi}$ is defined on the whole X_i^* since Φ is defined.

Firstly we prove the inclusion

$$\bigcup_{|\xi - \xi(i,j)| \le r} \mathcal{B}\left[\xi; \frac{\sqrt{2}}{2} h_g\right] \subset \mathcal{B}\left[\xi(i, j); rh_g + \frac{\sqrt{2}}{2} h_g\right]. \tag{10.49}$$

Indeed, for each ball from the left hand side $\mathcal{B}[\xi; \frac{\sqrt{2}}{2}h_g]$, and for each x belonging to it, we have

$$|x - \xi(i,j)| \leq |x - \xi| + |\xi(i,j) - \xi| \leq r + \frac{\sqrt{2}}{2}h_g < \varepsilon e^{0.51 n_2 h},$$

which means that $x \in \mathcal{B}\left[\xi(i,j); rh_g + \frac{\sqrt{2}}{2}h_g\right]$ and (10.49) is proved.

The definition of r in Step 2 and Lemma 10.6.6 imply that (10.48) holds on every

$$h^{-1}\left(\mathcal{B}\left[\xi(i,j); rh_g + \frac{\sqrt{2}}{2}h_g\right]\right).$$

Inclusion (10.49) and Step 3 of the algorithm guarantee that for all k, l, such that $m_{k,l}$ is set to 1 in the loop, (10.48) holds on $h^{-1}\left(\mathcal{B}[\xi(k,l); \frac{\sqrt{2}}{2}h_g]\right)$.

Therefore it remains to prove that

$$R \subset \bigcup_{\xi \in \mathcal{G}_R} \mathcal{B}\left[\xi; \frac{\sqrt{2}}{2}h_g\right].$$

The meshes of the grid are rhombuses with side h_g and a rhombus with a side of length a can be covered by the 4 closed balls with centers in its corners and a radius of $\frac{\sqrt{2}}{2}a$.

Now we notice that $B_i[\delta^u, \delta^s] = h_i^{-1}(X_i^*)$. This proves the lemma. □

Lemma 10.6.8. *Algorithm $Alg_2(i, j, h_g)$ successfully terminates for all i, j such that $a_{i,j} = 1$ in (10.19).*

Proof: The proof is given by straightforward numerical calculation.

The algorithms are implemented in C++ and programs can be downloaded from the site

http://phys.ucc.ie/~oll/lksplit/A.

Although the implementation of Alg_1 and Alg_2 is straightforward, the computations could not be performed on one PC within a reasonable time. Pilot experiments have shown that the mesh size h_g in the algorithm Alg_2 should be less then $2 \cdot 10^{-7}$. Fortunately, the algorithm Alg_2 is data parallelizable. To show split-hyperbolicity on a given set $Y \subset \cup_i Y_i$ it is enough to show it on each Y_i, which can be done independently on every Y_i.

Any reasonable task scheduler on a group of computers could be used to implement parallel computations since the demand for communications is small.

To avoid the problems of using non-homogeneous networks we have chosen the PVM library to perform network communications and to implement a simple scheduling. To simplify the parallelization, each of the original 8 sets X_i for which we need to prove split-hyperbolicity was subdivided into 100 smaller parallelograms.

To verify the statement of the lemma, we used a cluster of one hundred Pentium-3 1.2GHz machines, which belongs to the Boole Centre for Research in Informatics, University College, Cork, Ireland. All computations took approximately one day. □

10.7 Acknowledgements

The authors are grateful to John McInerney and Guillaume Huyet for helpful discussions. The first author was partially supported by the Enterprise Ireland Grant SC/2003/376. The second author was supported by IRCSET Grant PD/2004/26.

Bibliography

[1] D. V. ANOSOV, ed., *Dynamical Systems 9. Dynamical systems with hyperbolic behaviour*, Encyclopedia of Mathematical Sciences, 1, Springer–Verlag, Berlin, 1988.

[2] N. BOBYLEV, A. POKROVSKII, AND J. MCINERNEY, *On Positive Definiteness of Interval Homogeneous Forms*, Preprints of Institute for Nonlinear Science, 4, National University of Ireland, University College, Cork, 2000.

[3] M. BROKATE AND A. V. POKROVSKII, *Asymptotically stable oscillations in systems with hysteresis nonlinearities*, J. Differential Equations, 150 (1998), pp. 98–123.

[4] *Conley Index Theory*. Papers from the workshop held in Warsaw, June 1997, K. Mischaikow, M. Mrozek and P. Zgliczyński, eds., Banach Center Publications, 47, Polish Academy of Sciences, Institute of Mathematics, Warsaw, 1999.

[5] K. DEIMLING, *Nonlinear Functional Analysis*, Springer-Verlag, Berlin,1980.

[6] P. DIAMOND, P. E. KLOEDEN, V. S. KOZYAKIN, AND A. V. POKROVSKII, *Semi-hyperbolic mappings*, J. Nonlinear Sci., 5 (1995), pp. 419–431.

[7] P. DIAMOND, P. E. KLOEDEN, M. A. KRASNOSEL'SKII, AND A. V. POKROVSKII, *Chaotic dynamics in nonsmooth perturbations of bishadowing system*, Arab J. Math. Sci., 6(1) (2000), pp. 41–74.

[8] F. DAĬMOND, P. KLOEDEN, V. S. KOZYAKIN, M. A. KRASNOSEL'SKII, AND A. V. POKROVSKII, *Structural stability of the trajectories of dynamical systems with respect to hysteresis perturbations* (in Russian, MR96h:47078), Dokl. Akad. Nauk, 343(1) (1995), pp. 25–27.

[9] T. T. GEORGIOU AND M. C. SMITH, *Robustness of a relaxation oscillator*, George Zames commemorative issue. Internat. J. Robust Nonlinear Control, 10(11–12) (2000), pp. 1005–1024.

[10] G. H. GOLDSZTEIN, F. BRONER, AND S. H. STROGATZ, *Dynamical hysteresis without static hysteresis: scaling laws and asymptotic expansions*, SIAM J. Appl. Math., 57(4) (1997), pp. 1163–1187.

[11] E. HAIRER, S. P. NORSETT, AND G. WANNER, *Solving Ordinary Differential Equations. I. Nonstiff Problems*, Springer-Verlag Series in Computational Mathematics, 8. Springer-Verlag, Berlin-New York, 1987.

[12] R. E. HARTWIG AND M. NEUMANN, *Bounds on the exponent of primitivity which depend on the spectrum and the minimal polynomial*, Linear Algebra Appl., 184 (1993), pp. 103–122.

[13] D. HERTZ, *The extreme eigenvalues and stability of real symmetric interval matrices*, IEEE Trans. Automat. Control, 37(4) (1992), pp. 532–535.

[14] A. J. HOMBURG AND H. WEISS, *A geometric criterion for positive topological entropy. II. Homoclinic tangencies*, Comm. Math. Phys., 208(2) (1999), pp. 267–273.

[15] G. HUYET, J. K. WHITE, A. J. KENT, S. P. HEGARTY, J. V. MOLONEY, AND J. G. MCINERNEY, *Dynamics of a semiconductor laser with optical feedback*, Phys. Rev. A60 (1999), pp. 1534–1537.

[16] A. E. KATOK AND B. HASSELBLATT, *Introduction to the Modern Theory of Dynamical Systems*, Cambridge University Press, Cambridge, 1995.

[17] M. A. KRASNOSEL'SKII AND A. V. POKROVSKII, *Systems with Hysteresis*, Springer-Verlag, Berlin, 1989.

[18] M. A. KRASNOSEL'SKII AND P. P. ZABREIKO, *Geometrical Methods of Nonlinear Analysis*, Springer-Verlag, Berlin – Heidelberg – New York – Tokyo, 1984.

[19] P. KREJCI, *Hysteresis, Convexity and Dissipation in Hyperbolic Equations*, Gakkotosho, Tokyo, 1996.

[20] R. LANG AND K. KOBAYASHI, *External optical feedback effects on semiconductor injection laser properties*, IEEE J. Quantum Electron., QE-16 (1980), pp. 347–355.

[21] T. Y. LI AND J. A. YORKE, *Period three implies chaos*, Amer. Math. Monthly, 82 (1975), pp. 985–992.

[22] H. LI, J. YE, AND J. G. MCINERNEY, *Detailed analysis of coherence collapse in semiconductor lasers*, IEEE J. Quantum Electron., QE-29 (1993), pp. 2421–2432

[23] J. MOERK AND B. TROMBORG, *Stability analysis and the route to chaos for laser diodes with optical feedback*, IEEE Phot. Tech. Lett., 2 (1990), pp. 21–23.

[24] S. NEUFELD AND J. SHEN, *Some results on generalized exponents*, J. Graph Theory, 28(4) (1998), pp. 215–225.

[25] W. H. PRESS, S. A. TEUKOLSKY, W. T. VETTERLING, AND B. P. FLANNERY, *Numerical Recipies in C*, Cambridge University Press, Chapter 1, pp. 21–31, 1992.

[26] A. V. POKROVSKII, *Topological shadowing and split-hyperbolicity*, *J. for Difference and Differential Equations*, special issue dedicated to M. A. Krasnosel'skii, 4(3–4) (1997), pp. 335–360.

[27] A. V. POKROSKII, S. J. SZYBKAM, AND J. G. MCINERNEY, *Topological degree in locating homoclinic structures for discrete dynamical systems*, National University of Ireland, University College. Preprint INS, 01-001, Cork, 2001.

[28] A. V. POKROVSKII AND O. A. RASSKAZOV, *Methods of topological degree theory in the analysis of broken orbits*, National University of Ireland, University College. Preprint INS, Cork, 2000.

[29] O. RASSKAZOV, *Geometrical methods of nonlinear analysis with application to a model of a semiconductor laser*, Ph.D. thesis, National University of Ireland, University College, Cork, 2003.

[30] ——, *Forward and backward stable sets of split-hyperbolic mappings*, Izvestiya of RAEN, Series MMMIU, 5(1–2) (2001), pp. 185–205.

[31] O. RASSKAZOV, G. HUYET, J. MCINERNEY, AND A. POKROVSKII, *Rigorous Analysis of Complicated Behaviour in a Truncated Lang-Kobayashi Model*, National University of Ireland, University College. Reports of INS, 01-011, Cork, 2001.

[32] D. RUELLE, *Elements of Differentiable Dynamics and Bifurcation Theory*, Academic Press, Boston, 1989.

[33] G. R. SELL, *Lectures on Topological Dynamics and Differential Equations*, Van Nostrand–Reinbold, London, 1971.

[34] P. ZGLICZYŃSKI, *Fixed point index for iterations of mappings, topological horseshoe and chaos*, Topol. Methods Nonlinear Anal., 8(1) (1996), pp. 169–177.

[35] P. ZGLICZYŃSKI, *Computer assisted proof of the horseshoe dynamics in the Henon map*, Random Comput. Dynam., 5(1) (1997), pp. 1–17.

Index